A Concise Course in Algebraic Topology

Chicago Lectures in Mathematics Series
Robert J. Zimmer, series editor
J. P. May, Spencer J. Bloch, Norman R. Lebovitz, and Carlos Kenig, editors

Other *Chicago Lectures in Mathematics* titles available from the University of Chicago Press:

Simplicial Objects in Algebraic Topology, by J. Peter May (1967)
Fields and Rings, Second Edition, by Irving Kaplansky (1969, 1972)
Lie Algebras and Locally Compact Groups, by Irving Kaplansky (1971)
Several Complex Variables, by Raghavan Narasimhan (1971)
Torsion-Free Modules, by Eben Matlis (1973)
Stable Homotopy and Generalised Homology, by J. F. Adams (1974)
Rings with Involution, by I. N. Herstein (1976)
Theory of Unitary Group Representation, by George V. Mackey (1976)
Commutative Semigroup Rings, by Robert Gilmer (1984)
Infinite-Dimensional Optimization and Convexity, by Ivar Ekeland and Thomas Turnbull (1983)
Navier-Stokes Equations, by Peter Constantin and Ciprian Foias (1988)
Essential Results of Functional Analysis, by Robert J. Zimmer (1990)
Fuchsian Groups, by Svetlana Katok (1992)
Unstable Modules over the Steenrod Algebra and Sullivan's Fixed Point Set Conjecture, by Lionel Schwartz (1994)
Topological Classification of Stratified Spaces, by Shmuel Weinberger (1994)
Lectures on Exceptional Lie Groups, by J. F. Adams (1996)
Geometry of Nonpositively Curved Manifolds, by Patrick B. Eberlein (1996)
Dimension Theory in Dynamical Systems: Contemporary Views and Applications, by Yakov B. Pesin (1997)
Harmonic Analysis and Partial Differential Equations: Essays in Honor of Alberto Calderón, edited by Michael Christ, Carlos Kenig, and Cora Sadosky (1999)

J. P. May

A CONCISE COURSE IN ALGEBRAIC TOPOLOGY

The University of Chicago Press
Chicago and London

J. P. May is professor of mathematics at the University of Chicago. He is author or coauthor of many books, including *Simplicial Objects in Algebraic Topology* and *Equivariant Homotopy and Cohomology Theory*.

The University of Chicago Press, Chicago 60637
The University of Chicago Press, Ltd., London
©1999 by The University of Chicago
All rights reserved. Published 1999
08 07 06 05 04 03 02 01 00   2 3 4 5

ISBN: 0-226-51182-0 (cloth)
ISBN: 0-226-51183-9 (paper)

LIBRARY OF CONGRESS CATALOGING-IN-PUBLICATION DATA

May, J. Peter.
  A concise course in algebraic topology / J. P. May.
    p.   cm.— (Chicago lectures in mathematics series)
  Includes bibliographical references and index.
  ISBN 0-226-51182-0 (cloth : alk. paper). — ISBN 0-226-51183-9 (pbk. : alk. paper)
  1. Algebraic topology.   I. Title.   II. Series.
  QA612.M387 1999
  514'.2—dc21   99-29613
                                                                                  CIP

⊚The paper used in this publication meets the minimum requirements of the American National Standard for Information Sciences—Permanence of Paper for Printed Library Materials, ANSI Z39.48–1992.

Contents

Introduction	1
Chapter 1. The fundamental group and some of its applications	5
1. What is algebraic topology?	5
2. The fundamental group	6
3. Dependence on the basepoint	7
4. Homotopy invariance	7
5. Calculations: $\pi_1(\mathbb{R}) = 0$ and $\pi_1(S^1) = \mathbb{Z}$	8
6. The Brouwer fixed point theorem	10
7. The fundamental theorem of algebra	10
Chapter 2. Categorical language and the van Kampen theorem	13
1. Categories	13
2. Functors	13
3. Natural transformations	14
4. Homotopy categories and homotopy equivalences	14
5. The fundamental groupoid	15
6. Limits and colimits	16
7. The van Kampen theorem	17
8. Examples of the van Kampen theorem	19
Chapter 3. Covering spaces	21
1. The definition of covering spaces	21
2. The unique path lifting property	22
3. Coverings of groupoids	22
4. Group actions and orbit categories	23
5. The classification of coverings of groupoids	25
6. The construction of coverings of groupoids	27
7. The classification of coverings of spaces	28
8. The construction of coverings of spaces	29
Chapter 4. Graphs	33
1. The definition of graphs	33
2. Edge paths and trees	33
3. The homotopy types of graphs	34
4. Covers of graphs and Euler characteristics	35
5. Applications to groups	35
Chapter 5. Compactly generated spaces	37
1. The definition of compactly generated spaces	37
2. The category of compactly generated spaces	38

Chapter 6. Cofibrations	41
1. The definition of cofibrations	41
2. Mapping cylinders and cofibrations	42
3. Replacing maps by cofibrations	43
4. A criterion for a map to be a cofibration	43
5. Cofiber homotopy equivalence	44
Chapter 7. Fibrations	47
1. The definition of fibrations	47
2. Path lifting functions and fibrations	47
3. Replacing maps by fibrations	48
4. A criterion for a map to be a fibration	49
5. Fiber homotopy equivalence	50
6. Change of fiber	51
Chapter 8. Based cofiber and fiber sequences	55
1. Based homotopy classes of maps	55
2. Cones, suspensions, paths, loops	55
3. Based cofibrations	56
4. Cofiber sequences	57
5. Based fibrations	59
6. Fiber sequences	59
7. Connections between cofiber and fiber sequences	61
Chapter 9. Higher homotopy groups	63
1. The definition of homotopy groups	63
2. Long exact sequences associated to pairs	63
3. Long exact sequences associated to fibrations	64
4. A few calculations	64
5. Change of basepoint	66
6. n-Equivalences, weak equivalences, and a technical lemma	67
Chapter 10. CW complexes	71
1. The definition and some examples of CW complexes	71
2. Some constructions on CW complexes	72
3. HELP and the Whitehead theorem	73
4. The cellular approximation theorem	74
5. Approximation of spaces by CW complexes	75
6. Approximation of pairs by CW pairs	76
7. Approximation of excisive triads by CW triads	77
Chapter 11. The homotopy excision and suspension theorems	81
1. Statement of the homotopy excision theorem	81
2. The Freudenthal suspension theorem	83
3. Proof of the homotopy excision theorem	84
Chapter 12. A little homological algebra	89
1. Chain complexes	89
2. Maps and homotopies of maps of chain complexes	89
3. Tensor products of chain complexes	90
4. Short and long exact sequences	91

Chapter 13. Axiomatic and cellular homology theory   93
1. Axioms for homology   93
2. Cellular homology   94
3. Verification of the axioms   98
4. The cellular chains of products   99
5. Some examples: T, K, and $\mathbb{R}P^n$ 101

Chapter 14. Derivations of properties from the axioms   105
1. Reduced homology; based versus unbased spaces   105
2. Cofibrations and the homology of pairs   106
3. Suspension and the long exact sequence of pairs   107
4. Axioms for reduced homology   108
5. Mayer-Vietoris sequences   110
6. The homology of colimits   112

Chapter 15. The Hurewicz and uniqueness theorems   115
1. The Hurewicz theorem   115
2. The uniqueness of the homology of CW complexes   117

Chapter 16. Singular homology theory   121
1. The singular chain complex   121
2. Geometric realization   122
3. Proofs of the theorems   123
4. Simplicial objects in algebraic topology   124
5. Classifying spaces and $K(\pi,n)$s   126

Chapter 17. Some more homological algebra   129
1. Universal coefficients in homology   129
2. The Künneth theorem   130
3. Hom functors and universal coefficients in cohomology   131
4. Proof of the universal coefficient theorem   133
5. Relations between $\otimes$ and Hom   133

Chapter 18. Axiomatic and cellular cohomology theory   135
1. Axioms for cohomology   135
2. Cellular and singular cohomology   136
3. Cup products in cohomology   137
4. An example: $\mathbb{R}P^n$ and the Borsuk-Ulam theorem   138
5. Obstruction theory   140

Chapter 19. Derivations of properties from the axioms   143
1. Reduced cohomology groups and their properties   143
2. Axioms for reduced cohomology   144
3. Mayer-Vietoris sequences in cohomology   145
4. Lim^1 and the cohomology of colimits   146
5. The uniqueness of the cohomology of CW complexes   147

Chapter 20. The Poincaré duality theorem   149
1. Statement of the theorem   149
2. The definition of the cap product   151
3. Orientations and fundamental classes   153

4. The proof of the vanishing theorem	155
5. The proof of the Poincaré duality theorem	158
6. The orientation cover	161

Chapter 21. The index of manifolds; manifolds with boundary — 163
1. The Euler characteristic of compact manifolds — 163
2. The index of compact oriented manifolds — 164
3. Manifolds with boundary — 166
4. Poincaré duality for manifolds with boundary — 167
5. The index of manifolds that are boundaries — 169

Chapter 22. Homology, cohomology, and $K(\pi,n)$s — 171
1. $K(\pi,n)$s and homology — 171
2. $K(\pi,n)$s and cohomology — 173
3. Cup and cap products — 175
4. Postnikov systems — 178
5. Cohomology operations — 180

Chapter 23. Characteristic classes of vector bundles — 183
1. The classification of vector bundles — 183
2. Characteristic classes for vector bundles — 185
3. Stiefel-Whitney classes of manifolds — 187
4. Characteristic numbers of manifolds — 189
5. Thom spaces and the Thom isomorphism theorem — 190
6. The construction of the Stiefel-Whitney classes — 192
7. Chern, Pontryagin, and Euler classes — 193
8. A glimpse at the general theory — 196

Chapter 24. An introduction to K-theory — 199
1. The definition of K-theory — 199
2. The Bott periodicity theorem — 202
3. The splitting principle and the Thom isomorphism — 204
4. The Chern character; almost complex structures on spheres — 207
5. The Adams operations — 209
6. The Hopf invariant one problem and its applications — 211

Chapter 25. An introduction to cobordism — 215
1. The cobordism groups of smooth closed manifolds — 215
2. Sketch proof that $\mathcal{N}_*$ is isomorphic to $\pi_*(TO)$ — 216
3. Prespectra and the algebra $H_*(TO;\mathbb{Z}_2)$ — 219
4. The Steenrod algebra and its coaction on $H_*(TO)$ — 222
5. The relationship to Stiefel-Whitney numbers — 224
6. Spectra and the computation of $\pi_*(TO) = \pi_*(MO)$ — 226
7. An introduction to the stable category — 228

Suggestions for further reading — 231
1. A classic book and historical references — 231
2. Textbooks in algebraic topology and homotopy theory — 231
3. Books on CW complexes — 232
4. Differential forms and Morse theory — 232
5. Equivariant algebraic topology — 233

6.	Category theory and homological algebra	233
7.	Simplicial sets in algebraic topology	233
8.	The Serre spectral sequence and Serre class theory	233
9.	The Eilenberg-Moore spectral sequence	233
10.	Cohomology operations	234
11.	Vector bundles	234
12.	Characteristic classes	234
13.	K-theory	235
14.	Hopf algebras; the Steenrod algebra, Adams spectral sequence	235
15.	Cobordism	236
16.	Generalized homology theory and stable homotopy theory	236
17.	Quillen model categories	236
18.	Localization and completion; rational homotopy theory	237
19.	Infinite loop space theory	237
20.	Complex cobordism and stable homotopy theory	238
21.	Follow-ups to this book	238

Index   239

Introduction

The first year graduate program in mathematics at the University of Chicago consists of three three-quarter courses, in analysis, algebra, and topology. The first two quarters of the topology sequence focus on manifold theory and differential geometry, including differential forms and, usually, a glimpse of de Rham cohomology. The third quarter focuses on algebraic topology. I have been teaching the third quarter off and on since around 1970. Before that, the topologists, including me, thought that it would be impossible to squeeze a serious introduction to algebraic topology into a one quarter course, but we were overruled by the analysts and algebraists, who felt that it was unacceptable for graduate students to obtain their PhDs without having some contact with algebraic topology.

This raises a conundrum. A large number of students at Chicago go into topology, algebraic and geometric. The introductory course should lay the foundations for their later work, but it should also be viable as an introduction to the subject suitable for those going into other branches of mathematics. These notes reflect my efforts to organize the foundations of algebraic topology in a way that caters to both pedagogical goals. There are evident defects from both points of view. A treatment more closely attuned to the needs of algebraic geometers and analysts would include Čech cohomology on the one hand and de Rham cohomology and perhaps Morse homology on the other. A treatment more closely attuned to the needs of algebraic topologists would include spectral sequences and an array of calculations with them. In the end, the overriding pedagogical goal has been the introduction of basic ideas and methods of thought.

Our understanding of the foundations of algebraic topology has undergone subtle but serious changes since I began teaching this course. These changes reflect in part an enormous internal development of algebraic topology over this period, one which is largely unknown to most other mathematicians, even those working in such closely related fields as geometric topology and algebraic geometry. Moreover, this development is poorly reflected in the textbooks that have appeared over this period.

Let me give a small but technically important example. The study of generalized homology and cohomology theories pervades modern algebraic topology. These theories satisfy the excision axiom. One constructs most such theories homotopically, by constructing representing objects called spectra, and one must then prove that excision holds. There is a way to do this in general that is no more difficult than the standard verification for singular homology and cohomology. I find this proof far more conceptual and illuminating than the standard one even when specialized to singular homology and cohomology. (It is based on the approximation of excisive triads by weakly equivalent CW triads.) This should by now be a standard approach. However, to the best of my knowledge, there exists no rigorous exposition of this approach in the literature, at any level.

More centrally, there now exist axiomatic treatments of large swaths of homotopy theory based on Quillen's theory of closed model categories. While I do not think that a first course should introduce such abstractions, I do think that the exposition should give emphasis to those features that the axiomatic approach shows to be fundamental. For example, this is one of the reasons, although by no means the only one, that I have dealt with cofibrations, fibrations, and weak equivalences much more thoroughly than is usual in an introductory course.

Some parts of the theory are dealt with quite classically. The theory of fundamental groups and covering spaces is one of the few parts of algebraic topology that has probably reached definitive form, and it is well treated in many sources. Nevertheless, this material is far too important to all branches of mathematics to be omitted from a first course. For variety, I have made more use of the fundamental groupoid than in standard treatments,[1] and my use of it has some novel features. For conceptual interest, I have emphasized different categorical ways of modeling the topological situation algebraically, and I have taken the opportunity to introduce some ideas that are central to equivariant algebraic topology.

Poincaré duality is also too fundamental to omit. There are more elegant ways to treat this topic than the classical one given here, but I have preferred to give the theory in a quick and standard fashion that reaches the desired conclusions in an economical way. Thus here I have not presented the truly modern approach that applies to generalized homology and cohomology theories.[2]

The reader is warned that this book is not designed as a textbook, although it could be used as one in exceptionally strong graduate programs. Even then, it would be impossible to cover all of the material in detail in a quarter, or even in a year. There are sections that should be omitted on a first reading and others that are intended to whet the student's appetite for further developments. In practice, when teaching, my lectures are regularly interrupted by (purposeful) digressions, most often directly prompted by the questions of students. These introduce more advanced topics that are not part of the formal introductory course: cohomology operations, characteristic classes, K-theory, cobordism, etc., are often first introduced earlier in the lectures than a linear development of the subject would dictate.

These digressions have been expanded and written up here as sketches without complete proofs, in a logically coherent order, in the last four chapters. These are topics that I feel must be introduced in some fashion in any serious graduate level introduction to algebraic topology. A defect of nearly all existing texts is that they do not go far enough into the subject to give a feel for really substantial applications: the reader sees spheres and projective spaces, maybe lens spaces, and applications accessible with knowledge of the homology and cohomology of such spaces. That is not enough to give a real feeling for the subject. I am aware that this treatment suffers the same defect, at least before its sketchy last chapters.

Most chapters end with a set of problems. Most of these ask for computations and applications based on the material in the text, some extend the theory and introduce further concepts, some ask the reader to furnish or complete proofs omitted in the text, and some are essay questions which implicitly ask the reader

---

[1] But see R. Brown's book cited in §2 of the suggestions for further reading.

[2] That approach derives Poincaré duality as a consequence of Spanier-Whitehead and Atiyah duality, via the Thom isomorphism for oriented vector bundles.

to seek answers in other sources. Problems marked ∗ are more difficult or more peripheral to the main ideas. Most of these problems are included in the weekly problem sets that are an integral part of the course at Chicago. In fact, doing the problems is the heart of the course. (There are no exams and no grades; students are strongly encouraged to work together, and more work is assigned than a student can reasonably be expected to complete working alone.) *The reader is urged to try most of the problems: this is the way to learn the material.* The lectures focus on the ideas; their assimilation requires more calculational examples and applications than are included in the text.

I have ended with a brief and idiosyncratic guide to the literature for the reader interested in going further in algebraic topology.

These notes have evolved over many years, and I claim no originality for most of the material. In particular, many of the problems, especially in the more classical chapters, are the same as, or are variants of, problems that appear in other texts. Perhaps this is unavoidable: interesting problems that are doable at an early stage of the development are few and far between. I am especially aware of my debts to earlier texts by Massey, Greenberg and Harper, Dold, and Gray.

I am very grateful to John Greenlees for his careful reading and suggestions, especially of the last three chapters. I am also grateful to Igor Kriz for his suggestions and for trying out the book at the University of Michigan. By far my greatest debt, a cumulative one, is to several generations of students, far too numerous to name. They have caught countless infelicities and outright blunders, and they have contributed quite a few of the details. You know who you are. Thank you.

CHAPTER 1

The fundamental group and some of its applications

We introduce algebraic topology with a quick treatment of standard material about the fundamental groups of spaces, embedded in a geodesic proof of the Brouwer fixed point theorem and the fundamental theorem of algebra.

1. What is algebraic topology?

A topological space X is a set in which there is a notion of nearness of points. Precisely, there is given a collection of "open" subsets of X which is closed under finite intersections and arbitrary unions. It suffices to think of metric spaces. In that case, the open sets are the arbitrary unions of finite intersections of neighborhoods $U_\varepsilon(x) = \{y | d(x, y) < \varepsilon\}$.

A function $p : X \longrightarrow Y$ is continuous if it takes nearby points to nearby points. Precisely, $p^{-1}(U)$ is open if U is open. If X and Y are metric spaces, this means that, for any $x \in X$ and $\varepsilon > 0$, there exists $\delta > 0$ such that $p(U_\delta(x)) \subset U_\varepsilon(p(x))$.

Algebraic topology assigns discrete algebraic invariants to topological spaces and continuous maps. More narrowly, one wants the algebra to be invariant with respect to continuous deformations of the topology. Typically, one associates a group $A(X)$ to a space X and a homomorphism $A(p) : A(X) \longrightarrow A(Y)$ to a map $p : X \longrightarrow Y$; one usually writes $A(p) = p_*$.

A "homotopy" $h : p \simeq q$ between maps $p, q : X \longrightarrow Y$ is a continuous map $h : X \times I \longrightarrow Y$ such that $h(x, 0) = p(x)$ and $h(x, 1) = q(x)$, where I is the unit interval $[0, 1]$. We usually want $p_* = q_*$ if $p \simeq q$, or some invariance property close to this.

In oversimplified outline, the way homotopy theory works is roughly this.

1. One defines some algebraic construction A and proves that it is suitably homotopy invariant.
2. One computes A on suitable spaces and maps.
3. One takes the problem to be solved and deforms it to the point that step 2 can be used to solve it.

The further one goes in the subject, the more elaborate become the constructions A and the more horrendous become the relevant calculational techniques. This chapter will give a totally self-contained paradigmatic illustration of the basic philosophy. Our construction A will be the "fundamental group." We will calculate A on the circle S^1 and on some maps from S^1 to itself. We will then use the computation to prove the "Brouwer fixed point theorem" and the "fundamental theorem of algebra."

2. The fundamental group

Let X be a space. Two paths $f, g : I \longrightarrow X$ from x to y are equivalent if they are homotopic through paths from x to y. That is, there must exist a homotopy $h : I \times I \longrightarrow X$ such that

$$h(s,0) = f(s), \quad h(s,1) = g(s), \quad h(0,t) = x, \quad \text{and} \quad h(1,t) = y$$

for all $s, t \in I$. Write $[f]$ for the equivalence class of f. We say that f is a loop if $f(0) = f(1)$. Define $\pi_1(X, x)$ to be the set of equivalence classes of loops that start and end at x.

For paths $f : x \to y$ and $g : y \to z$, define $g \cdot f$ to be the path obtained by traversing first f and then g, going twice as fast on each:

$$(g \cdot f)(s) = \begin{cases} f(2s) & \text{if } 0 \leq s \leq 1/2 \\ g(2s-1) & \text{if } 1/2 \leq s \leq 1. \end{cases}$$

Define f^{-1} to be f traversed the other way around: $f^{-1}(s) = f(1-s)$. Define c_x to the constant loop at x: $c_x(s) = x$. Composition of paths passes to equivalence classes via $[g][f] = [g \cdot f]$. It is easy to check that this is well defined. Moreover, after passage to equivalence classes, this composition becomes associative and unital. It is easy enough to write down explicit formulas for the relevant homotopies. It is more illuminating to draw a picture of the domain squares and to indicate schematically how the homotopies are to behave on it. In the following, we assume given paths

$$f : x \to y, \quad g : y \to z, \quad \text{and} \quad h : z \to w.$$

$$h \cdot (g \cdot f) \simeq (h \cdot g) \cdot f$$

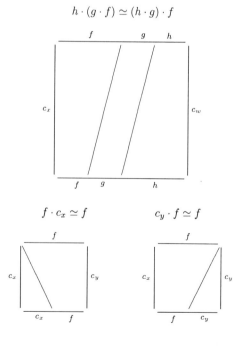

$$f \cdot c_x \simeq f \qquad\qquad c_y \cdot f \simeq f$$

Moreover, $[f^{-1} \cdot f] = [c_x]$ and $[f \cdot f^{-1}] = [c_y]$. For the first, we have the following schematic picture and corresponding formula. In the schematic picture,

$$f_t = f|[0,t] \quad \text{and} \quad f_t^{-1} = f^{-1}|[1-t, 1].$$

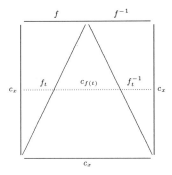

$$h(s,t) = \begin{cases} f(2s) & \text{if } 0 \le s \le t/2 \\ f(t) & \text{if } t/2 \le s \le 1 - t/2 \\ f(2-2s) & \text{if } 1 - t/2 \le s \le 1. \end{cases}$$

We conclude that $\pi_1(X, x)$ is a group with identity element $e = [c_x]$ and inverse elements $[f]^{-1} = [f^{-1}]$. It is called the fundamental group of X, or the first homotopy group of X. There are higher homotopy groups $\pi_n(X, x)$ defined in terms of maps $S^n \longrightarrow X$. We will get to them later.

3. Dependence on the basepoint

For a path $a : x \to y$, define $\gamma[a] : \pi_1(X, x) \longrightarrow \pi_1(X, y)$ by $\gamma[a][f] = [a \cdot f \cdot a^{-1}]$. It is easy to check that $\gamma[a]$ depends only on the equivalence class of a and is a homomorphism of groups. For a path $b : y \to z$, we see that $\gamma[b \cdot a] = \gamma[b] \circ \gamma[a]$. It follows that $\gamma[a]$ is an isomorphism with inverse $\gamma[a^{-1}]$. For a path $b : y \to x$, we have $\gamma[b \cdot a][f] = [b \cdot a][f][(b \cdot a)^{-1}]$. If the group $\pi_1(X, x)$ happens to be Abelian, which may or may not be the case, then this is just $[f]$. By taking $b = (a')^{-1}$ for another path $a' : x \to y$, we see that, when $\pi_1(X, x)$ is Abelian, $\gamma[a]$ is independent of the choice of the path class $[a]$. Thus, in this case, we have a canonical way to identify $\pi_1(X, x)$ with $\pi_1(X, y)$.

4. Homotopy invariance

For a map $p : X \longrightarrow Y$, define $p_* : \pi_1(X, x) \longrightarrow \pi_1(Y, p(x))$ by $p_*[f] = [p \circ f]$, where $p \circ f$ is the composite of p with the loop $f : I \longrightarrow X$. Clearly p_* is a homomorphism. The identity map id : $X \longrightarrow X$ induces the identity homomorphism. For a map $q : Y \longrightarrow Z$, $q_* \circ p_* = (q \circ p)_*$.

Now suppose given two maps $p, q : X \longrightarrow Y$ and a homotopy $h : p \simeq q$. We would like to conclude that $p_* = q_*$, but this doesn't quite make sense because homotopies needn't respect basepoints. However, the homotopy h determines the path $a : p(x) \to q(x)$ specified by $a(t) = h(x, t)$, and the next best thing happens.

PROPOSITION. *The following diagram is commutative:*

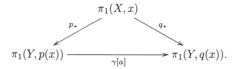

PROOF. Let $f : I \longrightarrow X$ be a loop at x. We must show that $q \circ f$ is equivalent to $a \cdot (p \circ f) \cdot a^{-1}$. It is easy to check that this is equivalent to showing that $c_{p(x)}$ is equivalent to $a^{-1} \cdot (q \circ f)^{-1} \cdot a \cdot (p \circ f)$. Define $j : I \times I \longrightarrow Y$ by $j(s,t) = h(f(s),t)$. Then

$$j(s,0) = (p \circ f)(s), \quad j(s,1) = (q \circ f)(s), \quad \text{and} \quad j(0,t) = a(t) = j(1,t).$$

Note that $j(0,0) = p(x)$. Schematically, on the boundary of the square, j is

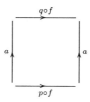

Thus, going counterclockwise around the boundary starting at $(0,0)$, we traverse $a^{-1} \cdot (q \circ f)^{-1} \cdot a \cdot (p \circ f)$. The map j induces a homotopy through loops between this composite and $c_{p(x)}$. Explicitly, a homotopy k is given by $k(s,t) = j(r_t(s))$, where $r_t : I \longrightarrow I \times I$ maps successive quarter intervals linearly onto the edges of the bottom left subsquare of $I \times I$ with edges of length t, starting at $(0,0)$:

□

5. Calculations: $\pi_1(\mathbb{R}) = 0$ and $\pi_1(S^1) = \mathbb{Z}$

Our first calculation is rather trivial. We take the origin 0 as a convenient basepoint for the real line $\mathbb{R}$.

LEMMA. $\pi_1(\mathbb{R}, 0) = 0$.

PROOF. Define $k : \mathbb{R} \times I \longrightarrow \mathbb{R}$ by $k(s,t) = (1-t)s$. Then k is a homotopy from the identity to the constant map at 0. For a loop $f : I \longrightarrow \mathbb{R}$ at 0, define $h(s,t) = k(f(s),t)$. The homotopy h shows that f is equivalent to c_0. □

Consider the circle S^1 to be the set of complex numbers $x = y + iz$ of norm 1, $y^2 + z^2 = 1$. Observe that S^1 is a group under multiplication of complex numbers. It is a topological group: multiplication is a continuous function. We take the identity element 1 as a convenient basepoint for S^1.

THEOREM. $\pi_1(S^1, 1) \cong \mathbb{Z}$.

PROOF. For each integer n, define a loop f_n in S^1 by $f_n(s) = e^{2\pi i n s}$. This is the composite of the map $I \longrightarrow S^1$ that sends s to $e^{2\pi i s}$ and the nth power map on S^1; if we identify the boundary points 0 and 1 of I, then the first map induces the evident identification of $I/\partial I$ with S^1. It is easy to check that $[f_m][f_n] = [f_{m+n}]$, and we define a homomorphism $i : \mathbb{Z} \longrightarrow \pi_1(S^1, 1)$ by $i(n) = [f_n]$. We claim that i is an isomorphism. The idea of the proof is to use the fact that, locally, S^1 looks just like $\mathbb{R}$.

Define $p : \mathbb{R} \longrightarrow S^1$ by $p(s) = e^{2\pi i s}$. Observe that p wraps each interval $[n, n+1]$ around the circle, starting at 1 and going counterclockwise. Since the exponential function converts addition to multiplication, we easily check that $f_n = p \circ \tilde{f}_n$, where $\tilde{f}_n$ is the path in $\mathbb{R}$ defined by $\tilde{f}_n(s) = sn$.

This lifting of paths works generally. For any path $f : I \longrightarrow S^1$ with $f(0) = 1$, there is a unique path $\tilde{f} : I \longrightarrow \mathbb{R}$ such that $\tilde{f}(0) = 0$ and $p \circ \tilde{f} = f$. To see this, observe that the inverse image in $\mathbb{R}$ of any small connected neighborhood in S^1 is a disjoint union of a copy of that neighborhood contained in each interval $(r+n, r+n+1)$ for some $r \in [0,1)$. Using the fact that I is compact, we see that we can subdivide I into finitely many closed subintervals such that f carries each subinterval into one of these small connected neighborhoods. Now, proceeding subinterval by subinterval, we obtain the required unique lifting of f by observing that the lifting on each subinterval is uniquely determined by the lifting of its initial point.

Define a function $j : \pi_1(S^1, 1) \longrightarrow \mathbb{Z}$ by $j[f] = \tilde{f}(1)$, the endpoint of the lifted path. This is an integer since $p(\tilde{f}(1)) = 1$. We must show that this integer is independent of the choice of f in its path class $[f]$. In fact, if we have a homotopy $h : f \simeq g$ through loops at 1, then the homotopy lifts uniquely to a homotopy $\tilde{h} : I \times I \longrightarrow \mathbb{R}$ such that $\tilde{h}(0,0) = 0$ and $p \circ \tilde{h} = h$. The argument is just the same as for $\tilde{f}$: we use the fact that $I \times I$ is compact to subdivide it into finitely many subsquares such that h carries each into a small connected neighborhood in S^1. We then construct the unique lift $\tilde{h}$ by proceeding subsquare by subsquare, starting at the lower left, say, and proceeding upward one row of squares at a time. By the uniqueness of lifts of paths, which works just as well for paths with any starting point, $c(t) = \tilde{h}(0,t)$ and $d(t) = \tilde{h}(1,t)$ specify constant paths since $h(0,t) = 1$ and $h(1,t) = 1$ for all t. Clearly c is constant at 0, so, again by the uniqueness of lifts of paths, we must have

$$\tilde{f}(s) = \tilde{h}(s, 0) \quad \text{and} \quad \tilde{g}(s) = \tilde{h}(s, 1).$$

But then our second constant path d starts at $\tilde{f}(1)$ and ends at $\tilde{g}(1)$.

Since $j[f_n] = n$ by our explicit formula for $\tilde{f}_n$, the composite $j \circ i : \mathbb{Z} \longrightarrow \mathbb{Z}$ is the identity. It suffices to check that the function j is one-to-one, since then both i and j will be one-to-one and onto. Thus suppose that $j[f] = j[g]$. This means that $\tilde{f}(1) = \tilde{g}(1)$. Therefore $\tilde{g}^{-1} \cdot \tilde{f}$ is a loop at 0 in $\mathbb{R}$. By the lemma, $[\tilde{g}^{-1} \cdot \tilde{f}] = [c_0]$. It follows upon application of p_* that

$$[g^{-1}][f] = [g^{-1} \cdot f] = [c_1].$$

Therefore $[f] = [g]$ and the proof is complete. □

6. The Brouwer fixed point theorem

Let D^2 be the unit disk $\{y + iz | y^2 + z^2 \leq 1\}$. Its boundary is S^1, and we let $i : S^1 \longrightarrow D^2$ be the inclusion. Exactly as for $\mathbb{R}$, we see that $\pi_1(D^2) = 0$ for any choice of basepoint.

PROPOSITION. *There is no continuous map* $r : D^2 \longrightarrow S^1$ *such that* $r \circ i = \mathrm{id}$.

PROOF. If there were such a map r, then the composite homomorphism

$$\pi_1(S^1, 1) \xrightarrow{i_*} \pi_1(D^2, 1) \xrightarrow{r_*} \pi_1(S^1, 1)$$

would be the identity. Since the identity homomorphism of $\mathbb{Z}$ does not factor through the zero group, this is impossible. □

THEOREM (Brouwer fixed point theorem). *Any continuous map*

$$f : D^2 \longrightarrow D^2$$

has a fixed point.

PROOF. Suppose that $f(x) \neq x$ for all x. Define $r(x) \in S^1$ to be the intersection with S^1 of the ray that starts at $f(x)$ and passes through x. Certainly $r(x) = x$ if $x \in S^1$. By writing an equation for r in terms of f, we see that r is continuous. This contradicts the proposition. □

7. The fundamental theorem of algebra

Let $\iota \in \pi_1(S^1, 1)$ be a generator. For a map $f : S^1 \longrightarrow S^1$, define an integer $\deg(f)$ by letting the composite

$$\pi_1(S^1, 1) \xrightarrow{f_*} \pi_1(S^1, f(1)) \xrightarrow{\gamma[a]} \pi_1(S^1, 1)$$

send ι to $\deg(f)\iota$. Here a is any path $f(1) \to 1$; $\gamma[a]$ is independent of the choice of $[a]$ since $\pi_1(S^1, 1)$ is Abelian. If $f \simeq g$, then $\deg(f) = \deg(g)$ by our homotopy invariance diagram and this independence of the choice of path. Conversely, our calculation of $\pi_1(S^1, 1)$ implies that if $\deg(f) = \deg(g)$, then $f \simeq g$, but we will not need that for the moment. It is clear that $\deg(f) = 0$ if f is the constant map at some point. It is also clear that if $f_n(x) = x^n$, then $\deg(f_n) = n$: we built that fact into our proof that $\pi_1(S^1, 1) = \mathbb{Z}$.

THEOREM (Fundamental theorem of algebra). *Let*

$$f(x) = x^n + c_1 x^{n-1} + \cdots + c_{n-1} x + c_n$$

be a polynomial with complex coefficients c_i, *where* $n > 0$. *Then there is a complex number* x *such that* $f(x) = 0$. *Therefore there are* n *such complex numbers (counted with multiplicities).*

PROOF. Using $f(x)/(x-c)$ for a root c, we see that the last statement will follow by induction from the first. We may as well assume that $f(x) \neq 0$ for $x \in S^1$. This allows us to define $\hat{f} : S^1 \longrightarrow S^1$ by $\hat{f}(x) = f(x)/|f(x)|$. We proceed to calculate $\deg(\hat{f})$. Suppose first that $f(x) \neq 0$ for all x such that $|x| \leq 1$. This allows us to define $h : S^1 \times I \longrightarrow S^1$ by $h(x,t) = f(tx)/|f(tx)|$. Then h is a homotopy from the constant map at $f(0)/|f(0)|$ to $\hat{f}$, and we conclude that $\deg(\hat{f}) = 0$. Suppose next

that $f(x) \neq 0$ for all x such that $|x| \geq 1$. This allows us to define $j : S^1 \times I \longrightarrow S^1$ by $j(x,t) = k(x,t)/|k(x,t)|$, where
$$k(x,t) = t^n f(x/t) = x^n + t(c_1 x^{n-1} + tc_2 x^{n-2} + \cdots + t^{n-1} c_n).$$
Then j is a homotopy from f_n to $\hat{f}$, and we conclude that $\deg(\hat{f}) = n$. One of our suppositions had better be false! □

It is to be emphasized how technically simple this is, requiring nothing remotely as deep as complex analysis. Nevertheless, homotopical proofs like this are relatively recent. Adequate language, elementary as it is, was not developed until the 1930s.

PROBLEMS

1. Let p be a polynomial function on $\mathbb{C}$ which has no root on S^1. Show that the number of roots of $p(z) = 0$ with $|z| < 1$ is the degree of the map $\hat{p} : S^1 \longrightarrow S^1$ specified by $\hat{p}(z) = p(z)/|p(z)|$.
2. Show that any map $f : S^1 \longrightarrow S^1$ such that $\deg(f) \neq 1$ has a fixed point.
3. Let G be a topological group and take its identity element e as its basepoint. Define the pointwise product of loops α and β by $(\alpha\beta)(t) = \alpha(t)\beta(t)$. Prove that $\alpha\beta$ is equivalent to the composition of paths $\beta \cdot \alpha$. Deduce that $\pi_1(G, e)$ is Abelian.

CHAPTER 2

Categorical language and the van Kampen theorem

We introduce categorical language and ideas and use them to prove the van Kampen theorem. This method of computing fundamental groups illustrates the general principle that calculations in algebraic topology usually work by piecing together a few pivotal examples by means of general constructions or procedures.

1. Categories

Algebraic topology concerns mappings from topology to algebra. Category theory gives us a language to express this. We just record the basic terminology, without being overly pedantic about it.

A category $\mathscr{C}$ consists of a collection of objects, a set $\mathscr{C}(A, B)$ of morphisms (also called maps) between any two objects, an identity morphism $\mathrm{id}_A \in \mathscr{C}(A, A)$ for each object A (usually abbreviated id), and a composition law

$$\circ : \mathscr{C}(B, C) \times \mathscr{C}(A, B) \longrightarrow \mathscr{C}(A, C)$$

for each triple of objects A, B, C. Composition must be associative, and identity morphisms must behave as their names dictate:

$$h \circ (g \circ f) = (h \circ g) \circ f, \quad \mathrm{id} \circ f = f, \quad \text{and} \quad f \circ \mathrm{id} = f$$

whenever the specified composites are defined. A category is "small" if it has a set of objects.

We have the category $\mathscr{S}$ of sets and functions, the category $\mathscr{U}$ of topological spaces and continuous functions, the category $\mathscr{G}$ of groups and homomorphisms, the category $\mathscr{A}b$ of Abelian groups and homomorphisms, and so on.

2. Functors

A functor $F : \mathscr{C} \longrightarrow \mathscr{D}$ is a map of categories. It assigns an object $F(A)$ of $\mathscr{D}$ to each object A of $\mathscr{C}$ and a morphism $F(f) : F(A) \longrightarrow F(B)$ of $\mathscr{D}$ to each morphism $f : A \longrightarrow B$ of $\mathscr{C}$ in such a way that

$$F(\mathrm{id}_A) = \mathrm{id}_{F(A)} \quad \text{and} \quad F(g \circ f) = F(g) \circ F(f).$$

More precisely, this is a covariant functor. A contravariant functor F reverses the direction of arrows, so that F sends $f : A \longrightarrow B$ to $F(f) : F(B) \longrightarrow F(A)$ and satisfies $F(g \circ f) = F(f) \circ F(g)$. A category $\mathscr{C}$ has an opposite category $\mathscr{C}^{op}$ with the same objects and with $\mathscr{C}^{op}(A, B) = \mathscr{C}(B, A)$. A contravariant functor $F : \mathscr{C} \longrightarrow \mathscr{D}$ is just a covariant functor $\mathscr{C}^{op} \longrightarrow \mathscr{D}$.

For example, we have forgetful functors from spaces to sets and from Abelian groups to sets, and we have the free Abelian group functor from sets to Abelian groups.

3. Natural transformations

A natural transformation $\alpha : F \longrightarrow G$ between functors $\mathscr{C} \longrightarrow \mathscr{D}$ is a map of functors. It consists of a morphism $\alpha_A : F(A) \longrightarrow G(A)$ for each object A of $\mathscr{C}$ such that the following diagram commutes for each morphism $f : A \longrightarrow B$ of $\mathscr{C}$:

$$\begin{array}{ccc} F(A) & \xrightarrow{F(f)} & F(B) \\ \alpha_A \downarrow & & \downarrow \alpha_B \\ G(A) & \xrightarrow{G(f)} & G(B). \end{array}$$

Intuitively, the maps α_A are defined in the same way for every A.

For example, if $F : \mathscr{S} \longrightarrow \mathscr{A}b$ is the functor that sends a set to the free Abelian group that it generates and $U : \mathscr{A}b \longrightarrow \mathscr{S}$ is the forgetful functor that sends an Abelian group to its underlying set, then we have a natural inclusion of sets $S \longrightarrow UF(S)$. The functors F and U are left adjoint and right adjoint to each other, in the sense that we have a natural isomorphism

$$\mathscr{A}b(F(S), A) \cong \mathscr{S}(S, U(A))$$

for a set S and an Abelian group A. This just expresses the "universal property" of free objects: a map of sets $S \longrightarrow U(A)$ extends uniquely to a homomorphism of groups $F(S) \longrightarrow A$. Although we won't bother with a formal definition, the notion of an adjoint pair of functors will play an important role later on.

Two categories $\mathscr{C}$ and $\mathscr{D}$ are equivalent if there are functors $F : \mathscr{C} \longrightarrow \mathscr{D}$ and $G : \mathscr{D} \longrightarrow \mathscr{C}$ and natural isomorphisms $FG \longrightarrow \mathrm{Id}$ and $GF \longrightarrow \mathrm{Id}$, where the Id are the respective identity functors.

4. Homotopy categories and homotopy equivalences

Let $\mathscr{T}$ be the category of spaces X with a chosen basepoint $x \in X$; its morphisms are continuous maps $X \longrightarrow Y$ that carry the basepoint of X to the basepoint of Y. The fundamental group specifies a functor $\mathscr{T} \longrightarrow \mathscr{G}$, where $\mathscr{G}$ is the category of groups and homomorphisms.

When we have a (suitable) relation of homotopy between maps in a category $\mathscr{C}$, we define the homotopy category $h\mathscr{C}$ to be the category with the same objects as $\mathscr{C}$ but with morphisms the homotopy classes of maps. We have the homotopy category $h\mathscr{U}$ of unbased spaces. On $\mathscr{T}$, we require homotopies to map basepoint to basepoint at all times t, and we obtain the homotopy category $h\mathscr{T}$ of based spaces. The fundamental group is a homotopy invariant functor on $\mathscr{T}$, in the sense that it factors through a functor $h\mathscr{T} \longrightarrow \mathscr{G}$.

A homotopy equivalence in $\mathscr{U}$ is an isomorphism in $h\mathscr{U}$. Less mysteriously, a map $f : X \longrightarrow Y$ is a homotopy equivalence if there is a map $g : Y \longrightarrow X$ such that both $g \circ f \simeq \mathrm{id}$ and $f \circ g \simeq \mathrm{id}$. Working in $\mathscr{T}$, we obtain the analogous notion of a based homotopy equivalence. Functors carry isomorphisms to isomorphisms, so we see that a based homotopy equivalence induces an isomorphism of fundamental groups. The same is true, less obviously, for unbased homotopy equivalences.

PROPOSITION. *If* $f : X \longrightarrow Y$ *is a homotopy equivalence, then*

$$f_* : \pi_1(X, x) \longrightarrow \pi_1(Y, f(x))$$

is an isomorphism for all $x \in X$.

PROOF. Let $g : Y \longrightarrow X$ be a homotopy inverse of f. By our homotopy invariance diagram, we see that the composites

$$\pi_1(X, x) \xrightarrow{f_*} \pi_1(Y, f(x)) \xrightarrow{g_*} \pi_1(X, (g \circ f)(x))$$

and

$$\pi_1(Y, y) \xrightarrow{g_*} \pi_1(X, g(y)) \xrightarrow{f_*} \pi_1(Y, (f \circ g)(y))$$

are isomorphisms determined by paths between basepoints given by chosen homotopies $g \circ f \simeq \mathrm{id}$ and $f \circ g \simeq \mathrm{id}$. Therefore, in each displayed composite, the first map is a monomorphism and the second is an epimorphism. Taking $y = f(x)$ in the second composite, we see that the second map in the first composite is an isomorphism. Therefore so is the first map. □

A space X is said to be contractible if it is homotopy equivalent to a point.

COROLLARY. *The fundamental group of a contractible space is zero.*

5. The fundamental groupoid

While algebraic topologists often concentrate on connected spaces with chosen basepoints, it is valuable to have a way of studying fundamental groups that does not require such choices. For this purpose, we define the "fundamental groupoid" $\Pi(X)$ of a space X to be the category whose objects are the points of X and whose morphisms $x \longrightarrow y$ are the equivalence classes of paths from x to y. Thus the set of endomorphisms of the object x is exactly the fundamental group $\pi_1(X, x)$.

The term "groupoid" is used for a category all morphisms of which are isomorphisms. The idea is that a group may be viewed as a groupoid with a single object. Taking morphisms to be functors, we obtain the category $\mathscr{GP}$ of groupoids. Then we may view Π as a functor $\mathscr{U} \longrightarrow \mathscr{GP}$.

There is a useful notion of a skeleton $sk\mathscr{C}$ of a category $\mathscr{C}$. This is a "full" subcategory with one object from each isomorphism class of objects of $\mathscr{C}$, "full" meaning that the morphisms between two objects of $sk\mathscr{C}$ are all of the morphisms between these objects in $\mathscr{C}$. The inclusion functor $J : sk\mathscr{C} \longrightarrow \mathscr{C}$ is an equivalence of categories. An inverse functor $F : \mathscr{C} \longrightarrow sk\mathscr{C}$ is obtained by letting $F(A)$ be the unique object in $sk\mathscr{C}$ that is isomorphic to A, choosing an isomorphism $\alpha_A : A \longrightarrow F(A)$, and defining $F(f) = \alpha_B \circ f \circ \alpha_A^{-1} : F(A) \longrightarrow F(B)$ for a morphism $f : A \longrightarrow B$ in $\mathscr{C}$. We choose α to be the identity morphism if A is in $sk\mathscr{C}$, and then $FJ = \mathrm{Id}$; the α_A specify a natural isomorphism $\alpha : \mathrm{Id} \longrightarrow JF$.

A category $\mathscr{C}$ is said to be connected if any two of its objects can be connected by a sequence of morphisms. For example, a sequence $A \longleftarrow B \longrightarrow C$ connects A to C, although there need be no morphism $A \longrightarrow C$. However, a groupoid $\mathscr{C}$ is connected if and only if any two of its objects are isomorphic. The group of endomorphisms of any object C is then a skeleton of $\mathscr{C}$. Therefore the previous paragraph specializes to give the following relationship between the fundamental group and the fundamental groupoid of a path connected space X.

PROPOSITION. *Let X be a path connected space. For each point $x \in X$, the inclusion $\pi_1(X, x) \longrightarrow \Pi(X)$ is an equivalence of categories.*

PROOF. We are regarding $\pi_1(X, x)$ as a category with a single object x, and it is a skeleton of $\Pi(X)$. □

6. Limits and colimits

Let $\mathscr{D}$ be a small category and let $\mathscr{C}$ be any category. A $\mathscr{D}$-shaped diagram in $\mathscr{C}$ is a functor $F : \mathscr{D} \longrightarrow \mathscr{C}$. A morphism $F \longrightarrow F'$ of $\mathscr{D}$-shaped diagrams is a natural transformation, and we have the category $\mathscr{D}[\mathscr{C}]$ of $\mathscr{D}$-shaped diagrams in $\mathscr{C}$. Any object C of $\mathscr{C}$ determines the constant diagram $\underline{C}$ that sends each object of $\mathscr{D}$ to C and sends each morphism of $\mathscr{D}$ to the identity morphism of C.

The colimit, $\operatorname{colim} F$, of a $\mathscr{D}$-shaped diagram F is an object of $\mathscr{C}$ together with a morphism of diagrams $\iota : F \longrightarrow \underline{\operatorname{colim} F}$ that is initial among all such morphisms. This means that if $\eta : F \longrightarrow \underline{A}$ is a morphism of diagrams, then there is a unique map $\tilde{\eta} : \operatorname{colim} F \longrightarrow A$ in $\mathscr{C}$ such that $\tilde{\eta} \circ \iota = \eta$. Diagrammatically, this property is expressed by the assertion that, for each map $d : D \longrightarrow D'$ in $\mathscr{D}$, we have a commutative diagram

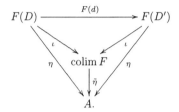

The limit of F is defined by reversing arrows: it is an object $\lim F$ of $\mathscr{C}$ together with a morphism of diagrams $\pi : \underline{\lim F} \longrightarrow F$ that is terminal among all such morphisms. This means that if $\varepsilon : \underline{A} \longrightarrow F$ is a morphism of diagrams, then there is a unique map $\tilde{\varepsilon} : A \longrightarrow \lim F$ in $\mathscr{C}$ such that $\pi \circ \tilde{\varepsilon} = \varepsilon$. Diagrammatically, this property is expressed by the assertion that, for each map $d : D \longrightarrow D'$ in $\mathscr{D}$, we have a commutative diagram

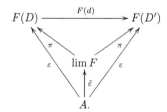

If $\mathscr{D}$ is a set regarded as a discrete category (only identity morphisms), then colimits and limits indexed on $\mathscr{D}$ are coproducts and products indexed on the set $\mathscr{D}$. Coproducts are disjoint unions in $\mathscr{S}$ or $\mathscr{U}$, wedges (or one-point unions) in $\mathscr{T}$, free products in $\mathscr{G}$, and direct sums in $\mathscr{A}b$. Products are Cartesian products in all of these categories; more precisely, they are Cartesian products of underlying sets, with additional structure. If $\mathscr{D}$ is the category displayed schematically as

$$e \longleftarrow d \longrightarrow f \quad \text{or} \quad d \rightrightarrows d',$$

where we have displayed all objects and all non-identity morphisms, then the colimits indexed on $\mathscr{D}$ are called pushouts or coequalizers, respectively. Similarly, if $\mathscr{D}$ is displayed schematically as

$$e \longrightarrow d \longleftarrow f \quad \text{or} \quad d \rightrightarrows d',$$

then the limits indexed on $\mathscr{D}$ are called pullbacks or equalizers, respectively.

A given category may or may not have all colimits, and it may have some but not others. A category is said to be cocomplete if it has all colimits, complete if it has all limits. The categories $\mathscr{S}$, $\mathscr{U}$, $\mathscr{T}$, $\mathscr{G}$, and $\mathscr{A}b$ are complete and cocomplete. If a category has coproducts and coequalizers, then it is cocomplete, and similarly for completeness. The proof is a worthwhile exercise.

7. The van Kampen theorem

The following is a modern dress treatment of the van Kampen theorem. I should admit that, in lecture, it may make more sense not to introduce the fundamental groupoid and to go directly to the fundamental group statement. The direct proof is shorter, but not as conceptual. However, as far as I know, the deduction of the fundamental group version of the van Kampen theorem from the fundamental groupoid version does not appear in the literature in full generality. The proof well illustrates how to manipulate colimits formally. We have used the van Kampen theorem as an excuse to introduce some basic categorical language, and we shall use that language heavily in our treatment of covering spaces in the next chapter.

THEOREM (van Kampen). *Let $\mathscr{O} = \{U\}$ be a cover of a space X by path connected open subsets such that the intersection of finitely many subsets in $\mathscr{O}$ is again in $\mathscr{O}$. Regard $\mathscr{O}$ as a category whose morphisms are the inclusions of subsets and observe that the functor Π, restricted to the spaces and maps in $\mathscr{O}$, gives a diagram*

$$\Pi|\mathscr{O} : \mathscr{O} \longrightarrow \mathscr{GP}$$

of groupoids. The groupoid $\Pi(X)$ is the colimit of this diagram. In symbols,

$$\Pi(X) \cong \mathrm{colim}_{U \in \mathscr{O}} \Pi(U).$$

PROOF. We must verify the universal property. For a groupoid $\mathscr{C}$ and a map $\eta : \Pi|\mathscr{O} \longrightarrow \mathscr{C}$ of $\mathscr{O}$-shaped diagrams of groupoids, we must construct a map $\tilde{\eta} : \Pi(X) \longrightarrow \mathscr{C}$ of groupoids that restricts to η_U on $\Pi(U)$ for each $U \in \mathscr{O}$. On objects, that is on points of X, we must define $\tilde{\eta}(x) = \eta_U(x)$ for $x \in U$. This is independent of the choice of U since $\mathscr{O}$ is closed under finite intersections. If a path $f : x \to y$ lies entirely in a particular U, then we must define $\tilde{\eta}[f] = \eta([f])$. Again, since $\mathscr{O}$ is closed under finite intersections, this specification is independent of the choice of U if f lies entirely in more than one U. Any path f is the composite of finitely many paths f_i, each of which does lie in a single U, and we must define $\tilde{\eta}[f]$ to be the composite of the $\tilde{\eta}[f_i]$. Clearly this specification will give the required unique map $\tilde{\eta}$, provided that $\tilde{\eta}$ so specified is in fact well defined. Thus suppose that f is equivalent to g. The equivalence is given by a homotopy $h : f \simeq g$ through paths $x \to y$. We may subdivide the square $I \times I$ into subsquares, each of which is mapped into one of the U. We may choose the subdivision so that the resulting subdivision of $I \times \{0\}$ refines the subdivision used to decompose f as the composite of paths f_i, and similarly for g and the resulting subdivision of $I \times \{1\}$. We see that the relation $[f] = [g]$ in $\Pi(X)$ is a consequence of a finite number of relations, each of which holds in one of the $\Pi(U)$. Therefore $\tilde{\eta}([f]) = \tilde{\eta}([g])$. This verifies the universal property and proves the theorem. □

The fundamental group version of the van Kampen theorem "follows formally." That is, it is an essentially categorical consequence of the version just proved. Arguments like this are sometimes called proof by categorical nonsense.

THEOREM (van Kampen). *Let X be path connected and choose a basepoint $x \in X$. Let $\mathscr{O}$ be a cover of X by path connected open subsets such that the intersection of finitely many subsets in $\mathscr{O}$ is again in $\mathscr{O}$ and x is in each $U \in \mathscr{O}$. Regard $\mathscr{O}$ as a category whose morphisms are the inclusions of subsets and observe that the functor $\pi_1(-, x)$, restricted to the spaces and maps in $\mathscr{O}$, gives a diagram*

$$\pi_1 | \mathscr{O} : \mathscr{O} \longrightarrow \mathscr{G}$$

of groups. The group $\pi_1(X, x)$ is the colimit of this diagram. In symbols,

$$\pi_1(X, x) \cong \mathrm{colim}_{U \in \mathscr{O}} \, \pi_1(U, x).$$

We proceed in two steps.

LEMMA. *The van Kampen theorem holds when the cover $\mathscr{O}$ is finite.*

PROOF. This step is based on the nonsense above about skeleta of categories. We must verify the universal property, this time in the category of groups. For a group G and a map $\eta : \pi_1 | \mathscr{O} \longrightarrow \underline{G}$ of $\mathscr{O}$-shaped diagrams of groups, we must show that there is a unique homomorphism $\tilde{\eta} : \pi_1(X, x) \longrightarrow G$ that restricts to η_U on $\pi_1(U, x)$. Remember that we think of a group as a groupoid with a single object and with the elements of the group as the morphisms. The inclusion of categories $J : \pi_1(X, x) \longrightarrow \Pi(X)$ is an equivalence. An inverse equivalence $F : \Pi(X) \longrightarrow \pi_1(X, x)$ is determined by a choice of path classes $x \longrightarrow y$ for $y \in X$; we choose c_x when $y = x$ and so ensure that $F \circ J = \mathrm{Id}$. Because the cover $\mathscr{O}$ is finite and closed under finite intersections, we can choose our paths inductively so that the path $x \longrightarrow y$ lies entirely in U whenever y is in U. This ensures that the chosen paths determine compatible inverse equivalences $F_U : \Pi(U) \longrightarrow \pi_1(U, x)$ to the inclusions $J_U : \pi_1(U, x) \longrightarrow \Pi(U)$. Thus the functors

$$\Pi(U) \xrightarrow{F_U} \pi_1(U, x) \xrightarrow{\eta_U} G$$

specify an $\mathscr{O}$-shaped diagram of groupoids $\Pi | \mathscr{O} \longrightarrow \underline{G}$. By the fundamental groupoid version of the van Kampen theorem, there is a unique map of groupoids

$$\xi : \Pi(X) \longrightarrow G$$

that restricts to $\eta_U \circ F_U$ on $\Pi(U)$ for each U. The composite

$$\pi_1(X, x) \xrightarrow{J} \Pi(X) \xrightarrow{\xi} G$$

is the required homomorphism $\tilde{\eta}$. It restricts to η_U on $\pi_1(U, x)$ by a little "diagram chase" and the fact that $F_U \circ J_U = \mathrm{Id}$. It is unique because ξ is unique. In fact, if we are given $\tilde{\eta} : \pi_1(X, x) \longrightarrow G$ that restricts to η_U on each $\pi_1(U, x)$, then $\tilde{\eta} \circ F : \Pi(X) \longrightarrow G$ restricts to $\eta_U \circ F_U$ on each $\Pi(U)$; therefore $\xi = \tilde{\eta} \circ F$ and thus $\xi \circ J = \tilde{\eta}$. □

PROOF OF THE VAN KAMPEN THEOREM. We deduce the general case from the case just proved. Let $\mathscr{F}$ be the set of those finite subsets of the cover $\mathscr{O}$ that are closed under finite intersection. For $\mathscr{S} \in \mathscr{F}$, let $U_{\mathscr{S}}$ be the union of the U in $\mathscr{S}$. Then $\mathscr{S}$ is a cover of $U_{\mathscr{S}}$ to which the lemma applies. Thus

$$\mathrm{colim}_{U \in \mathscr{S}} \, \pi_1(U, x) \cong \pi_1(U_{\mathscr{S}}, x).$$

Regard $\mathscr{F}$ as a category with a morphism $\mathscr{S} \longrightarrow \mathscr{T}$ whenever $U_{\mathscr{S}} \subset U_{\mathscr{T}}$. We claim first that

$$\mathrm{colim}_{\mathscr{S} \in \mathscr{F}} \, \pi_1(U_{\mathscr{S}}, x) \cong \pi_1(X, x).$$

In fact, by the usual subdivision argument, any loop $I \longrightarrow X$ and any equivalence $h : I \times I \longrightarrow X$ between loops has image in some $U_{\mathscr{S}}$. This implies directly that $\pi_1(X, x)$, together with the homomorphisms $\pi_1(U_{\mathscr{S}}, x) \longrightarrow \pi_1(X, x)$, has the universal property that characterizes the claimed colimit. We claim next that

$$\operatorname{colim}_{U \in \mathscr{O}} \pi_1(U, x) \cong \operatorname{colim}_{\mathscr{S} \in \mathscr{F}} \pi_1(U_{\mathscr{S}}, x),$$

and this will complete the proof. Substituting in the colimit on the right, we have

$$\operatorname{colim}_{\mathscr{S} \in \mathscr{F}} \pi_1(U_{\mathscr{S}}, x) \cong \operatorname{colim}_{\mathscr{S} \in \mathscr{F}} \operatorname{colim}_{U \in \mathscr{S}} \pi_1(U, x).$$

By a comparison of universal properties, this iterated colimit is isomorphic to the single colimit

$$\operatorname{colim}_{(U, \mathscr{S}) \in (\mathscr{O}, \mathscr{F})} \pi_1(U, x).$$

Here the indexing category $(\mathscr{O}, \mathscr{F})$ has objects the pairs $(U, \mathscr{S})$ with $U \in \mathscr{S}$; there is a morphism $(U, \mathscr{S}) \longrightarrow (V, \mathscr{T})$ whenever both $U \subset V$ and $U_{\mathscr{S}} \subset U_{\mathscr{T}}$. A moment's reflection on the relevant universal properties should convince the reader of the claimed identification of colimits: the system on the right differs from the system on the left only in that the homomorphisms $\pi_1(U, x) \longrightarrow \pi_1(V, x)$ occur many times in the system on the right, each appearance making the same contribution to the colimit. If we assume known a priori that colimits of groups exist, we can formalize this as follows. We have a functor $\mathscr{O} \longrightarrow \mathscr{F}$ that sends U to the singleton set $\{U\}$ and thus a functor $\mathscr{O} \longrightarrow (\mathscr{O}, \mathscr{F})$ that sends U to $(U, \{U\})$. The functor $\pi_1(-, x): \mathscr{O} \longrightarrow \mathscr{G}$ factors through $(\mathscr{O}, \mathscr{F})$, hence we have an induced map of colimits

$$\operatorname{colim}_{U \in \mathscr{O}} \pi_1(U, x) \longrightarrow \operatorname{colim}_{(U, \mathscr{S}) \in (\mathscr{O}, \mathscr{F})} \pi_1(U, x).$$

Projection to the first coordinate gives a functor $(\mathscr{O}, \mathscr{F}) \longrightarrow \mathscr{O}$. Its composite with $\pi_1(-, x): \mathscr{O} \longrightarrow \mathscr{G}$ defines the colimit on the right, hence we have an induced map of colimits

$$\operatorname{colim}_{(U, \mathscr{S}) \in (\mathscr{O}, \mathscr{F})} \pi_1(U, x) \longrightarrow \operatorname{colim}_{U \in \mathscr{O}} \pi_1(U, x).$$

These maps are inverse isomorphisms. $\square$

8. Examples of the van Kampen theorem

So far, we have only computed the fundamental groups of the circle and of contractible spaces. The van Kampen theorem lets us extend these calculations. We now drop notation for the basepoint, writing $\pi_1(X)$ instead of $\pi_1(X, x)$.

PROPOSITION. *Let X be the wedge of a set of path connected based spaces X_i, each of which contains a contractible neighborhood V_i of its basepoint. Then $\pi_1(X)$ is the coproduct (= free product) of the groups $\pi_1(X_i)$.*

PROOF. Let U_i be the union of X_i and the V_j for $j \neq i$. We apply the van Kampen theorem with $\mathscr{O}$ taken to be the U_i and their finite intersections. Since any intersection of two or more of the U_i is contractible, the intersections make no contribution to the colimit and the conclusion follows. $\square$

COROLLARY. *The fundamental group of a wedge of circles is a free group with one generator for each circle.*

Any compact surface is homeomorphic to a sphere, or to a connected sum of tori $T^2 = S^1 \times S^1$, or to a connected sum of projective planes $\mathbb{R}P^2 = S^2/\mathbb{Z}_2$ (where we write $\mathbb{Z}_2 = \mathbb{Z}/2\mathbb{Z}$). We shall see shortly that $\pi_1(\mathbb{R}P^2) = \mathbb{Z}_2$. We also have the following observation, which is immediate from the universal property of products. Using this information, it is an exercise to compute the fundamental group of any compact surface from the van Kampen theorem.

LEMMA. *For based spaces X and Y, $\pi_1(X \times Y) \cong \pi_1(X) \times \pi_1(Y)$.*

We shall later use the following application of the van Kampen theorem to prove that any group is the fundamental group of some space. We need a definition.

DEFINITION. A space X is said to be simply connected if it is path connected and satisfies $\pi_1(X) = 0$.

PROPOSITION. *Let $X = U \cup V$, where U, V, and $U \cap V$ are path connected open neighborhoods of the basepoint of X and V is simply connected. Then $\pi_1(U) \longrightarrow \pi_1(X)$ is an epimorphism whose kernel is the smallest normal subgroup of $\pi_1(U)$ that contains the image of $\pi_1(U \cap V)$.*

PROOF. Let N be the cited kernel and consider the diagram

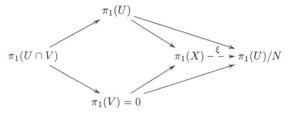

The universal property gives rise to the map ξ, and ξ is an isomorphism since, by an easy algebraic inspection, $\pi_1(U)/N$ is the pushout in the category of groups of the homomorphisms $\pi_1(U \cap V) \longrightarrow \pi_1(U)$ and $\pi_1(U \cap V) \longrightarrow 0$. □

PROBLEMS

1. Compute the fundamental group of the two-holed torus (the compact surface of genus 2 obtained by sewing together two tori along the boundaries of an open disk removed from each).
2. The Klein bottle K is the quotient space of $S^1 \times I$ obtained by identifying $(z, 0)$ with $(z^{-1}, 1)$ for $z \in S^1$. Compute $\pi_1(K)$.
3. * Let $X = \{(p,q) | p \neq -q\} \subset S^n \times S^n$. Define a map $f : S^n \longrightarrow X$ by $f(p) = (p,p)$. Prove that f is a homotopy equivalence.
4. Let $\mathscr{C}$ be a category that has all coproducts and coequalizers. Prove that $\mathscr{C}$ is cocomplete (has all colimits). Deduce formally, by use of opposite categories, that a category that has all products and equalizers is complete.

CHAPTER 3

Covering spaces

We run through the theory of covering spaces and their relationship to fundamental groups and fundamental groupoids. This is standard material, some of the oldest in algebraic topology. However, I know of no published source for the use that we shall make of the orbit category $\mathcal{O}(\pi_1(B,b))$ in the classification of coverings of a space B. This point of view gives us the opportunity to introduce some ideas that are central to equivariant algebraic topology, the study of spaces with group actions. In any case, this material is far too important to all branches of mathematics to omit.

1. The definition of covering spaces

While the reader is free to think about locally contractible spaces, weaker conditions are appropriate for the full generality of the theory of covering spaces. A space X is said to be locally path connected if for any $x \in X$ and any neighborhood U of x, there is a smaller neighborhood V of x each of whose points can be connected to x by a path in U. This is equivalent to the seemingly more stringent requirement that the topology of X have a basis consisting of path connected open sets. In fact, if X is locally path connected and U is an open neighborhood of a point x, then the set

$$V = \{y \,|\, y \text{ can be connected to } x \text{ by a path in } U\}$$

is a path connected open neighborhood of x that is contained in U. Observe that if X is connected and locally path connected, then it is path connected. Throughout this chapter, we assume that all given spaces are connected and locally path connected.

DEFINITION. A map $p: E \longrightarrow B$ is a covering (or cover, or covering space) if it is surjective and if each point $b \in B$ has an open neighborhood V such that each component of $p^{-1}(V)$ is open in E and is mapped homeomorphically onto V by p. We say that a path connected open subset V with this property is a fundamental neighborhood of B. We call E the total space, B the base space, and $F_b = p^{-1}(b)$ a fiber of the covering p.

Any homeomorphism is a cover. A product of covers is a cover. The projection $\mathbb{R} \longrightarrow S^1$ is a cover. Each $f_n : S^1 \longrightarrow S^1$ is a cover. The projection $S^n \longrightarrow \mathbb{R}P^n$ is a cover, where the real projective space $\mathbb{R}P^n$ is obtained from S^n by identifying antipodal points. If $f : A \longrightarrow B$ is a map (where A is connected and locally path connected) and D is a component of the pullback of f along p, then $p : D \longrightarrow A$ is a cover.

2. The unique path lifting property

The following result is abstracted from what we saw in the case of the particular cover $\mathbb{R} \longrightarrow S^1$. It describes the behavior of p with respect to path classes and fundamental groups.

THEOREM (Unique path lifting). *Let $p : E \longrightarrow B$ be a covering, let $b \in B$, and let $e, e' \in F_b$.*
  (i) *A path $f : I \longrightarrow B$ with $f(0) = b$ lifts uniquely to a path $g : I \longrightarrow E$ such that $g(0) = e$ and $p \circ g = f$.*
  (ii) *Equivalent paths $f \simeq f' : I \longrightarrow B$ that start at b lift to equivalent paths $g \simeq g' : I \longrightarrow E$ that start at e, hence $g(1) = g'(1)$.*
  (iii) $p_* : \pi_1(E, e) \longrightarrow \pi_1(B, b)$ *is a monomorphism.*
  (iv) $p_*(\pi_1(E, e'))$ *is conjugate to* $p_*(\pi_1(E, e))$.
  (v) *As e' runs through F_b, the groups $p_*(\pi_1(E, e'))$ run through all conjugates of $p_*(\pi_1(E, e))$ in $\pi_1(B, b)$.*

PROOF. For (i), subdivide I into subintervals each of which maps to a fundamental neighborhood under f, and lift f to g inductively by use of the prescribed homeomorphism property of fundamental neighborhoods. For (ii), let $h : I \times I \longrightarrow B$ be a homotopy $f \simeq f'$ through paths $b \longrightarrow b'$. Subdivide the square into subsquares each of which maps to a fundamental neighborhood under f. Proceeding inductively, we see that h lifts uniquely to a homotopy $H : I \times I \longrightarrow E$ such that $H(0,0) = e$ and $p \circ H = h$. By uniqueness, H is a homotopy $g \simeq g'$ through paths $e \longrightarrow e'$, where $g(1) = e' = g'(1)$. Parts (iii)–(v) are formal consequences of (i) and (ii), as we shall see in the next section. □

DEFINITION. A covering $p : E \longrightarrow B$ is regular if $p_*(\pi_1(E, e))$ is a normal subgroup of $\pi_1(B, b)$. It is universal if E is simply connected.

As we shall explain in §4, for a universal cover $p : E \longrightarrow B$, the elements of F_b are in bijective correspondence with the elements of $\pi_1(B, b)$. We illustrate the force of this statement.

EXAMPLE. For $n \geq 2$, S^n is a universal cover of $\mathbb{R}P^n$. Therefore $\pi_1(\mathbb{R}P^n)$ has only two elements. There is a unique group with two elements, and this proves our earlier claim that $\pi_1(\mathbb{R}P^n) = \mathbb{Z}_2$.

3. Coverings of groupoids

Much of the theory of covering spaces can be recast conceptually in terms of fundamental groupoids. This point of view separates the essentials of the topology from the formalities and gives a convenient language in which to describe the algebraic classification of coverings.

DEFINITION. (i) Let $\mathscr{C}$ be a category and x be an object of $\mathscr{C}$. The category $x \backslash \mathscr{C}$ of objects under x has objects the maps $f : x \longrightarrow y$ in $\mathscr{C}$; for objects $f : x \longrightarrow y$ and $g : x \longrightarrow z$, the morphisms $\gamma : f \longrightarrow g$ in $x \backslash \mathscr{C}$ are the morphisms $\gamma : y \longrightarrow z$ in $\mathscr{C}$ such that $\gamma \circ f = g : x \longrightarrow z$. Composition and identity maps are given by composition and identity maps in $\mathscr{C}$. When $\mathscr{C}$ is a groupoid, $\gamma = g \circ f^{-1}$, and the objects of $x \backslash \mathscr{C}$ therefore determine the category.

(ii) Let $\mathscr{C}$ be a small groupoid. Define the star of x, denoted $St(x)$ or $St_\mathscr{C}(x)$, to be the set of objects of $x \backslash \mathscr{C}$, that is, the set of morphisms of $\mathscr{C}$ with source x. Write $\mathscr{C}(x, x) = \pi(\mathscr{C}, x)$ for the group of automorphisms of the object x.

(iii) Let $\mathscr{E}$ and $\mathscr{B}$ be small connected groupoids. A covering $p: \mathscr{E} \longrightarrow \mathscr{B}$ is a functor that is surjective on objects and restricts to a bijection
$$p: St(e) \longrightarrow St(p(e))$$
for each object e of $\mathscr{E}$. For an object b of $\mathscr{B}$, let F_b denote the set of objects of $\mathscr{E}$ such that $p(e) = b$. Then $p^{-1}(St(b))$ is the disjoint union over $e \in F_b$ of $St(e)$.

Parts (i) and (ii) of the unique path lifting theorem can be restated as follows.

PROPOSITION. *If $p: E \longrightarrow B$ is a covering of spaces, then the induced functor $\Pi(p): \Pi(E) \longrightarrow \Pi(B)$ is a covering of groupoids.*

Parts (iii), (iv), and (v) of the unique path lifting theorem are categorical consequences that apply to any covering of groupoids, where they read as follows.

PROPOSITION. *Let $p: \mathscr{E} \longrightarrow \mathscr{B}$ be a covering of groupoids, let b be an object of $\mathscr{B}$, and let e and e' be objects of F_b.*
  (i) *$p: \pi(\mathscr{E}, e) \longrightarrow \pi(\mathscr{B}, b)$ is a monomorphism.*
  (ii) *$p(\pi(\mathscr{E}, e'))$ is conjugate to $p(\pi(\mathscr{E}, e))$.*
  (iii) *As e' runs through F_b, the groups $p(\pi(E, e'))$ run through all conjugates of $p(\pi(\mathscr{E}, e))$ in $\pi(\mathscr{B}, b)$.*

PROOF. For (i), if $g, g' \in \pi(\mathscr{E}, e)$ and $p(g) = p(g')$, then $g = g'$ by the injectivity of p on $St(e)$. For (ii), there is a map $g: e \longrightarrow e'$ since $\mathscr{E}$ is connected. Conjugation by g gives a homomorphism $\pi(\mathscr{E}, e) \longrightarrow \pi(\mathscr{E}, e')$ that maps under p to conjugation of $\pi(\mathscr{B}, b)$ by its element $p(g)$. For (iii), the surjectivity of p on $St(e)$ gives that any $f \in \pi(\mathscr{B}, b)$ is of the form $p(g)$ for some $g \in St(e)$. If e' is the target of g, then $p(\pi(\mathscr{E}, e'))$ is the conjugate of $p(\pi(\mathscr{E}, e))$ by f. □

The fibers F_b of a covering of groupoids are related by translation functions.

DEFINITION. Let $p: \mathscr{E} \longrightarrow \mathscr{B}$ be a covering of groupoids. Define the fiber translation functor $T = T(p): \mathscr{B} \longrightarrow \mathscr{S}$ as follows. For an object b of $\mathscr{B}$, $T(b) = F_b$. For a morphism $f: b \longrightarrow b'$ of $\mathscr{B}$, $T(f): F_b \longrightarrow F_{b'}$ is specified by $T(f)(e) = e'$, where e' is the target of the unique g in $St(e)$ such that $p(g) = f$.

It is an exercise from the definition of a covering of a groupoid to verify that T is a well defined functor. For a covering space $p: E \longrightarrow B$ and a path $f: b \longrightarrow b'$, $T(f): F_b \longrightarrow F_{b'}$ is given by $T(f)(e) = g(1)$ where g is the path in E that starts at e and covers f.

PROPOSITION. *Any two fibers F_b and $F_{b'}$ of a covering of groupoids have the same cardinality. Therefore any two fibers of a covering of spaces have the same cardinality.*

PROOF. For $f: b \longrightarrow b'$, $T(f): F_b \longrightarrow F_{b'}$ is a bijection with inverse $T(f^{-1})$. □

4. Group actions and orbit categories

The classification of coverings is best expressed in categorical language that involves actions of groups and groupoids on sets.

A (left) action of a group G on a set S is a function $G \times S \longrightarrow S$ such that $es = s$ (where e is the identity element) and $(g'g)s = g'(gs)$ for all $s \in S$. The *isotropy group* G_s of a point s is the subgroup $\{g|gs = s\}$ of G. An action is *free* if $gs = s$ implies $g = e$, that is, if $G_s = e$ for every $s \in S$.

The orbit generated by a point s is $\{gs|g \in G\}$. An action is *transitive* if for every pair s, s' of elements of S, there is an element g of G such that $gs = s'$. Equivalently, S consists of a single orbit. If H is a subgroup of G, the set G/H of cosets gH is a transitive G-set. When G acts transitively on a set S, we obtain an isomorphism of G-sets between S and the G-set G/G_s for any fixed $s \in S$ by sending gs to the coset gG_s.

The following lemma describes the group of automorphisms of a transitive G-set S. For a subgroup H of G, let NH denote the normalizer of H in G and define $WH = NH/H$. Such quotient groups WH are sometimes called Weyl groups.

LEMMA. *Let G act transitively on a set S, choose $s \in S$, and let $H = G_s$. Then WH is isomorphic to the group $\mathrm{Aut}_G(S)$ of automorphisms of the G-set S.*

PROOF. For $n \in NH$ with image $\bar{n} \in WH$, define an automorphism $\phi(\bar{n})$ of S by $\phi(\bar{n})(gs) = gns$. For an automorphism ϕ of S, we have $\phi(s) = ns$ for some $n \in G$. For $h \in H$, $hns = \phi(hs) = \phi(s) = ns$, hence $n^{-1}hn \in G_s = H$ and $n \in NH$. Clearly $\phi = \phi(\bar{n})$, and it is easy to check that this bijection between WH and $\mathrm{Aut}_G(S)$ is an isomorphism of groups. □

We shall also need to consider G-maps between different G-sets G/H.

LEMMA. *A G-map $\alpha : G/H \longrightarrow G/K$ has the form $\alpha(gH) = g\gamma K$, where the element $\gamma \in G$ satisfies $\gamma^{-1}h\gamma \in K$ for all $h \in H$.*

PROOF. If $\alpha(eH) = \gamma K$, then the relation
$$\gamma K = \alpha(eH) = \alpha(hH) = h\alpha(eH) = h\gamma K$$
implies that $\gamma^{-1}h\gamma \in K$ for $h \in H$. □

DEFINITION. The category $\mathscr{O}(G)$ of canonical orbits has objects the G-sets G/H and morphisms the G-maps of G-sets.

The previous lemmas give some feeling for the structure of $\mathscr{O}(G)$ and lead to the following alternative description.

LEMMA. *The category $\mathscr{O}(G)$ is isomorphic to the category $\mathscr{G}$ whose objects are the subgroups of G and whose morphisms are the distinct subconjugacy relations $\gamma^{-1}H\gamma \subset K$ for $\gamma \in G$.*

If we regard G as a category with a single object, then a (left) action of G on a set S is the same thing as a covariant functor $G \longrightarrow \mathscr{S}$. (A right action is the same thing as a contravariant functor.) If $\mathscr{B}$ is a small groupoid, it is therefore natural to think of a covariant functor $T : \mathscr{B} \longrightarrow \mathscr{S}$ as a generalization of a group action. For each object b of $\mathscr{B}$, T restricts to an action of $\pi(\mathscr{B}, b)$ on $T(b)$. We say that the functor T is *transitive* if this group action is transitive for each object b. If $\mathscr{B}$ is connected, this holds for all objects b if it holds for any one object b.

For example, for a covering of groupoids $p : \mathscr{E} \longrightarrow \mathscr{B}$, the fiber translation functor T restricts to give an action of $\pi(\mathscr{B}, b)$ on the set F_b. For $e \in F_b$, the isotropy group of e is precisely $p(\pi(\mathscr{E}, e))$. That is, $T(f)(e) = e$ if and only if the lift of f to an element of $St(e)$ is an automorphism of e. Moreover, the action

is transitive since there is an isomorphism in $\mathscr{E}$ connecting any two points of F_b. Therefore, as a $\pi(\mathscr{B},b)$-set,
$$F_b \cong \pi(\mathscr{B},b)/p(\pi(\mathscr{E},e)).$$

DEFINITION. A covering $p : \mathscr{E} \longrightarrow \mathscr{B}$ of groupoids is regular if $p(\pi(\mathscr{E},e))$ is a normal subgroup of $\pi(\mathscr{B},b)$. It is universal if $p(\pi(\mathscr{E},e)) = \{e\}$. Clearly a covering space is regular or universal if and only if its associated covering of fundamental groupoids is regular or universal.

A covering of groupoids is universal if and only if $\pi(\mathscr{B},b)$ acts freely on F_b, and then F_b is isomorphic to $\pi(\mathscr{B},b)$ as a $\pi(\mathscr{B},b)$-set. Specializing to covering spaces, this sharpens our earlier claim that the elements of F_b and $\pi_1(B,b)$ are in bijective correspondence.

5. The classification of coverings of groupoids

Fix a small connected groupoid $\mathscr{B}$ throughout this section and the next. We explain the classification of coverings of $\mathscr{B}$. This gives an algebraic prototype for the classification of coverings of spaces. We begin with a result that should be called the fundamental theorem of covering groupoid theory. We assume once and for all that all given groupoids are small and connected.

THEOREM. Let $p : \mathscr{E} \longrightarrow \mathscr{B}$ be a covering of groupoids, let $\mathscr{X}$ be a groupoid, and let $f : \mathscr{X} \longrightarrow \mathscr{B}$ be a functor. Choose a base object $x_0 \in \mathscr{X}$, let $b_0 = f(x_0)$, and choose $e_0 \in F_{b_0}$. Then there exists a functor $g : \mathscr{X} \longrightarrow \mathscr{E}$ such that $g(x_0) = e_0$ and $p \circ g = f$ if and only if
$$f(\pi(\mathscr{X},x_0)) \subset p(\pi(\mathscr{E},e_0))$$
in $\pi(\mathscr{B},b_0)$. When this condition holds, there is a unique such functor g.

PROOF. If g exists, its properties directly imply that $\text{im}(f) \subset \text{im}(p)$. For an object x of $\mathscr{X}$ and a map $\alpha : x_0 \longrightarrow x$ in $\mathscr{X}$, let $\tilde{\alpha}$ be the unique element of $St(e_0)$ such that $p(\tilde{\alpha}) = f(\alpha)$. If g exists, $g(\alpha)$ must be $\tilde{\alpha}$ and therefore $g(x)$ must be the target $T(f(\alpha))(e_0)$ of $\tilde{\alpha}$. The inclusion $f(\pi(\mathscr{X},x_0)) \subset p(\pi(\mathscr{E},e_0))$ ensures that $T(f(\alpha))(e_0)$ is independent of the choice of α, so that g so specified is a well defined functor. In fact, given another map $\alpha' : x_0 \longrightarrow x$, $\alpha^{-1} \circ \alpha'$ is an element of $\pi(\mathscr{X},x_0)$. Therefore
$$f(\alpha)^{-1} \circ f(\alpha') = f(\alpha^{-1} \circ \alpha') = p(\beta)$$
for some $\beta \in \pi(\mathscr{E},e_0)$. Thus
$$p(\tilde{\alpha} \circ \beta) = f(\alpha) \circ p(\beta) = f(\alpha) \circ f(\alpha)^{-1} \circ f(\alpha') = f(\alpha').$$
This means that $\tilde{\alpha} \circ \beta$ is the unique element $\tilde{\alpha}'$ of $St(e_0)$ such that $p(\tilde{\alpha}') = f(\alpha')$, and its target is the target of $\tilde{\alpha}$, as required. □

DEFINITION. A map $g : \mathscr{E} \longrightarrow \mathscr{E}'$ of coverings of $\mathscr{B}$ is a functor g such that the following diagram of functors is commutative:

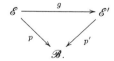

Let Cov($\mathscr{B}$) denote the category of coverings of $\mathscr{B}$; when $\mathscr{B}$ is understood, we write Cov($\mathscr{E}, \mathscr{E}'$) for the set of maps $\mathscr{E} \longrightarrow \mathscr{E}'$ of coverings of $\mathscr{B}$.

LEMMA. *A map $g : \mathscr{E} \longrightarrow \mathscr{E}'$ of coverings is itself a covering.*

PROOF. The functor g is surjective on objects since, if $e' \in \mathscr{E}'$ and we choose an object $e \in \mathscr{E}$ and a map $f : g(e) \longrightarrow e'$ in $\mathscr{E}'$, then $e' = g(T(p'(f))(e))$. The map $g : St_{\mathscr{E}}(e) \longrightarrow St_{\mathscr{E}'}(g(e))$ is a bijection since its composite with the bijection $p' : St_{\mathscr{E}'}(g(e)) \longrightarrow St_{\mathscr{B}}(p'(g(e)))$ is the bijection $p : St_{\mathscr{E}}(e) \longrightarrow St_{\mathscr{B}}(p(e))$. $\square$

The fundamental theorem immediately determines all maps of coverings of $\mathscr{B}$ in terms of group level data.

THEOREM. *Let $p : \mathscr{E} \longrightarrow \mathscr{B}$ and $p' : \mathscr{E}' \longrightarrow \mathscr{B}$ be coverings and choose base objects $b \in \mathscr{B}$, $e \in \mathscr{E}$, and $e' \in \mathscr{E}'$ such that $p(e) = b = p'(e')$. There exists a map $g : \mathscr{E} \longrightarrow \mathscr{E}'$ of coverings with $g(e) = e'$ if and only if*

$$p(\pi(\mathscr{E}, e)) \subset p'(\pi(\mathscr{E}', e')),$$

and there is then only one such g. In particular, two maps of covers $g, g' : \mathscr{E} \longrightarrow \mathscr{E}'$ coincide if $g(e) = g'(e)$ for any one object $e \in \mathscr{E}$. Moreover, g is an isomorphism if and only if the displayed inclusion of subgroups of $\pi(\mathscr{B}, b)$ is an equality. Therefore $\mathscr{E}$ and $\mathscr{E}'$ are isomorphic if and only if $p(\pi(\mathscr{E}, e))$ and $p'(\pi(\mathscr{E}', e'))$ are conjugate whenever $p(e) = p'(e')$.

COROLLARY. *If it exists, the universal cover of $\mathscr{B}$ is unique up to isomorphism and covers any other cover.*

That the universal cover does exist will be proved in the next section. It is useful to recast the previous theorem in terms of actions on fibers.

THEOREM. *Let $p : \mathscr{E} \longrightarrow \mathscr{B}$ and $p' : \mathscr{E}' \longrightarrow \mathscr{B}$ be coverings, choose a base object $b \in \mathscr{B}$, and let $G = \pi(\mathscr{B}, b)$. If $g : \mathscr{E} \longrightarrow \mathscr{E}'$ is a map of coverings, then g restricts to a map $F_b \longrightarrow F'_b$ of G-sets, and restriction to fibers specifies a bijection between $\mathrm{Cov}(\mathscr{E}, \mathscr{E}')$ and the set of G-maps $F_b \longrightarrow F'_b$.*

PROOF. Let $e \in F_b$ and $f \in \pi(\mathscr{B}, b)$. By definition, fe is the target of the map $\tilde{f} \in St_{\mathscr{E}}(e)$ such that $p(\tilde{f}) = f$. Clearly $g(fe)$ is the target of $g(\tilde{f}) \in St_{\mathscr{E}'}(g(e))$ and $p'(g(\tilde{f})) = p(\tilde{f}) = f$. Again by definition, this gives $g(fe) = fg(e)$. The previous theorem shows that restriction to fibers is an injection on $\mathrm{Cov}(\mathscr{E}, \mathscr{E}')$. To show surjectivity, let $\alpha : F_b \longrightarrow F'_b$ be a G-map. Choose $e \in F_b$ and let $e' = \alpha(e)$. Since α is a G-map, the isotropy group $p(\pi(\mathscr{E}, e))$ of e is contained in the isotropy group $p'(\pi(\mathscr{E}', e'))$ of e'. Therefore the previous theorem ensures the existence of a covering map g that restricts to α on fibers. $\square$

DEFINITION. Let $\mathrm{Aut}(\mathscr{E}) \subset \mathrm{Cov}(\mathscr{E}, \mathscr{E})$ denote the group of automorphisms of a cover $\mathscr{E}$. Note that, since it is possible to have conjugate subgroups H and H' of a group G such that H is a proper subgroup of H', it is possible to have a map of covers $g : \mathscr{E} \longrightarrow \mathscr{E}$ such that g is not an isomorphism.

COROLLARY. *Let $p : \mathscr{E} \longrightarrow \mathscr{B}$ be a covering and choose objects $b \in \mathscr{B}$ and $e \in F_b$. Write $G = \pi(\mathscr{B}, b)$ and $H = p(\pi(\mathscr{E}, e))$. Then $\mathrm{Aut}(\mathscr{E})$ is isomorphic to the group of automorphisms of the G-set F_b and therefore to the group WH. If p is regular, then $\mathrm{Aut}(\mathscr{E}) \cong G/H$. If p is universal, then $\mathrm{Aut}(\mathscr{E}) \cong G$.*

6. The construction of coverings of groupoids

We have given an algebraic classification of all possible covers of $\mathscr{B}$: there is at most one isomorphism class of covers corresponding to each conjugacy class of subgroups of $\pi(\mathscr{B}, b)$. We show that all of these possibilities are actually realized. Since this algebraic result is not needed in the proof of its topological analogue, we shall not give complete details.

THEOREM. *Choose a base object b of $\mathscr{B}$ and let $G = \pi(\mathscr{B}, b)$. There is a functor*

$$\mathscr{E}(-) : \mathscr{O}(G) \longrightarrow \mathrm{Cov}(\mathscr{B})$$

that is an equivalence of categories. For each subgroup H of G, the covering $p : \mathscr{E}(G/H) \longrightarrow \mathscr{B}$ has a canonical base object e in its fiber over b such that

$$p(\pi(\mathscr{E}(G/H), e)) = H.$$

Moreover, $F_b = G/H$ as a G-set and, for a G-map $\alpha : G/H \longrightarrow G/K$ in $\mathscr{O}(G)$, the restriction of $\mathscr{E}(\alpha) : \mathscr{E}(G/H) \longrightarrow \mathscr{E}(G/K)$ to fibers over b coincides with α.

PROOF. The idea is that, up to bijection, $St_{\mathscr{E}(G/H)}(e)$ must be the same set for each H, but the nature of its points can differ with H. At one extreme, $\mathscr{E}(G/G) = \mathscr{B}$, $p = \mathrm{id}$, $e = b$, and the set of morphisms from b to any other object b' is a copy of $\pi(\mathscr{B}, b)$. At the other extreme, $\mathscr{E}(G/e)$ is a universal cover of $\mathscr{B}$ and there is just one morphism from e to any other object e'. In general, the set of objects of $\mathscr{E}(G/H)$ is defined to be $St_{\mathscr{B}}(b)/H$, the coset of the identity morphism being e. Here G and hence its subgroup H act from the right on $St_{\mathscr{B}}(b)$ by composition in $\mathscr{B}$. We define $p : \mathscr{E}(G/H) \longrightarrow \mathscr{B}$ on objects by letting $p(fH)$ be the target of f, which is independent of the coset representative f. We define morphism sets by

$$\mathscr{E}(G/H)(fH, f'H) = \{f' \circ h \circ f^{-1} | h \in H\} \subset \mathscr{B}(p(fH), p(f'H)).$$

Again, this is independent of the choices of coset representatives f and f'. Composition and identities are inherited from those of $\mathscr{B}$, and p is given on morphisms by the displayed inclusions. It is easy to check that $p : \mathscr{E}(G/H) \longrightarrow \mathscr{B}$ is a covering, and it is clear that $p(\pi(\mathscr{E}(G/H), e)) = H$.

This defines the object function of the functor $\mathscr{E} : \mathscr{O}(G) \longrightarrow \mathrm{Cov}(\mathscr{B})$. To define $\mathscr{E}$ on morphisms, consider $\alpha : G/H \longrightarrow G/K$. If $\alpha(eH) = gK$, then $g^{-1}Hg \subset K$ and $\alpha(fH) = fgK$. The functor $\mathscr{E}(\alpha) : \mathscr{E}(G/H) \longrightarrow \mathscr{E}(G/K)$ sends the object fH to the object $\alpha(fH) = fgK$ and sends the morphism $f' \circ h \circ f^{-1}$ to the same morphism of $\mathscr{B}$ regarded as $f'g \circ g^{-1}hg \circ g^{-1}f^{-1}$. It is easily checked that each $\mathscr{E}(\alpha)$ is a well defined functor, and that $\mathscr{E}$ is functorial in α.

To show that the functor $\mathscr{E}(-)$ is an equivalence of categories, it suffices to show that it maps the morphism set $\mathscr{O}(G)(G/H, G/K)$ bijectively onto the morphism set $\mathrm{Cov}(\mathscr{E}(G/H), \mathscr{E}(G/K))$ and that every covering of $\mathscr{B}$ is isomorphic to one of the coverings $\mathscr{E}(G/H)$. These statements are immediate from the results of the previous section. □

The following remarks place the orbit category $\mathscr{O}(\pi(\mathscr{B}, b))$ in perspective by relating it to several other equivalent categories.

REMARK. Consider the category $\mathscr{S}^{\mathscr{B}}$ of functors $T : \mathscr{B} \longrightarrow \mathscr{S}$ and natural transformations. Let $G = \pi(\mathscr{B}, b)$. Regarding G as a category with one object b, it is a skeleton of $\mathscr{B}$, hence the inclusion $G \subset \mathscr{B}$ is an equivalence of categories. Therefore, restriction of functors T to G-sets $T(b)$ gives an equivalence of categories from $\mathscr{S}^{\mathscr{B}}$ to the category of G-sets. This restricts to an equivalence between the respective subcategories of transitive objects. We have chosen to focus on transitive objects since we prefer to insist that coverings be connected. The inclusion of the orbit category $\mathscr{O}(G)$ in the category of transitive G-sets is an equivalence of categories because $\mathscr{O}(G)$ is a full subcategory that contains a skeleton. We could shrink $\mathscr{O}(G)$ to a skeleton by choosing one H in each conjugacy class of subgroups of G, but the resulting equivalent subcategory is a less natural mathematical object.

7. The classification of coverings of spaces

In this section and the next, we shall classify covering spaces and their maps by arguments precisely parallel to those for covering groupoids in the previous sections. In fact, applied to the associated coverings of fundamental groupoids, some of the algebraic results directly imply their topological analogues. We begin with the following result, which deserves to be called the fundamental theorem of covering space theory and has many other applications. It asserts that the fundamental group gives the only "obstruction" to solving a certain lifting problem. Recall our standing assumption that all given spaces are connected and locally path connected.

THEOREM. *Let $p : E \longrightarrow B$ be a covering and let $f : X \longrightarrow B$ be a continuous map. Choose $x \in X$, let $b = f(x)$, and choose $e \in F_b$. There exists a map $g : X \longrightarrow E$ such that $g(x) = e$ and $p \circ g = f$ if and only if*

$$f_*(\pi_1(X,x)) \subset p_*(\pi_1(E,e))$$

in $\pi_1(B, b)$. When this condition holds, there is a unique such map g.

PROOF. If g exists, its properties directly imply that $\text{im}(f_*) \subset \text{im}(p_*)$. Thus assume that $\text{im}(f_*) \subset \text{im}(p_*)$. Applied to the covering $\Pi(p) : \Pi(E) \longrightarrow \Pi(B)$, the analogue for groupoids gives a functor $\Pi(X) \longrightarrow \Pi(E)$ that restricts on objects to the unique map $g : X \longrightarrow E$ of sets such that $g(x) = e$ and $p \circ g = f$. We need only check that g is continuous, and this holds because p is a local homeomorphism. In detail, if $y \in X$ and $g(y) \in U$, where U is an open subset of E, then there is a smaller open neighborhood U' of $g(y)$ that p maps homeomorphically onto an open subset V of B. If W is any path connected neighborhood of y such that $f(W) \subset V$, then $g(W) \subset U'$ by inspection of the definition of g. □

DEFINITION. A map $g : E \longrightarrow E'$ of coverings over B is a map g such that the following diagram is commutative:

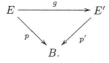

Let $\text{Cov}(B)$ denote the category of coverings of the space B; when B is understood, we write $\text{Cov}(E, E')$ for the set of maps $E \longrightarrow E'$ of coverings of B.

LEMMA. *A map $g : E \longrightarrow E'$ of coverings is itself a covering.*

PROOF. The map g is surjective by the algebraic analogue. The fundamental neighborhoods for g are the components of the inverse images in E' of the neighborhoods of B which are fundamental for both p and p'. □

The following remarkable theorem is an immediate consequence of the fundamental theorem of covering space theory.

THEOREM. *Let $p : E \longrightarrow B$ and $p' : E' \longrightarrow B$ be coverings and choose $b \in B$, $e \in E$, and $e' \in E'$ such that $p(e) = b = p'(e')$. There exists a map $g : E \longrightarrow E'$ of coverings with $g(e) = e'$ if and only if*
$$p_*(\pi_1(E, e)) \subset p'_*(\pi_1(E', e')),$$
and there is then only one such g. In particular, two maps of covers $g, g' : E \longrightarrow E'$ coincide if $g(e) = g'(e)$ for any one $e \in E$. Moreover, g is a homeomorphism if and only if the displayed inclusion of subgroups of $\pi_1(B, b)$ is an equality. Therefore E and E' are homeomorphic if and only if $p_*(\pi_1(E, e))$ and $p'_*(\pi_1(E', e'))$ are conjugate whenever $p(e) = p'(e')$.

COROLLARY. *If it exists, the universal cover of B is unique up to isomorphism and covers any other cover.*

Under a necessary additional hypothesis on B, we shall prove in the next section that the universal cover does exist.

We hasten to add that the theorem above is atypical of algebraic topology. It is not usually the case that algebraic invariants like the fundamental group totally determine the existence and uniqueness of maps of topological spaces with prescribed properties. The following immediate implication of the theorem gives one explanation.

COROLLARY. *The fundamental groupoid functor induces a bijection*
$$\mathrm{Cov}(E, E') \longrightarrow \mathrm{Cov}(\Pi(E), \Pi(E')).$$

Just as for groupoids, we can recast the theorem in terms of fibers. In fact, via the previous corollary, the following result is immediate from its analogue for groupoids.

THEOREM. *Let $p : E \longrightarrow B$ and $p' : E' \longrightarrow B$ be coverings, choose a basepoint $b \subset B$, and let $G = \pi_1(B, b)$. If $g : E \longrightarrow E'$ is a map of coverings, then g restricts to a map $F_b \longrightarrow F'_b$ of G-sets, and restriction to fibers specifies a bijection between $\mathrm{Cov}(E, E')$ and the set of G-maps $F_b \longrightarrow F'_b$.*

DEFINITION. Let $\mathrm{Aut}(E) \subset \mathrm{Cov}(E, E)$ denote the group of automorphisms of a cover E. Again, just as for groupoids, it is possible to have a map of covers $g : E \longrightarrow E$ such that g is not an isomorphism.

COROLLARY. *Let $p : E \longrightarrow B$ be a covering and choose $b \in B$ and $e \in F_b$. Write $G = \pi_1(B, b)$ and $H = p_*(\pi_1(E, e))$. Then $\mathrm{Aut}(E)$ is isomorphic to the group of automorphisms of the G-set F_b and therefore to the group WH. If p is regular, then $\mathrm{Aut}(E) \cong G/H$. If p is universal, then $\mathrm{Aut}(E) \cong G$.*

8. The construction of coverings of spaces

We have now given an algebraic classification of all possible covers of B: there is at most one isomorphism class of covers corresponding to each conjugacy class of subgroups of $\pi_1(B, b)$. We show here that all of these possibilities are actually realized. We shall first construct universal covers and then show that the existence of universal covers implies the existence of all other possible covers. Again, while it suffices to think in terms of locally contractible spaces, appropriate generality demands a weaker hypothesis. We say that a space B is semi-locally simply connected if every point $b \in B$ has a neighborhood U such that $\pi_1(U, b) \longrightarrow \pi_1(B, b)$ is the trivial homomorphism.

THEOREM. *If B is connected, locally path connected, and semi-locally simply connected, then B has a universal cover.*

PROOF. Fix a basepoint $b \in B$. We turn the properties of paths that must hold in a universal cover into a construction. Define E to be the set of equivalence classes of paths f in B that start at b and define $p : E \longrightarrow B$ by $p[f] = f(1)$. Of course, the equivalence relation is homotopy through paths from b to a given endpoint, so that p is well defined. Thus, as a set, E is just $St_{\Pi(B)}(b)$, exactly as in the construction of the universal cover of $\Pi(B)$. The topology of B has a basis consisting of path connected open subsets U such that $\pi_1(U, u) \longrightarrow \pi_1(B, u)$ is trivial for all $u \in U$. Since every loop in U is equivalent in B to the trivial loop, any two paths $u \longrightarrow u'$ in such a U are equivalent in B. We shall topologize E so that p is a cover with these U as fundamental neighborhoods. For a path f in B that starts at b and ends in U, define a subset $U[f]$ of E by

$$U[f] = \{[g] \mid [g] = [c \cdot f] \text{ for some } c : I \longrightarrow U\}.$$

The set of all such $U[f]$ is a basis for a topology on E since if $U[f]$ and $U'[f']$ are two such sets and $[g]$ is in their intersection, then

$$W[g] \subset U[f] \cap U'[f']$$

for any open set W of B such that $p[g] \in W \subset U \cap U'$. For $u \in U$, there is a unique $[g]$ in each $U[f]$ such that $p[g] = u$. Thus p maps $U[f]$ homeomorphically onto U and, if we choose a basepoint u in each U, then $p^{-1}(U)$ is the disjoint union of those $U[f]$ such that f ends at u. It only remains to show that E is connected, locally path connected, and simply connected, and the second of these is clear. Give E the basepoint $e = [c_b]$. For $[f] \in E$, define a path $\tilde{f} : I \longrightarrow E$ by $\tilde{f}(s) = [f_s]$, where $f_s(t) = f(st)$; $\tilde{f}$ is continuous since each $\tilde{f}^{-1}(U[g])$ is open by the definition of $U[g]$ and the continuity of f. Since $\tilde{f}$ starts at e and ends at $[f]$, E is path connected. Since $f_s(1) = f(s)$, $p \circ \tilde{f} = f$. Thus, by definition,

$$T[f](e) = [\tilde{f}(1)] = [f].$$

Restricting attention to loops f, we see that $T[f](e) = e$ if and only if $[f] = e$ as an element of $\pi_1(B, b)$. Thus the action of $\pi_1(B, b)$ on F_b is free and the isotropy group $p_*(\pi_1(E, e))$ is trivial. □

We shall construct general covers by passage to orbit spaces from the universal cover, and we need some preliminaries.

DEFINITION. A G-space X is a space X that is a G-set with continuous action map $G \times X \longrightarrow X$. Define the orbit space X/G to be the set of orbits $\{Gx \mid x \in X\}$ with its topology as a quotient space of X.

The definition makes sense for general topological groups G. However, our interest here is in discrete groups G, for which the continuity condition just means that action by each element of G is a homeomorphism. The functoriality on $\mathscr{O}(G)$ of our construction of general covers will be immediate from the following observation.

LEMMA. *Let X be a G-space. Then passage to orbit spaces defines a functor $X/(-) : \mathscr{O}(G) \longrightarrow \mathscr{U}$.*

PROOF. The functor sends G/H to X/H and sends a map $\alpha : G/H \longrightarrow G/K$ to the map $X/H \longrightarrow X/K$ that sends the coset Hx to the coset $K\gamma^{-1}x$, where α is given by the subconjugacy relation $\gamma^{-1}H\gamma \subset K$. □

The starting point of the construction of general covers is the following description of regular covers and in particular of the universal cover.

PROPOSITION. *Let $p : E \longrightarrow B$ be a cover such that $\mathrm{Aut}(E)$ acts transitively on F_b. Then the cover p is regular and $E/\mathrm{Aut}(E)$ is homeomorphic to B.*

PROOF. For any points $e, e' \in F_b$, there exists $g \in \mathrm{Aut}(E)$ such that $g(e) = e'$ and thus $p_*(\pi_1(E, e)) = p_*(\pi_1(E, e'))$. Therefore all conjugates of $p_*(\pi_1(E, e))$ are equal to $p_*(\pi_1(E, e))$ and $p_*(\pi_1(E, e))$ is a normal subgroup of $\pi_1(B, b)$. The homeomorphism is clear since, locally, both p and passage to orbits identify the different components of the inverse images of fundamental neighborhoods. □

THEOREM. *Choose a basepoint $b \in B$ and let $G = \pi_1(B, b)$. There is a functor*

$$E(-) : \mathscr{O}(G) \longrightarrow \mathrm{Cov}(B)$$

that is an equivalence of categories. For each subgroup H of G, the covering $p : E(G/H) \longrightarrow B$ has a canonical basepoint e in its fiber over b such that

$$p_*(\pi_1(E(G/H), e)) = H.$$

Moreover, $F_b \cong G/H$ as a G-set and, for a G-map $\alpha : G/H \longrightarrow G/K$ in $\mathscr{O}(G)$, the restriction of $E(\alpha) : E(G/H) \longrightarrow E(G/K)$ to fibers over b coincides with α.

PROOF. Let $p : E \longrightarrow B$ be the universal cover of B and fix $e \in E$ such that $p(e) = b$. We have the isomorphism $\mathrm{Aut}(E) \cong \pi_1(B, b)$ given by mapping $g : E \longrightarrow E$ to the path class $[f] \in G$ such that $g(e) = T(f)(e)$, where $T(f)(e)$ is the endpoint of the path $\tilde{f}$ that starts at e and lifts f. We identify subgroups of G with subgroups of $\mathrm{Aut}(E)$ via this isomorphism. We define $E(G/H)$ to be the orbit space E/H and we let $q : E \longrightarrow E/H$ be the quotient map. We may identify B with $E/\mathrm{Aut}(E)$, and inclusion of orbits specifies a map $p' : E/H \longrightarrow B$ such that $p' \circ q = p : E \longrightarrow B$. If $U \subset B$ is a fundamental neighborhood for p and V is a component of $p^{-1}(U) \subset E$, then

$$p^{-1}(U) = \coprod\nolimits_{g \in \mathrm{Aut}(E)} gV.$$

Passage to orbits over H simply identifies some of these components, and we see immediately that both p' and q are covers. If $e' = q(e)$, then p'_* maps $\pi_1(E/H, e')$ isomorphically onto H since, by construction, the isotropy group of e' under the action of $\pi_1(B, b)$ is precisely H. Rewriting $p' = p$ and $e' = e$ generically, this gives the stated properties of the coverings $E(G/H)$. The functoriality on $\mathscr{O}(G)$ follows directly from the previous lemma.

The functor $E(-)$ is an equivalence of categories since the results of the previous section imply that it maps the morphism set $\mathscr{O}(G)(G/H, G/K)$ bijectively onto the morphism set $\mathrm{Cov}(E(G/H), E(G/K))$ and that every covering of B is isomorphic to one of the coverings $E(G/H)$. □

The classification theorems for coverings of spaces and coverings of groupoids are nicely related. In fact, the following diagram of functors commutes up to natural isomorphism:

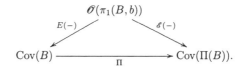

COROLLARY. $\Pi : \mathrm{Cov}(B) \longrightarrow \mathrm{Cov}(\Pi(B))$ *is an equivalence of categories.*

PROBLEMS

In the following two problems, let G be a connected and locally path connected topological group with identity element e, let $p : H \longrightarrow G$ be a covering, and fix $f \in H$ such that $p(f) = e$. Prove the following. (Hint: Make repeated use of the fundamental theorem for covering spaces.)

1. (a) H has a unique continuous product $H \times H \longrightarrow H$ with identity element f such that p is a homomorphism.
   (b) H is a topological group under this product, and H is Abelian if G is.
2. (a) The kernel K of p is a discrete normal subgroup of H.
   (b) In general, any discrete normal subgroup K of a connected topological group H is contained in the center of H.
   (c) For $k \in K$, define $t(k) : H \longrightarrow H$ by $t(k)(h) = kh$. Then $k \longrightarrow t(k)$ specifies an isomorphism between K and the group $\mathrm{Aut}(H)$.

Let X and Y be connected, locally path connected, and Hausdorff. A map $f : X \longrightarrow Y$ is said to be a local homeomorphism if every point of X has an open neighborhood that maps homeomorphically onto an open set in Y.

3. Give an example of a surjective local homeomorphism that is not a covering.
4. * Let $f : X \longrightarrow Y$ be a local homeomorphism, where X is compact. Prove that f is a (surjective!) covering with finite fibers.

Let X be a G-space, where G is a (discrete) group. For a subgroup H of G, define
$$X^H = \{x | hx = x \text{ for all } h \in H\} \subset X;$$
X^H is the H-fixed point subspace of X. Topologize the set of functions $G/H \longrightarrow X$ as the product of copies of X indexed on the elements of G/H, and give the set of G-maps $G/H \longrightarrow X$ the subspace topology.

5. Show that the space of G-maps $G/H \longrightarrow X$ is naturally homeomorphic to X^H. In particular, $\mathscr{O}(G/H, G/K) \cong (G/K)^H$.
6. Let X be a G-space. Show that passage to fixed point spaces, $G/H \longmapsto X^H$, is the object function of a *contravariant* functor $X^{(-)} : \mathscr{O}(G) \longrightarrow \mathscr{U}$.

CHAPTER 4

Graphs

We define graphs, describe their homotopy types, and use them to show that a subgroup of a free group is free and that any group is the fundamental group of some space.

1. The definition of graphs

We give the definition in a form that will later make it clear that a graph is exactly a one-dimensional CW complex. Note that the zero-sphere S^0 is a discrete space with two points. We think of S^0 as the boundary of I and so label the points 0 and 1.

DEFINITION. A graph X is a space that is obtained from a (discrete) set X^0 of points, called vertices, and a (discrete) set J of functions $j : S^0 \longrightarrow X^0$ as the quotient space of the disjoint union $X^0 \amalg (J \times I)$ that is obtained by identifying $(j, 0)$ with $j(0)$ and $(j, 1)$ with $j(1)$. The images of the intervals $\{j\} \times I$ are called edges. A graph is finite if it has only finitely many vertices and edges or, equivalently, if it is a compact space. A graph is locally finite if each vertex is a boundary point of only finitely many edges or, equivalently, if it is a locally compact space. A subgraph A of X is a graph $A \subset X$ with $A^0 \subset X^0$. That is, A is the union of some of the vertices and edges of X.

Observe that a graph is a locally contractible space: any neighborhood of any point contains a contractible neighborhood of that point. Therefore a connected graph has all possible covers.

2. Edge paths and trees

An oriented edge $k : I \longrightarrow X$ in a graph X is the traversal of an edge in either the forward or backward direction. An edge path is a finite composite of oriented edges k_n with $k_{n+1}(0) = k_n(1)$. Such a path is reduced if it is never the case that k_{n+1} is k_n with the opposite orientation. An edge path is closed if it starts and ends at the same vertex (and is thus a loop).

DEFINITION. A tree is a connected graph with no closed reduced edge paths.

A subspace A of a space X is a deformation retract if there is a homotopy $h : X \times I \longrightarrow X$ such that $h(x, 0) = x$, $h(a, t) = a$, and $h(x, 1) \in A$ for all $x \in X$, $a \in A$, and $t \in I$. Such a homotopy is called a deformation of X onto A.

LEMMA. *Any vertex v_0 of a tree T is a deformation retract of T.*

PROOF. This is true by induction on the number of edges when T is finite since we can prune the last branch. For the general case, observe that each vertex v lies in some finite connected subtree $T(v)$ that also contains v_0. Choose an edge path

$a(v) : I \longrightarrow T(v)$ connecting v to v_0. For an edge j from v to v', $T(v) \cup T(v') \cup j$ is a finite connected subtree of T. On the square $j \times I$, we define

$$h : j \times I \longrightarrow T(v) \cup T(v') \cup j$$

by requiring $h = a(v)$ on $\{v\} \times I$, $h = a(v')$ on $\{v'\} \times I$, $h(x,0) = x$ and $h(x,1) = v_0$ for all $x \in j$, and extending over the interior of the square by use of the simple connectivity of $T(v) \cup T(v') \cup j$. As j runs over the edges, these homotopies glue together to specify a deformation h of T onto v_0. □

A subtree of a graph X is maximal if it is contained in no strictly larger tree.

LEMMA. *If a tree T is a subgraph of a graph X, then T is contained in a maximal tree. If X is connected, then a tree in X is maximal if and only if it contains all vertices of X.*

PROOF. Since the union of an increasing family of trees in X is a tree, the first statement holds by Zorn's lemma. If X is connected, then a tree containing all vertices is maximal since addition of an edge would result in a subgraph that contains a closed reduced edge path and, conversely, a tree T that does not contain all vertices is not maximal since a vertex not in T can be connected to a vertex in T by a reduced edge path consisting of edges not in T. □

3. The homotopy types of graphs

Graph theory is a branch of combinatorics. The homotopy theory of graphs is essentially trivial, by the following result.

THEOREM. *Let X be a connected graph with maximal tree T. Then the quotient space X/T is the wedge of one circle for each edge of X not in T, and the quotient map $q : X \longrightarrow X/T$ is a homotopy equivalence.*

PROOF. The first clause is evident. The second is a direct consequence of a later result (that will be left as an exercise): for a suitably nice inclusion, called a "cofibration," of a contractible space T in a space X, the quotient map $X \longrightarrow X/T$ is a homotopy equivalence. A direct proof in the present situation is longer and uglier. With the notation in our proof that a vertex v_0 is a deformation retract of T via a deformation h, define a loop $b_j = a(v') \cdot j \cdot a(v)^{-1}$ at v_0 for each edge $j : v \longrightarrow v'$ not in T. The b_j together specify a map b from $X/T \cong \bigvee_j S^1$ to X. The composite $q \circ b : X/T \longrightarrow X/T$ is the wedge over j of copies of the loop $c_{v_0} \cdot \mathrm{id} \cdot c_{v_0}^{-1} : S^1 \longrightarrow S^1$ and is therefore homotopic to the identity. To prove that $b \circ q$ is homotopic to the identity, observe that h is a homotopy $\mathrm{id} \simeq b \circ q$ on T. This homotopy extends to a homotopy $H : \mathrm{id} \simeq b \circ q$ on all of X. To see this, we need only construct H on $j \times I$ for an edge $j : v \longrightarrow v'$ not in T. The following schematic description of the prescribed behavior on the boundary of the square makes it clear that H exists:

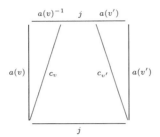

4. Covers of graphs and Euler characteristics

Define the Euler characteristic $\chi(X)$ of a finite graph X to be $V - E$, where V is the number of vertices of X and E is the number of edges. By induction on the number of edges, $\chi(T) = 1$ for any finite tree. The determination of the homotopy types of graphs has the following immediate implication.

COROLLARY. *If X is a connected graph, then $\pi_1(X)$ is a free group with one generator for each edge not in a given maximal tree. If X is finite, then $\pi_1(X)$ is free on $1 - \chi(X)$ generators; in particular, $\chi(X) \leq 1$, with equality if and only if X is a tree.*

THEOREM. *If B is a connected graph with vertex set B^0 and $p : E \longrightarrow B$ is a covering, then E is a connected graph with vertex set $E^0 = p^{-1}(B^0)$ and with one edge for each edge j of B and point $e \in F_{j(0)}$. Therefore, if B is finite and p is a finite cover whose fibers have cardinality n, then E is finite and $\chi(E) = n\chi(B)$.*

PROOF. Regard an edge j of B as a path $I \longrightarrow B$ and let $k(e) : I \longrightarrow E$ be the unique path such that $p \circ k = j$ and $k(e)(0) = e$, where $e \in F_{j(0)}$. We claim that E is a graph with E^0 as vertex set and the $k(e)$ as edges. An easy path lifting argument shows that each point of $E - E^0$ is an interior point of exactly one edge, hence we have a continuous bijection from the graph $E^0 \amalg (K \times I)/(\sim)$ to E, where K is the evident set of "attaching maps" $S^0 \longrightarrow E^0$ for the specified edges. This map is a homeomorphism since it is a local homeomorphism over B. □

5. Applications to groups

The following purely algebraic result is most simply proved by topology.

THEOREM. *A subgroup H of a free group G is free. If G is free on k generators and H has finite index n in G, then H is free on $1 - n + nk$ generators.*

PROOF. Realize G as $\pi_1(B)$, where B is the wedge of one circle for each generator of G in a given free basis. Construct a covering $p : E \longrightarrow B$ such that $p_*(\pi_1(E)) = H$. Since E is a graph, H must be free. If G has k generators, then $\chi(B) = 1 - k$. If $[G : H] = n$, then F_b has cardinality n and $\chi(E) = n\chi(B)$. Therefore $1 - \chi(E) = 1 - n + nk$. □

We can extend the idea to realize any group as the fundamental group of some connected space.

THEOREM. *For any group G, there is a connected space X such that $\pi_1(X)$ is isomorphic to G.*

PROOF. We may write $G = F/N$ for some free group F and normal subgroup N. As above, we may realize the inclusion of N in F by passage to fundamental groups from a cover $p : E \longrightarrow B$. Define the (unreduced) cone on E to be $CE = (E \times I)/(E \times \{1\})$ and define
$$X = B \cup_p CE/(\sim),$$
where $(e, 0) \sim p(e)$. Let U and V be the images in X of $B \amalg (E \times [0, 3/4))$ and $E \times (1/4, 1]$, respectively, and choose a basepoint in $E \times \{1/2\}$. Since U and $U \cap V$ are homotopy equivalent to B and E via evident deformations and V is contractible, a consequence of the van Kampen theorem gives the conclusion. $\square$

The space X constructed in the proof is called the "homotopy cofiber" of the map p. It is an important general construction to which we shall return shortly.

PROBLEMS

1. Let F be a free group on two generators a and b. How many subgroups of F have index 2? Specify generators for each of these subgroups.
2. Prove that a non-trivial normal subgroup N with infinite index in a free group F cannot be finitely generated.
3. * Essay: Describe a necessary and sufficient condition for a graph to be embeddable in the plane.

CHAPTER 5

Compactly generated spaces

We briefly describe the category of spaces in which algebraic topologists customarily work. The ordinary category of spaces allows pathology that obstructs a clean development of the foundations. The homotopy and homology groups of spaces are supported on compact subspaces, and it turns out that if one assumes a separation property that is a little weaker than the Hausdorff property, then one can refine the point-set topology of spaces to eliminate such pathology without changing these invariants. We shall leave the proofs to the reader, but the wise reader will simply take our word for it, at least on a first reading: we do not want to overemphasize this material, the importance of which can only become apparent in retrospect.

1. The definition of compactly generated spaces

We shall understand compact spaces to be both compact and Hausdorff, following Bourbaki. A space X is said to be "weak Hausdorff" if $g(K)$ is closed in X for every map $g : K \longrightarrow X$ from a compact space K into X. When this holds, the image $g(K)$ is Hausdorff and is therefore a compact subspace of X. This separation property lies between T_1 (points are closed) and Hausdorff, but it is not much weaker than the latter.

A subspace A of X is said to be "compactly closed" if $g^{-1}(A)$ is closed in K for any map $g : K \longrightarrow X$ from a compact space K into X. When X is weak Hausdorff, this holds if and only if the intersection of A with each compact subset of X is closed. A space X is a "k-space" if every compactly closed subspace is closed.

A space X is "compactly generated" if it is a weak Hausdorff k-space. For example, any locally compact space and any weak Hausdorff space that satisfies the first axiom of countability (every point has a countable neighborhood basis) is compactly generated. We have expressed the definition in a form that should make the following statement clear.

LEMMA. *If X is a compactly generated space and Y is any space, then a function $f : X \longrightarrow Y$ is continuous if and only if its restriction to each compact subspace K of X is continuous.*

We can make a space X into a k-space by giving it a new topology in which a space is closed if and only if it is compactly closed in the original topology. We call the resulting space kX. Clearly the identity function $kX \longrightarrow X$ is continuous. If X is weak Hausdorff, then so is kX, hence kX is compactly generated. Moreover, X and kX then have exactly the same compact subsets.

Write $X \times_c Y$ for the product of X and Y with its usual topology and write $X \times Y = k(X \times_c Y)$. If X and Y are weak Hausdorff, then $X \times Y = kX \times kY$. If X is locally compact and Y is compactly generated, then $X \times Y = X \times_c Y$.

37

By definition, a space X is Hausdorff if the diagonal subspace $\Delta X = \{(x,x)\}$ is closed in $X \times_c X$. The weak Hausdorff property admits a similar characterization.

LEMMA. *If X is a k-space, then X is weak Hausdorff if and only if ΔX is closed in $X \times X$.*

2. The category of compactly generated spaces

One major source of point-set level pathology can be passage to quotient spaces. Use of compactly generated topologies alleviates this.

PROPOSITION. *If X is compactly generated and $\pi : X \longrightarrow Y$ is a quotient map, then Y is compactly generated if and only if $(\pi \times \pi)^{-1}(\Delta Y)$ is closed in $X \times X$.*

The interpretation is that a quotient space of a compactly generated space by a "closed equivalence relation" is compactly generated. We are particularly interested in the following consequence.

PROPOSITION. *If X and Y are compactly generated spaces, A is a closed subspace of X, and $f : A \longrightarrow Y$ is any continuous map, then the pushout $Y \cup_f X$ is compactly generated.*

Another source of pathology is passage to colimits over sequences of maps $X_i \longrightarrow X_{i+1}$. When the given maps are inclusions, the colimit is the union of the sets X_i with the "topology of the union;" a set is closed if and only if its intersection with each X_i is closed.

PROPOSITION. *If $\{X_i\}$ is a sequence of compactly generated spaces and inclusions $X_i \longrightarrow X_{i+1}$ with closed images, then $\operatorname{colim} X_i$ is compactly generated.*

We now adopt a more categorical point of view. We redefine $\mathscr{U}$ to be the category of compactly generated spaces and continuous maps, and we redefine $\mathscr{T}$ to be its subcategory of based spaces and based maps.

Let $w\mathscr{U}$ be the category of weak Hausdorff spaces. We have the functor $k : w\mathscr{U} \longrightarrow \mathscr{U}$, and we have the forgetful functor $j : \mathscr{U} \longrightarrow w\mathscr{U}$, which embeds $\mathscr{U}$ as a full subcategory of $w\mathscr{U}$. Clearly

$$\mathscr{U}(X, kY) \cong w\mathscr{U}(jX, Y)$$

for $X \in \mathscr{U}$ and $Y \in w\mathscr{U}$ since the identity map $kY \longrightarrow Y$ is continuous and continuity of maps defined on compactly generated spaces is compactly determined. Thus k is right adjoint to j.

We can construct colimits and limits of spaces by performing these constructions on sets: they inherit topologies that give them the universal properties of colimits and limits in the classical category of spaces. Limits of weak Hausdorff spaces are weak Hausdorff, but limits of k-spaces need not be k-spaces. We construct limits of compactly generated spaces by applying the functor k to their limits as spaces. It is a categorical fact that functors which are right adjoints preserve limits, so this does give categorical limits in $\mathscr{U}$. This is how we defined $X \times Y$, for example.

Point-set level colimits of weak Hausdorff spaces need not be weak Hausdorff. However, if a point-set level colimit of compactly generated spaces is weak Hausdorff, then it is a k-space and therefore compactly generated. We shall only be interested in colimits in those cases where this holds. The propositions above give examples. In such cases, these constructions give categorical colimits in $\mathscr{U}$.

From here on, we agree that all given spaces are to be compactly generated, and we agree to redefine any construction on spaces by applying the functor k to it. For example, for spaces X and Y in $\mathscr{U}$, we understand the function space $\mathrm{Map}(X,Y) = Y^X$ to mean the set of continuous maps from X to Y with the k-ification of the standard compact-open topology; the latter topology has as basis the finite intersections of the subsets of the form $\{f | f(K) \subset U\}$ for some compact subset K of X and open subset U of Y. This leads to the following adjointness homeomorphism, which holds without restriction when we work in the category of compactly generated spaces.

PROPOSITION. *For spaces X, Y, and Z in $\mathscr{U}$, the canonical bijection*
$$Z^{(X \times Y)} \cong (Z^Y)^X$$
is a homeomorphism.

Observe in particular that a homotopy $X \times I \longrightarrow Y$ can equally well be viewed as a map $X \longrightarrow Y^I$. These adjoint, or "dual," points of view will play an important role in the next two chapters.

PROBLEMS

1. (a) Any subspace of a weak Hausdorff space is weak Hausdorff.
   (b) Any closed subspace of a k-space is a k-space.
   (c) An open subset U of a compactly generated space X is compactly generated if each point has an open neighborhood in X with closure contained in U.
2. * A Tychonoff (or completely regular) space X is a T_1-space (points are closed) such that for each point $x \in X$ and each closed subset A such that $x \notin A$, there is a function $f : X \longrightarrow I$ such that $f(x) = 0$ and $f(a) = 1$ if $a \in A$. Prove the following (e.g., Kelley, *General Topology*).
   (a) A space is Tychonoff if and only if it can be embedded in a cube (a product of copies of I).
   (b) There are Tychonoff spaces that are not k-spaces, but every cube is a compact Hausdorff space.
3. Brief essay: In view of Problems 1 and 2, what should we mean by a "subspace" of a compactly generated space. (We do *not* want to restrict the allowable set of subsets.)

CHAPTER 6

Cofibrations

Exact sequences that feature in the study of homotopy, homology, and cohomology groups all can be derived homotopically from the theory of cofiber and fiber sequences that we present in this and the following two chapters. Abstractions of these ideas are at the heart of modern axiomatic treatments of homotopical algebra and of the foundations of algebraic K-theory.

The theories of cofiber and fiber sequences illustrate an important, but informal, duality theory, known as Eckmann-Hilton duality. It is based on the adjunction between Cartesian products and function spaces. Our standing hypothesis that all spaces in sight are compactly generated allows the theory to be developed without further restrictions on the given spaces. We discuss "cofibrations" here and the "dual" notion of "fibrations" in the next chapter.

1. The definition of cofibrations

DEFINITION. A map $i : A \longrightarrow X$ is a cofibration if it satisfies the homotopy extension property (HEP). This means that if $h \circ i_0 = f \circ i$ in the diagram

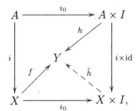

then there exists $\tilde{h}$ that makes the diagram commute.

Here $i_0(x) = (x, 0)$. We do not require $\tilde{h}$ to be unique, and it usually isn't. Using our alternative way of writing homotopies, we see that the "test diagram" displayed in the definition can be rewritten in the equivalent form

$$\begin{array}{ccc} A & \xrightarrow{h} & Y^I \\ {\scriptstyle i}\downarrow & {\scriptstyle \tilde{h}}\nearrow & \downarrow{\scriptstyle p_0} \\ X & \xrightarrow{f} & Y, \end{array}$$

where $p_0(\xi) = \xi(0)$.

Pushouts of cofibrations are cofibrations, in the sense of the following result. We generally write $B \cup_g X$ for the pushout of a given cofibration $i : A \longrightarrow X$ and a map $g : A \longrightarrow B$.

LEMMA. *If* $i : A \longrightarrow X$ *is a cofibration and* $g : A \longrightarrow B$ *is any map, then the induced map* $B \longrightarrow B \cup_g X$ *is a cofibration.*

PROOF. Notice that $(B \cup_g X) \times I \cong (B \times I) \cup_{g \times \mathrm{id}} (X \times I)$ and consider a typical test diagram for the HEP. The proof is a formal chase of the following diagram:

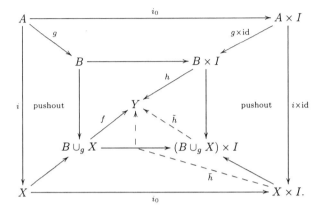

We first use that $A \longrightarrow X$ is a cofibration to obtain a homotopy $\bar{h} : X \times I \longrightarrow Y$ and then use the right-hand pushout to see that $\bar{h}$ and h induce the required homotopy $\tilde{h}$. □

2. Mapping cylinders and cofibrations

Although the HEP is expressed in terms of general test diagrams, there is a certain universal test diagram. Namely, we can let Y in our original test diagram be the "mapping cylinder"
$$Mi \equiv X \cup_i (A \times I),$$
which is the pushout of i and i_0. Indeed, suppose that we can construct a map r that makes the following diagram commute:

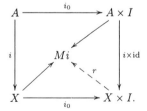

By the universal property of pushouts, the given maps f and h in our original test diagram induce a map $Mi \longrightarrow Y$, and its composite with r gives a homotopy $\tilde{h}$ that makes the test diagram commute.

A map r that makes the previous diagram commute satisfies $r \circ j = \mathrm{id}$, where $j : Mi \longrightarrow X \times I$ is the map that restricts to i_0 on X and to $i \times \mathrm{id}$ on $A \times I$. As a matter of point-set topology, left as an exercise, it follows that a cofibration is an inclusion with closed image.

3. Replacing maps by cofibrations

We can use the mapping cylinder construction to decompose an arbitrary map $f : X \longrightarrow Y$ as the composite of a cofibration and a homotopy equivalence. That is, up to homotopy, any map can be replaced by a cofibration. To see this, recall that $Mf = Y \cup_f (X \times I)$ and observe that f coincides with the composite

$$X \xrightarrow{j} Mf \xrightarrow{r} Y,$$

where $j(x) = (x, 1)$ and where $r(y) = y$ on Y and $r(x, s) = f(x)$ on $X \times I$. If $i : Y \longrightarrow Mf$ is the inclusion, then $r \circ i = \mathrm{id}$ and $\mathrm{id} \simeq i \circ r$. In fact, we can define a deformation $h : Mf \times I \longrightarrow Mf$ of Mf onto $i(Y)$ by setting

$$h(y, t) = y \quad \text{and} \quad h((x, s), t) = (x, (1-t)s).$$

It is not hard to check directly that $j : X \longrightarrow Mf$ satisfies the HEP, and this will also follow from the general criterion for a map to be a cofibration to which we turn next.

4. A criterion for a map to be a cofibration

We want a criterion that allows us to recognize cofibrations when we see them. We shall often consider pairs (X, A) consisting of a space X and a subspace A. Cofibration pairs will be those pairs that "behave homologically" just like the associated quotient spaces X/A.

DEFINITION. A pair (X, A) is an NDR-pair (= neighborhood deformation retract pair) if there is a map $u : X \longrightarrow I$ such that $u^{-1}(0) = A$ and a homotopy $h : X \times I \longrightarrow X$ such that $h_0 = \mathrm{id}$, $h(a, t) = a$ for $a \in A$ and $t \in I$, and $h(x, 1) \in A$ if $u(x) < 1$; (X, A) is a DR-pair if $u(x) < 1$ for all $x \in X$, in which case A is a deformation retract of X.

LEMMA. If (h, u) and (j, v) represent (X, A) and (Y, B) as NDR-pairs, then (k, w) represents the "product pair" $(X \times Y, X \times B \cup A \times Y)$ as an NDR-pair, where $w(x, y) = \min(u(x), v(y))$ and

$$k(x, y, t) = \begin{cases} (h(x, t), j(y, tu(x)/v(y))) & \text{if } v(y) \geq u(x) \\ (h(x, tv(y)/u(x)), j(y, t)) & \text{if } u(x) \geq v(y). \end{cases}$$

If (X, A) or (Y, B) is a DR-pair, then so is $(X \times Y, X \times B \cup A \times Y)$.

PROOF. If $v(y) = 0$ and $v(y) \geq u(x)$, then $u(x) = 0$ and both $y \in B$ and $x \in A$; therefore we can and must understand $k(x, y, t)$ to be (x, y). It is easy to check from this and the symmetric observation that k is a well defined continuous homotopy as desired. □

THEOREM. Let A be a closed subspace of X. Then the following are equivalent:
(i) (X, A) is an NDR-pair.
(ii) $(X \times I, X \times \{0\} \cup A \times I)$ is a DR-pair.
(iii) $X \times \{0\} \cup A \times I$ is a retract of $X \times I$.
(iv) The inclusion $i : A \longrightarrow X$ is a cofibration.

PROOF. The lemma gives that (i) implies (ii), (ii) trivially implies (iii), and we have already seen that (iii) and (iv) are equivalent. Assume given a retraction

$r: X \times I \longrightarrow X \times \{0\} \cup A \times I$. Let $\pi_1: X \times I \longrightarrow X$ and $\pi_2: X \times I \longrightarrow I$ be the projections and define $u: X \longrightarrow I$ by

$$u(x) = \sup\{t - \pi_2 r(x,t) | t \in I\}$$

and $h: X \times I \longrightarrow X$ by

$$h(x,t) = \pi_1 r(x,t).$$

Then (h, u) represents (X, A) as an NDR-pair. Here $u^{-1}(0) = A$ since $u(x) = 0$ implies that $r(x,t) \in A \times I$ for $t > 0$ and thus also for $t = 0$ since $A \times I$ is closed in $X \times I$. □

5. Cofiber homotopy equivalence

It is often important to work in the category of spaces under a given space A, and we shall later need a basic result about homotopy equivalences in this category. We shall also need a generalization concerning homotopy equivalences of pairs. The reader is warned that the results of this section, although easy enough to understand, have fairly lengthy and unilluminating proofs.

A space under A is a map $i: A \longrightarrow X$. A map of spaces under A is a commutative diagram

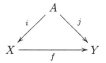

A homotopy between maps under A is a homotopy that at each time t is a map under A. We then write $h: f \simeq f'$ rel A and have $h(i(a),t) = j(a)$ for all $a \in A$ and $t \in I$. There results a notion of a homotopy equivalence under A. Such an equivalence is called a "cofiber homotopy equivalence." The name is suggested by the following result, whose proof illustrates a more substantial use of the HEP than we have seen before.

PROPOSITION. *Let $i: A \longrightarrow X$ and $j: A \longrightarrow Y$ be cofibrations and let $f: X \longrightarrow Y$ be a map such that $f \circ i = j$. Suppose that f is a homotopy equivalence. Then f is a cofiber homotopy equivalence.*

PROOF. It suffices to find a map $g: Y \longrightarrow X$ under A and a homotopy $g \circ f \simeq \text{id}$ rel A. Indeed, g will then be a homotopy equivalence, and we can repeat the argument to obtain $f': X \longrightarrow Y$ such that $f' \circ g \simeq \text{id}$ rel A; it will follow formally that $f' \simeq f$ rel A. By hypothesis, there is a map $g'': Y \longrightarrow X$ that is a homotopy inverse to f. Since $g'' \circ f \simeq \text{id}$, $g'' \circ j \simeq i$. Since j satisfies the HEP, it follows directly that g'' is homotopic to a map g' such that $g' \circ j = i$. It suffices to prove that $g' \circ f: X \longrightarrow X$ has a left homotopy inverse $e: X \longrightarrow X$ under A, since $g = e \circ g'$ will then satisfy $g \circ f \simeq \text{id}$ rel A. Replacing our original map f with $g' \circ f$, we see that it suffices to obtain a left homotopy inverse under A to a map $f: X \longrightarrow X$ such that $f \circ i = i$ and $f \simeq \text{id}$. Choose a homotopy $h: f \simeq \text{id}$. Since $h_0 \circ i = f \circ i = i$ and $h_1 = \text{id}$, we can apply the HEP to $h \circ (i \times \text{id}): A \times I \longrightarrow X$ and the identity map of X to obtain a homotopy $k: \text{id} \simeq k_1 \equiv e$ such that $k \circ (i \times \text{id}) = h \circ (i \times \text{id})$. Certainly $e \circ i = i$. Now apply the HEP to the following diagram:

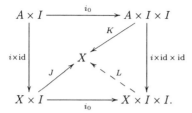

Here J is the homotopy $e \circ f \simeq \text{id}$ specified by

$$J(x,s) = \begin{cases} k(f(x), 1-2s) & \text{if } s \leq 1/2 \\ h(x, 2s-1) & \text{if } 1/2 \leq s. \end{cases}$$

The homotopy between homotopies K is specified by

$$K(a,s,t) = \begin{cases} k(i(a), 1-2s(1-t)) & \text{if } s \leq 1/2 \\ h(i(a), 1-2(1-s)(1-t)) & \text{if } s \geq 1/2. \end{cases}$$

Traversal of L around the three faces of $I \times I$ other than that specified by J gives a homotopy

$$e \circ f = J_0 = L_{0,0} \simeq L_{0,1} \simeq L_{1,1} \simeq L_{1,0} = J_1 = \text{id rel } A. \quad \square$$

The proposition applies to the following previously encountered situation.

EXAMPLE. Let $i : A \longrightarrow X$ be a cofibration. We then have the commutative diagram

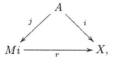

where $j(a) = (a, 1)$. The obvious homotopy inverse $\iota : X \longrightarrow Mi$ has $\iota(x) = (x, 0)$ and is thus very far from being a map under A. The proposition ensures that ι is homotopic to a map under A that is homotopy inverse to r under A.

The following generalization asserts that, for inclusions that are cofibrations, a pair of homotopy equivalences is a homotopy equivalence of pairs. It is often used implicitly in setting up homology and cohomology theories on pairs of spaces.

PROPOSITION. *Assume given a commutative diagram*

$$\begin{array}{ccc} A & \xrightarrow{d} & B \\ i \downarrow & & \downarrow j \\ X & \xrightarrow{f} & Y \end{array}$$

in which i and j are cofibrations and d and f are homotopy equivalences. Then $(f, d) : (X, A) \longrightarrow (Y, B)$ is a homotopy equivalence of pairs.

PROOF. The statement means that there are homotopy inverses e of d and g of f such that $g \circ j = i \circ e$ together with homotopies $H : g \circ f \simeq \mathrm{id}$ and $K : f \circ g \simeq \mathrm{id}$ that extend homotopies $h : e \circ d \simeq \mathrm{id}$ and $k : d \circ e \simeq \mathrm{id}$. Choose any homotopy inverse e to d, together with homotopies $h : e \circ d \simeq \mathrm{id}$ and $\ell : d \circ e \simeq \mathrm{id}$. By HEP for j, there is a homotopy inverse g' for f such that $g' \circ j = i \circ e$. Then, by HEP for i, there is a homotopy m of $g' \circ f$ such that $m \circ (i \times \mathrm{id}) = i \circ h$. Let $\phi = m_1$. Then $\phi \circ i = i$ and ϕ is a cofiber homotopy equivalence by the previous result. Let $\psi : X \longrightarrow X$ be a homotopy inverse under i and let $n : \psi \circ \phi \simeq \mathrm{id}$ be a homotopy under i. Define $g = \psi \circ g'$. Clearly $g \circ j = i \circ e$. Using that the pairs $(I \times I, I \times \{0\})$ and $(I \times I, I \times \{0\} \cup \partial I \times I)$ are homeomorphic, we can construct a homotopy between homotopies Λ by applying HEP to the diagram

$$\begin{array}{ccc}
(A \times I \times 0) \cup (A \times \partial I \times I) & \xrightarrow{\subset} & A \times I \times I \\
{\scriptstyle i \times \mathrm{id}} \downarrow & \searrow^{\Gamma} \quad X \quad \nwarrow^{\Lambda} & \downarrow {\scriptstyle i \times \mathrm{id}} \\
(X \times I \times 0) \cup (X \times \partial I \times I) & \xrightarrow[\subset]{\gamma} & X \times I \times I.
\end{array}$$

Here

$$\gamma(x,s,0) = \begin{cases} \psi(m(x,2s)) & \text{if } s \leq 1/2 \\ n(x, 2s-1) & \text{if } s \geq 1/2, \end{cases}$$
$$\gamma(x,0,t) = (g \circ f)(x) = (\psi \circ g' \circ f)(x),$$

and

$$\gamma(x,1,t) = x,$$

while

$$\Gamma(a,s,t) = \begin{cases} i(h(a, 2s/(1+t))) & \text{if } 2s \leq 1+t \\ i(a) & \text{if } 2s \geq 1+t \end{cases}$$

Define $H(x,s) = \Lambda(x,s,1)$. Then $H : g \circ f \simeq \mathrm{id}$ and $H \circ (i \times \mathrm{id}) = i \circ h$. Application of this argument with d and f replaced by e and g gives a left homotopy inverse f' to g and a homotopy $L : f' \circ g \simeq \mathrm{id}$ such that $f' \circ i = j \circ d$ and $L \circ (j \times \mathrm{id}) = j \circ \ell$. Adding homotopies by concentrating them on successive fractions of the unit interval and letting the negative of a homotopy be obtained by reversal of direction, define

$$k = (-\ell)(de \times \mathrm{id}) + dh(e \times \mathrm{id}) + \ell$$

and

$$K = (-L)(fg \times \mathrm{id}) + f'H(g \times \mathrm{id}) + L.$$

Then $K : f \circ g \simeq \mathrm{id}$ and $K \circ (j \times \mathrm{id}) = j \circ k$. $\square$

PROBLEMS

1. Show that a cofibration $i : A \longrightarrow X$ is an inclusion with closed image.
2. Let $i : A \longrightarrow X$ be a cofibration, where A is a contractible space. Prove that the quotient map $X \longrightarrow X/A$ is a homotopy equivalence.

CHAPTER 7

Fibrations

We "dualize" the definitions and theory of the previous chapter to the study of fibrations, which are "up to homotopy" generalizations of covering spaces.

1. The definition of fibrations

DEFINITION. A surjective map $p : E \longrightarrow B$ is a fibration if it satisfies the covering homotopy property (CHP). This means that if $h \circ i_0 = p \circ f$ in the diagram

$$\begin{array}{ccc} Y & \xrightarrow{f} & E \\ {\scriptstyle i_0}\downarrow & {\scriptstyle \tilde{h}}\nearrow & \downarrow {\scriptstyle p} \\ Y \times I & \xrightarrow{h} & B, \end{array}$$

then there exists $\tilde{h}$ that makes the diagram commute.

This notion of a fibration is due to Hurewicz. There is a more general notion of a Serre fibration, in which the test spaces Y are restricted to be cubes I^n. Serre fibrations are more appropriate for many purposes, but we shall make no use of them. The test diagram in the definition can be rewritten in the equivalent form

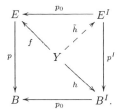

Here $p_0(\beta) = \beta(0)$ for $\beta \in B^I$. With this formulation, we can "dualize" the proof that pushouts of cofibrations are cofibrations to show that pullbacks of fibrations are fibrations. We often write $A \times_g E$ for the pullback of a given fibration $p : E \longrightarrow B$ and a map $g : A \longrightarrow B$.

LEMMA. If $p : E \longrightarrow B$ is a fibration and $g : A \longrightarrow B$ is any map, then the induced map $A \times_g E \longrightarrow A$ is a fibration.

2. Path lifting functions and fibrations

Although the CHP is expressed in terms of general test diagrams, there is a certain universal test diagram. Namely, we can let Y in our original test diagram be the "mapping path space"

$$Np \equiv E \times_p B^I = \{(e, \beta) | \beta(0) = p(e)\} \subset E \times B^I.$$

That is, Np is the pullback of p and p_0 in the second form of the test diagram and, with $Y = Np$, f and h in that diagram are the evident projections. A map $s : Np \longrightarrow E^I$ such that $k \circ s = \mathrm{id}$, where $k : E^I \longrightarrow Np$ has coordinates p_0 and p^I, is called a path lifting function. Thus
$$s(e, \beta)(0) = e \quad \text{and} \quad p \circ s(e, \beta) = \beta.$$
Given a general test diagram, there results a map $g : Y \longrightarrow Np$ determined by f and h, and we can take $\tilde{h} = s \circ g$.

In general, path lifting functions are not unique. In fact, we have already studied the special kinds of fibrations for which they are unique.

LEMMA. *If $p : E \longrightarrow B$ is a covering, then p is a fibration with a unique path lifting function s.*

PROOF. The unique lifts of paths with a given initial point specify s. □

Fibrations and cofibrations are related by the following useful observation.

LEMMA. *If $i : A \longrightarrow X$ is a cofibration and B is a space, then the induced map*
$$p = B^i : B^X \longrightarrow B^A$$
is a fibration.

PROOF. It is an easy matter to check that we have a homeomorphism
$$B^{Mi} = B^{X \times \{0\} \cup A \times I} \cong B^X \times_p (B^A)^I = Np.$$
If $r : X \times I \longrightarrow Mi$ is a retraction, then
$$B^r : Np \cong B^{Mi} \longrightarrow B^{X \times I} \cong (B^X)^I$$
is a path lifting function. □

3. Replacing maps by fibrations

We can use the mapping path space construction to decompose an arbitrary map $f : X \longrightarrow Y$ as the composite of a homotopy equivalence and a fibration. That is, up to homotopy, any map can be replaced by a fibration. To see this, recall that $Nf = X \times_f Y^I$ and observe that f coincides with the composite
$$X \xrightarrow{\nu} Nf \xrightarrow{\rho} Y,$$
where $\nu(x) = (x, c_{f(x)})$ and $\rho(x, \chi) = \chi(1)$. Let $\pi : Nf \longrightarrow X$ be the projection. Then $\pi \circ \nu = \mathrm{id}$ and $\mathrm{id} \simeq \nu \circ \pi$ since we can define a deformation $h : Nf \times I \longrightarrow Nf$ of Nf onto $\nu(X)$ by setting
$$h(x, \chi)(t) = (x, \chi_t), \quad \text{where } \chi_t(s) = \chi((1 - t)s).$$
We check directly that $\rho : Nf \longrightarrow Y$ satisfies the CHP. Consider a test diagram

$$\begin{array}{ccc} A & \xrightarrow{g} & Nf \\ {\scriptstyle i_0}\downarrow & \tilde{h}\nearrow & \downarrow{\scriptstyle \rho} \\ A \times I & \xrightarrow{h} & Y. \end{array}$$

We are given g and h such that $h \circ i_0 = \rho \circ g$ and must construct $\tilde{h}$ that makes the diagram commute. We write $g(a) = (g_1(a), g_2(a))$ and set
$$\tilde{h}(a, t) = (g_1(a), j(a, t)),$$

where
$$j(a,t)(s) = \begin{cases} g_2(a)(s+st) & \text{if } 0 \le s \le 1/(1+t) \\ h(a, s+ts-1) & \text{if } 1/(1+t) \le s \le 1. \end{cases}$$

4. A criterion for a map to be a fibration

Again, we want a criterion that allows us to recognize fibrations when we see them. Here the idea of duality fails, and we instead think of fibrations as generalizations of coverings. When restricted to the spaces U in a well chosen open cover $\mathscr{O}$ of the base space B, a covering is homeomorphic to the projection $U \times F \longrightarrow U$, where F is a fixed discrete set.

The obvious generalization of this is the notion of a bundle. A map $p : E \longrightarrow B$ is a bundle if, when restricted to the spaces U in a well chosen open cover $\mathscr{O}$ of B, there are homeomorphisms $\phi : U \times F \longrightarrow p^{-1}(U)$ such that $p \circ \phi = \pi_1$, where F is a fixed topological space.

We shall require of well chosen open covers that they be numerable. This means that there are continuous maps $\lambda_U : B \longrightarrow I$ such that $\lambda_U^{-1}(0,1] = U$ and that the cover is locally finite, in the sense that each $b \in B$ is a point of only finitely many $U \in \mathscr{O}$. Any open cover of a paracompact space has a numerable refinement. With this proviso on the open covers allowed in the definition of a bundle, the following result shows in particular that every bundle is a fibration.

THEOREM. *Let $p : E \longrightarrow B$ be a map and let $\mathscr{O}$ be a numerable open cover of B. Then p is a fibration if and only if $p : p^{-1}(U) \longrightarrow U$ is a fibration for every $U \in \mathscr{O}$.*

PROOF. Since pullbacks of fibrations are fibrations, necessity is obvious. Thus assume that $p|p^{-1}(U)$ is a fibration for each $U \in \mathscr{O}$. We shall construct a path lifting function for B by patching together path lifting functions for the $p|p^{-1}(U)$, but we first set up the scaffolding of the patching argument. Choose maps $\lambda_U : B \longrightarrow I$ such that $\lambda_U^{-1}(0,1] = U$. For a finite ordered subset $T = \{U_1, \ldots, U_n\}$ of sets in $\mathscr{O}$, define $c(T) = n$ and define $\lambda_T : B^I \longrightarrow I$ by
$$\lambda_T(\beta) = \inf\{(\lambda_{U_i} \circ \beta)(t) | (i-1)/n \le t \le i/n,\ 1 \le i \le n\}.$$
Let $W_T = \lambda_T^{-1}(0,1]$. Equivalently,
$$W_T = \{\beta | \beta(t) \in U_i \text{ if } t \in [(i-1)/n, i/n]\} \subset B^I.$$
The set $\{W_T\}$ is an open cover of B^I, but it need not be locally finite. However, $\{W_T | c(T) < n\}$ is locally finite for each fixed n. If $c(T) = n$, define $\gamma_T : B^I \longrightarrow I$ by
$$\gamma_T(\beta) = \max\{0, \lambda_T(\beta) - n\textstyle\sum_{c(S)<n} \lambda_S(\beta)\},$$
and define
$$V_T = \{\beta | \gamma_T(\beta) > 0\} \subset W_T.$$
Then $\{V_T\}$ is a locally finite open cover of B^I. We choose a total ordering of the set of all finite ordered subsets T of $\mathscr{O}$.

With this scaffolding in place, choose path lifting functions
$$s_U : p^{-1}(U) \times_p U^I \longrightarrow p^{-1}(U)^I$$
for $U \in \mathscr{O}$, so that $(p \circ s_U)(e, \beta) = \beta$ and $s_U(e, \beta)(0) = e$. For a given $T = \{U_1, \ldots, U_n\}$, consider paths $\beta \in V_T$. For $0 \le u < v \le 1$, let $\beta[u,v]$ be the restriction of β to $[u,v]$. If $u \in [(i-1)/n, i/n]$ and $v \in [(j-1)/n, j/n]$, where

$0 \leq i \leq j \leq n$, and if $e \in p^{-1}(\beta(u))$, define $s_T(e, \beta[u,v]) : [u,v] \longrightarrow E$ to be the path that starts at e and covers $\beta[u,v]$ that is obtained by applying s_{U_i} to lift over $[u, i/n]$ (or over $[u,v]$ if $i = j$), using $s_{U_{i+1}}$, starting at the point where the first lifted path ends, to lift over $[i/n, (i+1)/n]$ and so on inductively, ending with use of s_{U_j} to lift over $[(j-1)/n, v]$. (Technically, since we are lifting over partial intervals and the s_U lift paths defined on I to paths defined on I, this involves a rescaling: we must shrink I linearly onto our subinterval, then apply the relevant part of β, next lift the resulting path, and finally apply the result to the linear expansion of our subinterval onto I.) For a point (e, β) in Np, define $s(e, \beta)$ to be the concatenation of the paths $s_{T_j}(e_{j-1}, \beta[u_{j-1}, u_j])$, $1 \leq j \leq q$, where the T_i, in order, run through the set of all T such that $\beta \in V_T$, where $u_0 = 0$ and $u_j = \sum_{i=1}^{j} \gamma_{T_i}(\beta)$ for $1 \leq j \leq q$, and where $e_0 = e$ and e_j is the endpoint of $s_{T_j}(e_{j-1}, \beta[u_{j-1}, u_j])$ for $1 \leq j < q$. Certainly $s(e, \beta) = e$ and $(p \circ s)(e, \beta) = \beta$. It is not hard to check that s is well defined and continuous, hence it is a path lifting function for p. □

5. Fiber homotopy equivalence

It is often important to study fibrations over a given base space B, working in the category of spaces over B. A space over B is a map $p : E \longrightarrow B$. A map of spaces over B is a commutative diagram

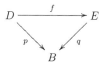

A homotopy between maps over B is a homotopy that at each time t is a map over B. There results a notion of a homotopy equivalence over B. Such an equivalence is called a "fiber homotopy equivalence." The name is suggested by the following result, whose proof is precisely dual to the corresponding result for cofibrations and is left as an exercise.

PROPOSITION. Let $p : D \longrightarrow B$ and $q : E \longrightarrow B$ be fibrations and let $f : D \longrightarrow E$ be a map such that $q \circ f = p$. Suppose that f is a homotopy equivalence. Then f is a fiber homotopy equivalence.

EXAMPLE. Let $p : E \longrightarrow B$ be a fibration. We then have the commutative diagram

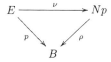

where $\nu(e) = (e, c_{p(e)})$ and $\rho(e, \chi) = \chi(1)$. The obvious homotopy inverse $\pi : Np \longrightarrow E$ is not a map over B, but the proposition ensures that it is homotopic to a map over B that is homotopy inverse to ν over B.

The result generalizes as follows, the proof again being dual to the proof of the corresponding result for cofibrations.

PROPOSITION. *Assume given a commutative diagram*

$$\begin{array}{ccc} D & \xrightarrow{f} & E \\ p\downarrow & & \downarrow q \\ A & \xrightarrow{d} & B \end{array}$$

in which p and q are fibrations and d and f are homotopy equivalences. Then $(f,d) : p \longrightarrow q$ is a homotopy equivalence of fibrations.

The statement means that there are homotopy inverses e of d and g of f such that $p \circ g = e \circ q$ together with homotopies $H : g \circ f \simeq \text{id}$ and $K : f \circ g \simeq \text{id}$ that cover homotopies $h : e \circ d \simeq \text{id}$ and $k : d \circ e \simeq \text{id}$.

6. Change of fiber

Translation of fibers along paths in the base space played a fundamental role in our study of covering spaces. Fibrations admit an up to homotopy version of that theory that well illustrates the use of the CHP and will be used later.

Let $p : E \longrightarrow B$ be a fibration with fiber F_b over $b \in B$ and let $i_b : F_b \longrightarrow E$ be the inclusion. For a path $\beta : I \longrightarrow B$ from b to b', the CHP gives a lift $\tilde{\beta}$ in the diagram

$$\begin{array}{ccc} F_b \times \{0\} & \xrightarrow{i_b} & E \\ \downarrow & \tilde{\beta} \nearrow & \downarrow p \\ F_b \times I & \xrightarrow{\pi_2} I \xrightarrow{\beta} & B. \end{array}$$

At time t, $\tilde{\beta}$ maps F_b to the fiber $F_{\beta(t)}$. In particular, at $t = 1$, this gives a map

$$\tau[\beta] \equiv [\tilde{\beta}_1] : F_b \longrightarrow F_{b'},$$

which we call the translation of fibers along the path class $[\beta]$.

We claim that, as indicated by our choice of notation, the homotopy class of the map $\tilde{\beta}_1$ is independent of the choice of β in its path class. Thus suppose that β and β' are equivalent paths from b to b', let $h : I \times I \longrightarrow B$ be a homotopy $\beta \simeq \beta'$ through paths from b to b', and let $\tilde{\beta}' : F_b \times I \longrightarrow E$ cover $\beta' \pi_2$. Observe that if

$$J^2 = I \times \partial I \cup \{0\} \times I \subset I^2,$$

then the pairs (I^2, J^2) and $(I \times I, I \times \{0\})$ are homeomorphic. Define $f : F_b \times J^2 \longrightarrow E$ to be $\tilde{\beta}$ on $F_b \times I \times \{0\}$, $\tilde{\beta}'$ on $F_b \times I \times \{1\}$, and $i_b \circ \pi_1$ on $F_b \times \{0\} \times I$. Then another application of the CHP gives a lift $\tilde{h}$ in the diagram

$$\begin{array}{ccc} F_b \times J^2 & \xrightarrow{f} & E \\ \downarrow & \tilde{h} \nearrow & \downarrow p \\ F_b \times I^2 & \xrightarrow{\pi_2} I^2 \xrightarrow{h} & B. \end{array}$$

Thus $\tilde{h} : \tilde{\beta} \simeq \tilde{\beta}'$ through maps $F_b \times I \longrightarrow E$, each of which starts at the inclusion of F_b in E. At time $t = 1$, this gives a homotopy $\tilde{\beta}_1 \simeq \tilde{\beta}'_1$. Thus $\tau[\beta] = [\tilde{\beta}_1]$ is a well defined *homotopy class* of maps $F_b \longrightarrow F_{b'}$.

We think of $\tau[\beta]$ as a map in the homotopy category $h\mathscr{U}$. It is clear that, in the homotopy category,

$$\tau[c_b] = [\text{id}] \quad \text{and} \quad \tau[\gamma \cdot \beta] = \tau[\gamma] \circ \tau[\beta]$$

if $\gamma(0) = \beta(1)$. It follows that $\tau[\beta]$ is an isomorphism with inverse $\tau[\beta^{-1}]$. This can be stated formally as follows.

THEOREM. *Lifting of equivalence classes of paths in B to homotopy classes of maps of fibers specifies a functor $\lambda : \Pi(B) \longrightarrow h\mathscr{U}$. Therefore, if B is path connected, then any two fibers of B are homotopy equivalent.*

Just as the fundamental group $\pi_1(B, b)$ of the base space of a covering acts on the fiber F_b, so the fundamental group $\pi_1(B, b)$ of the base space of a fibration acts "up to homotopy" on the fiber, in a sense made precise by the following corollary. For a space X, let $\pi_0(X)$ denote the set of path components of X. The set of homotopy equivalences of X is denoted $\text{Aut}(X)$ and is topologized as a subspace of the function space of maps $X \longrightarrow X$. The composite of homotopy equivalences is a homotopy equivalence, and composition defines a continuous product on $\text{Aut}(X)$. With this product, $\text{Aut}(X)$ is a "topological monoid," namely a space with a continuous and associative multiplication with a two-sided identity element, but it is not a group. However, the path components of $\text{Aut}(X)$ are the homotopy classes of homotopy equivalences of X, and these do form a group under composition.

COROLLARY. *Lifting of equivalence classes of loops specifies a homomorphism $\pi_1(B, b) \longrightarrow \pi_0(\text{Aut}(F_b))$.*

We have the following naturality statement with respect to maps of fibrations.

THEOREM. *Let p and q be fibrations in the commutative diagram*

$$\begin{array}{ccc} D & \xrightarrow{g} & E \\ q \downarrow & & \downarrow p \\ A & \xrightarrow{f} & B. \end{array}$$

For a path $\alpha : I \longrightarrow A$ from a to a', the following diagram commutes in $h\mathscr{U}$:

$$\begin{array}{ccc} F_a & \xrightarrow{g} & F_{f(a)} \\ \tau[\alpha] \downarrow & & \downarrow \tau[f \circ \alpha] \\ F_{a'} & \xrightarrow{g} & F_{f(a')}. \end{array}$$

If, further, $h : f \simeq f'$ and $H : g \simeq g'$ in the commutative diagram

$$\begin{array}{ccc} D \times I & \xrightarrow{H} & E \\ q \times \text{id} \downarrow & & \downarrow p \\ A \times I & \xrightarrow{h} & B, \end{array}$$

then the following diagram in $h\mathcal{U}$ also commutes, where $h(a)(t) = h(a,t)$:

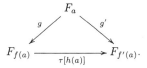

PROOF. Let $\tilde{\alpha} : F_a \times I \longrightarrow D$ lift α and $\tilde{\beta} : F_{f(a)} \times I \longrightarrow E$ lift $f \circ \alpha$. Define $j : F_a \times J^2 \longrightarrow E$ to be $g \circ \tilde{\alpha}$ on $F_a \times I \times \{0\}$, $\tilde{\beta} \circ (g \times \mathrm{id})$ on $F_a \times I \times \{1\}$, and $g \circ \pi_1$ on $F_a \times \{0\} \times I$. Define $k : I^2 \longrightarrow B$ to be the constant homotopy which at each time t is $f \circ \alpha$. Another application of the CHP gives a lift $\tilde{k}$ in the diagram:

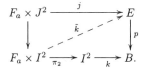

Here $\tilde{k}$ is a homotopy $g \circ \tilde{\alpha} \simeq \tilde{\beta} \circ (g \times \mathrm{id})$ through homotopies starting at $g \circ \pi_1 : F_a \times I \longrightarrow E$. This gives the diagram claimed in the first statement. For the second statement, define $\alpha : I \longrightarrow A \times I$ by $\alpha(t) = (a,t)$, so that $h(a) = h \circ \alpha$. Define $\tilde{\alpha} : F_a \longrightarrow F_a \times I$ by $\tilde{\alpha}(f) = (f,t)$. Then $\tilde{\alpha}$ lifts α and
$$\tau[\alpha] = [\mathrm{id}] : F_a = F_a \times \{0\} \longrightarrow F_a \times \{1\} = F_a.$$
We conclude that the second statement is a special case of the first. □

PROBLEM

1. Prove the proposition stated in §5.

CHAPTER 8

Based cofiber and fiber sequences

We use cofibrations and fibrations in the category $\mathscr{T}$ of based spaces to generate two "exact sequences of spaces" from a given map of based spaces. We shall write $*$ generically for the basepoints of based spaces. Much that we do for cofibrations can be done equally well in the unbased context of the previous chapter. However, the dual theory of fibration sequences only makes sense in the based context.

1. Based homotopy classes of maps

For based spaces X and Y, we let $[X, Y]$ denote the set of based homotopy classes of based maps $X \longrightarrow Y$. This set has a natural basepoint, namely the homotopy class of the constant map from X to the basepoint of Y.

The appropriate analogue of the Cartesian product in the category of based spaces is the "smash product" $X \wedge Y$ defined by

$$X \wedge Y = X \times Y / X \vee Y.$$

Here $X \vee Y$ is viewed as the subspace of $X \times Y$ consisting of those pairs (x, y) such that either x is the basepoint of X or y is the basepoint of Y.

The appropriate based analogue of the function space is the subspace $F(X, Y)$ of Y^X consisting of the based maps, with the constant based map as basepoint. With these definitions, we have a natural homeomorphism of based spaces

$$F(X \wedge Y, Z) \cong F(X, F(Y, Z))$$

for based spaces X and Y.

Recall that $\pi_0(X)$ denotes the set of path components of X. When X is based, so is this set, and we sometimes denote it by $\pi_0(X, *)$. Observe that $[X, Y]$ may be identified with $\pi_0(F(X, Y))$.

2. Cones, suspensions, paths, loops

Let X be a based space. We define the cone on X to be $CX = X \wedge I$, where I is given the basepoint 1. That is,

$$CX = X \times I / (\{*\} \times I \cup X \times \{1\}).$$

We view S^1 as $I/\partial I$, denote its basepoint by 1, and define the suspension of X to be $\Sigma X = X \wedge S^1$. That is,

$$\Sigma X = X \times S^1 / (\{*\} \times S^1 \cup X \times \{1\}).$$

These are sometimes called the reduced cone and suspension, to distinguish them from the unreduced constructions, in which the line $\{*\} \times I$ through the basepoint of X is not identified to a point. We shall make use of both constructions in our work, but we shall not distinguish them notationally.

Dually, we define the path space of X to be $PX = F(I, X)$, where I is given the basepoint 0. Thus the points of PX are the paths in X that start at the basepoint. We define the loop space of X to be $\Omega X = F(S^1, X)$. Its points are the loops at the basepoint.

We have the adjunction

$$F(\Sigma X, Y) \cong F(X, \Omega Y).$$

Passing to π_0, this gives that

$$[\Sigma X, Y] \cong [X, \Omega Y].$$

Composition of loops defines a multiplication on this set. Explicitly, for $f, g : \Sigma X \longrightarrow Y$, we write

$$(g + f)(x \wedge t) = (g(x) \cdot f(x))(t) = \begin{cases} f(x \wedge 2t) & \text{if } 0 \leq t \leq 1/2 \\ g(x \wedge (2t-1)) & \text{if } 1/2 \leq t \leq 1. \end{cases}$$

LEMMA. $[\Sigma X, Y]$ is a group and $[\Sigma^2 X, Y]$ is an Abelian group.

PROOF. The first statement is proved just as for the fundamental group. For the second, think of maps $f, g : \Sigma^2 X \longrightarrow Y$ as maps $S^2 \longrightarrow F(X, Y)$ and think of S^2 as the quotient $I^2/\partial I^2$. Then a homotopy between $g + f$ and $f + g$ can be pictured schematically as follows:

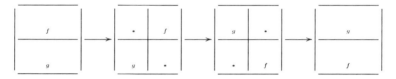

□

3. Based cofibrations

The definition of a cofibration has an evident based variant, in which all given and constructed maps in our test diagrams are required to be based. A based map $i : A \longrightarrow X$ that is a cofibration in the unbased sense is necessarily a cofibration in the based sense since the basepoint of X must lie in A.

We say that X is "nondegenerately based," or "well pointed," if the inclusion of its basepoint is a cofibration in the unbased sense. If A and X are nondegenerately based and $i : A \longrightarrow X$ is a based cofibration, then i is necessarily an unbased cofibration.

We refer to based cofibrations simply as cofibrations in the rest of this chapter.

Write Y_+ for the union of a space Y and a disjoint basepoint and observe that we can identify $X \wedge Y_+$ with $X \times Y/\{*\} \times Y$.

The space $X \wedge I_+$ is called the reduced cylinder on X, and a based homotopy $X \times I \longrightarrow Y$ is the same thing as a based map $X \wedge I_+ \longrightarrow Y$. We change notations and write Mf for the based mapping cylinder $Y \cup_f (X \wedge I_+)$ of a based map f.

As in the unbased case, we conclude that a based map $i : A \longrightarrow X$ is a cofibration if and only if Mi is a retract of $X \wedge I_+$.

4. Cofiber sequences

For a based map $f : X \longrightarrow Y$, define the "homotopy cofiber" Cf to be
$$Cf = Y \cup_f CX = Mf/j(X),$$
where $j : X \longrightarrow Mf$ sends x to $(x, 1)$. As in the unbased case, our original map f is the composite of the cofibration j and the evident retraction $r : Mf \longrightarrow Y$. Thus Cf is constructed by first replacing f by the cofibration j and then taking the associated quotient space.

Let $i : Y \longrightarrow Cf$ be the inclusion. It is a cofibration since it is the pushout of f and the cofibration $X \longrightarrow CX$ that sends x to $(x, 0)$. Let
$$\pi : Cf \longrightarrow Cf/Y \cong \Sigma X$$
be the quotient map. The sequence
$$X \xrightarrow{f} Y \xrightarrow{i} Cf \xrightarrow{\pi} \Sigma X \xrightarrow{-\Sigma f} \Sigma Y \xrightarrow{-\Sigma i} \Sigma Cf \xrightarrow{-\Sigma \pi} \Sigma^2 X \xrightarrow{\Sigma^2 f} \Sigma^2 Y \longrightarrow \cdots$$
is called the cofiber sequence generated by the map f; here
$$(-\Sigma f)(x \wedge t) = f(x) \wedge (1 - t).$$

These "long exact sequences of based spaces" give rise to long exact sequences of pointed sets, where a sequence
$$S' \xrightarrow{f} S \xrightarrow{g} S''$$
of pointed sets is said to be exact if $g(s) = *$ if and only if $s = f(s')$ for some s'.

THEOREM. *For any based space Z, the induced sequence*
$$\cdots \longrightarrow [\Sigma Cf, Z] \longrightarrow [\Sigma Y, Z] \longrightarrow [\Sigma X, Z] \longrightarrow [Cf, Z] \longrightarrow [Y, Z] \longrightarrow [X, Z]$$
is an exact sequence of pointed sets, or of groups to the left of $[\Sigma X, Z]$, or of Abelian groups to the left of $[\Sigma^2 X, Z]$.

Exactness is clear at the first stage, where we are considering the composite of $f : X \longrightarrow Y$ and the inclusion i of Y in the cofiber Cf. To see this, consider the diagram

$$\begin{array}{ccc} X \xrightarrow{f} Y \xrightarrow{i} Cf = Y \cup_f CX \\ {\scriptstyle g} \downarrow \quad \swarrow\, \tilde{g} = g \cup h \\ Z \end{array}$$

Here $h : g \circ f \simeq c_*$, and we view h as a map $CX \longrightarrow Z$. Thus we check exactness by using any given homotopy to extend g over the cofiber. We emphasize that this applies to any composite pair of maps of the form (f, i), where i is the inclusion of the target of f in the cofiber of f.

We claim that, up to homotopy equivalence, each consecutive pair of maps in our cofiber sequence is the composite of a map and the inclusion of its target in its cofiber. This will imply the theorem. We observe that, for any map f, interchange of the cone and suspension coordinate gives a homeomorphism
$$\Sigma Cf \cong C(\Sigma f)$$

such that the following diagram commutes:

$$\begin{array}{ccccccc}
\Sigma X & \xrightarrow{\Sigma f} & \Sigma Y & \xrightarrow{\Sigma i(f)} & \Sigma Cf & \xrightarrow{\Sigma \pi(f)} & \Sigma^2 X \\
\| & & \| & & \downarrow \cong & & \downarrow \tau \\
\Sigma X & \xrightarrow{\Sigma f} & \Sigma Y & \xrightarrow{i(\Sigma f)} & C(\Sigma f) & \xrightarrow{\pi(\Sigma f)} & \Sigma^2 X.
\end{array}$$

Here $\tau : \Sigma^2 X \longrightarrow \Sigma^2 X$ is the homeomorphism obtained by interchanging the two suspension coordinates; we shall see later, and leave as an exercise here, that τ is homotopic to $-$ id. We have written $i(f)$, $\pi(f)$, etc., to indicate the maps to which the generic constructions i and π are applied. Using this inductively, we see that we need only verify our claim for the two pairs of maps $(i(f), \pi(f))$ and $(\pi(f), -\Sigma f)$. The following two lemmas will imply the claim in these two cases. More precisely, they will imply the claim directly for the first pair and will imply that the second pair is equivalent to a pair of the same form as the first pair.

LEMMA. *If $i : A \longrightarrow X$ is a cofibration, then the quotient map*

$$\psi : Ci \longrightarrow Ci/CA \cong X/A$$

is a based homotopy equivalence.

PROOF. Since i is a cofibration, there is a retraction

$$r : X \wedge I_+ \longrightarrow Mi = X \cup_i (A \wedge I_+).$$

We embed X as $X \times \{1\}$ in the source and collapse out $A \times \{1\}$ from the target. The resulting composite $X \longrightarrow Ci$ maps A to $\{*\}$ and so induces a map $\phi : X/A \longrightarrow Ci$. The map r restricts to the identity on $A \wedge I_+$, and if we collapse out $A \wedge I_+$ from its source and target, then r becomes a homotopy id $\simeq \psi \circ \phi$. The map r on $X \wedge I_+$ glues together with the map $h : CA \wedge I_+ \longrightarrow CA$ specified by

$$h(a, s, t) = (a, \max(s, t))$$

to give a homotopy $Ci \wedge I_+ \longrightarrow Ci$ from the identity to $\phi \circ \psi$. $\square$

LEMMA. *The left triangle commutes and the right triangle commutes up to homotopy in the diagram*

$$X \xrightarrow{f} Y \xrightarrow{i(f)} Cf \xrightarrow{\pi(f)} \Sigma X \xrightarrow{-\Sigma f} \Sigma Y \longrightarrow \cdots$$

with $i(i(f))$, ψ, $\pi(i(f))$ through $Ci(f)$.

PROOF. Observe that $Ci(f)$ is obtained by gluing the cones CX and CY along their bases via the map $f : X \longrightarrow Y$. The left triangle commutes since collapsing out CY from $Ci(f)$ is the same as collapsing out Y from Cf. A homotopy $h : Ci(f) \wedge I_+ \longrightarrow \Sigma Y$ from π to $(-\Sigma f) \circ \psi$ is given by

$$h(x, s, t) = (f(x), t - st) \quad \text{on} \quad CX$$

and

$$h(y, s, t) = (y, s + t - st) \quad \text{on} \quad CY. \quad \square$$

5. Based fibrations

Similarly, the definition of a fibration has an evident based variant, in which all given and constructed maps in our test diagrams are required to be based. A based fibration $p : E \longrightarrow B$ is necessarily a fibration in the unbased sense, as we see by restricting to spaces of the form Y_+ in test diagrams and noting that $Y_+ \wedge I_+ \cong (Y \times I)_+$. Less obviously, if p is a based map that is an unbased fibration, then it satisfies the based CHP for test diagrams in which Y is nondegenerately based.

We refer to based fibrations simply as fibrations in the rest of this chapter.

Observe that a based homotopy $X \wedge I_+ \longrightarrow Y$ is the same thing as a based map $X \longrightarrow F(I_+, Y)$. Here $F(I_+, Y)$ is the same space as Y^I, but given a basepoint determined by the basepoint of Y. Therefore the based version of the mapping path space Nf of a based map $f : X \longrightarrow Y$ is the same space as the unbased version, but given a basepoint determined by the given basepoints of X and Y. However, because path spaces are always defined with I having basepoint 0 rather than 1, we find it convenient to redefine Nf correspondingly, setting

$$Nf = \{(x, \chi) | \chi(1) = f(x)\} \subset X \times Y^I.$$

As in the unbased case, we easily check that a based map $p : E \longrightarrow B$ is a fibration if and only if there is a based path lifting function

$$s : Np \longrightarrow F(I_+, E).$$

6. Fiber sequences

For a based map $f : X \longrightarrow Y$, define the "homotopy fiber" Ff to be

$$Ff = X \times_f PY = \{(x, \chi) | f(x) = \chi(1)\} \subset X \times PY.$$

Equivalently, Ff is the pullback displayed in the diagram

$$\begin{array}{ccc} Ff & \longrightarrow & PY \\ \pi \downarrow & & \downarrow p_1 \\ X & \xrightarrow{f} & Y, \end{array}$$

where $\pi(x, \chi) = x$. As a pullback of a fibration, π is a fibration.

If $\rho : Nf \longrightarrow Y$ is defined by $\rho(x, \chi) = \chi(0)$, then $f = \rho \circ \nu$, where $\nu(x) = (x, c_{f(x)})$, and Ff is the fiber $\rho^{-1}(*)$. Thus the homotopy fiber Ff is constructed by first replacing f by the fibration ρ and then taking the actual fiber.

Let $\iota : \Omega Y \longrightarrow Ff$ be the inclusion specified by $\iota(\chi) = (*, \chi)$. The sequence

$$\cdots \longrightarrow \Omega^2 X \xrightarrow{\Omega^2 f} \Omega^2 Y \xrightarrow{-\Omega \iota} \Omega Ff \xrightarrow{-\Omega \pi} \Omega X \xrightarrow{-\Omega f} \Omega Y \xrightarrow{\iota} Ff \xrightarrow{\pi} X \xrightarrow{f} Y$$

is called the fiber sequence generated by the map f; here

$$(-\Omega f)(\zeta)(t) = (f \circ \zeta)(1-t) \quad \text{for } \zeta \in \Omega X.$$

These "long exact sequences of based spaces" also give rise to long exact sequences of pointed sets, this time covariantly.

THEOREM. *For any based space Z, the induced sequence*

$$\cdots \longrightarrow [Z, \Omega Ff] \longrightarrow [Z, \Omega X] \longrightarrow [Z, \Omega Y] \longrightarrow [Z, Ff] \longrightarrow [Z, X] \longrightarrow [Z, Y]$$

is an exact sequence of pointed sets, or of groups to the left of $[Z, \Omega Y]$, or of Abelian groups to the left of $[Z, \Omega^2 Y]$.

Exactness is clear at the first stage. To see this, consider the diagram

$$\begin{array}{c} & & & Z \\ & \tilde{g}=g \times h \nearrow & & \downarrow g \\ Ff = X \times_f PY & \xrightarrow{\pi} & X & \xrightarrow{f} & Y \end{array}$$

Here $h : c_* \simeq f \circ g$, and we view h as a map $Z \longrightarrow PY$. Thus we check exactness by using any given homotopy to lift g to the fiber.

We claim that, up to homotopy equivalence, each consecutive pair of maps in our fiber sequence is the composite of a map and the projection from its fiber onto its source. This will imply the theorem. We observe that, for any map f, interchange of coordinates gives a homeomorphism

$$\Omega Ff \cong F(\Omega f)$$

such that the following diagram commutes:

$$\begin{array}{ccccccc} \Omega^2 Y & \xrightarrow{\Omega \iota(f)} & \Omega Ff & \xrightarrow{\Omega \pi(f)} & \Omega X & \xrightarrow{\Omega f} & \Omega Y \\ \tau \downarrow & & \downarrow \cong & & \| & & \| \\ \Omega^2 Y & \xrightarrow{\iota(\Omega f)} & F(\Omega f) & \xrightarrow{\pi(\Omega f)} & \Omega X & \xrightarrow{\Omega f} & \Omega Y. \end{array}$$

Here τ is obtained by interchanging the loop coordinates and is homotopic to $-\text{id}$. We have written $\iota(f)$, $\pi(f)$, etc., to indicate the maps to which the generic constructions ι and π are applied. Using this inductively, we see that we need only verify our claim for the two pairs of maps $(\iota(f), \pi(f))$ and $(-\Omega f, \iota(f))$. The following two lemmas will imply the claim in these two cases. More precisely, they will imply the claim directly for the first pair and will imply that the second pair is equivalent to a pair of the same form as the first pair. The proofs of the lemmas are left as exercises.

LEMMA. *If $p : E \longrightarrow B$ is a fibration, then the inclusion*

$$\phi : p^{-1}(*) \longrightarrow Fp$$

specified by $\phi(e) = (e, c_*)$ is a based homotopy equivalence.

LEMMA. *The right triangle commutes and the left triangle commutes up to homotopy in the diagram*

$$\begin{array}{ccccccccc} \cdots & \longrightarrow & \Omega X & \xrightarrow{-\Omega f} & \Omega Y & \xrightarrow{\iota(f)} & Ff & \xrightarrow{\pi(f)} & X & \xrightarrow{f} & Y. \\ & & & \searrow_{\iota(\pi(f))} & \downarrow \phi & \nearrow_{\pi(\pi(f))} & & & & & \\ & & & & F\pi(f) & & & & & & \end{array}$$

7. Connections between cofiber and fiber sequences

It is often useful to know that cofiber sequences and fiber sequences can be connected to one another. The adjunction between loops and suspension has "unit" and "counit" maps

$$\eta : X \longrightarrow \Omega\Sigma X \quad \text{and} \quad \varepsilon : \Sigma\Omega X \longrightarrow X.$$

Explicitly, $\eta(x)(t) = x \wedge t$ and $\varepsilon(\chi \wedge t) = \chi(t)$ for $x \in X$, $\chi \in \Omega X$, and $t \in S^1$. For a map $f : X \longrightarrow Y$, we define

$$\eta : Ff \longrightarrow \Omega Cf \quad \text{and} \quad \varepsilon : \Sigma Ff \longrightarrow Cf$$

by

$$\eta(x,\gamma)(t) = \varepsilon(x,\gamma,t) = \begin{cases} \gamma(2t) & \text{if } t \leq 1/2 \\ (x, 2t-1) & \text{if } t \geq 1/2 \end{cases}$$

for $x \in X$ and $\gamma \in PY$ such that $\gamma(1) = f(x)$. Thus ε is just the adjoint of η.

LEMMA. *Let $f : X \longrightarrow Y$ be a map of based spaces. Then the following diagram, in which the top row is the suspension of part of the fiber sequence of f and the bottom row is the loops on part of the cofiber sequence of f, is homotopy commutative:*

$$\begin{array}{ccccccccc}
\Sigma\Omega Ff & \xrightarrow{\Sigma\Omega p} & \Sigma\Omega X & \xrightarrow{\Sigma\Omega f} & \Sigma\Omega Y & \xrightarrow{\Sigma\iota} & \Sigma Ff & \xrightarrow{\Sigma p} & \Sigma X \\
\downarrow{\varepsilon} & & \downarrow{\varepsilon} & & \downarrow{\varepsilon} & & \downarrow{\varepsilon} & & \| \\
\Omega Y & \xrightarrow{\iota} & Ff & \xrightarrow{p} & X & \xrightarrow{f} & Y & \xrightarrow{i} & Cf & \xrightarrow{\pi} & \Sigma X \\
\| & & \downarrow{\eta} & & \downarrow{\eta} & & \downarrow{\eta} & & \downarrow{\eta} \\
\Omega Y & \xrightarrow{\Omega i} & \Omega Cf & \xrightarrow{\Omega\pi} & \Omega\Sigma X & \xrightarrow{\Omega\Sigma f} & \Omega\Sigma Y & \xrightarrow{\Omega\Sigma i} & \Omega\Sigma Cf.
\end{array}$$

PROOF. Four of the squares commute by naturality and the remaining four squares consist of two pairs that are adjoint to each other. To see that the two bottom left squares commute up to homotopy one need only write down the relevant maps explicitly. □

Another easily verified result along the same lines relates the quotient map $(Mf, X) \longrightarrow (Cf, *)$ to $\eta : Ff \longrightarrow \Omega Cf$. Here in the based context we let Mf be the reduced mapping cylinder, in which the line through the basepoint of X is collapsed to a point.

LEMMA. *Let $f : X \longrightarrow Y$ be a map of based spaces. Then the following diagram is homotopy commutative, where $j : X \longrightarrow Mf$ is the inclusion, $r : Mf \longrightarrow Y$ is the retraction, and π is induced by the quotient map $Mf \longrightarrow Cf$:*

$$\begin{array}{ccc}
Fj = X \times_j PMf & \xrightarrow{Fr = \text{id} \times Pr} & X \times_f PY = Ff \\
& \searrow{\pi} \quad \swarrow{\eta} & \\
& \Omega Cf. &
\end{array}$$

PROBLEM

1. Prove the two lemmas stated at the end of §6.

CHAPTER 9

Higher homotopy groups

The most basic invariants in algebraic topology are the homotopy groups. They are very easy to define, but very hard to compute. We give the basic properties of these groups here.

1. The definition of homotopy groups

For $n \geq 0$ and a based space X, define
$$\pi_n(X) = \pi_n(X, *) = [S^n, X],$$
the set of homotopy classes of based maps $S^n \longrightarrow X$. This is a group if $n \geq 1$ and an Abelian group if $n \geq 2$. When $n = 0$ and $n = 1$, this agrees with our previous definitions. Observe that
$$\pi_n(X) = \pi_{n-1}(\Omega X) = \cdots = \pi_0(\Omega^n X).$$

For $* \in A \subset X$, the (homotopy) fiber of the inclusion $A \longrightarrow X$ may be identified with the space $P(X; *, A)$ of paths in X that begin at the basepoint and end in A. For $n \geq 1$, define
$$\pi_n(X, A) = \pi_n(X, A, *) = \pi_{n-1} P(X; *, A).$$
This is a group if $n \geq 2$ and an Abelian group if $n \geq 3$. Again,
$$\pi_n(X, A) = \pi_0(\Omega^{n-1} P(X; *, A)).$$
These are called relative homotopy groups.

2. Long exact sequences associated to pairs

With $Fi = P(X; *, A)$, we have the fiber sequence
$$\cdots \longrightarrow \Omega^2 A \longrightarrow \Omega^2 X \longrightarrow \Omega Fi \longrightarrow \Omega A \longrightarrow \Omega X \xrightarrow{\iota} Fi \xrightarrow{p_1} A \xrightarrow{i} X$$
associated to the inclusion $i : A \longrightarrow X$, where p_1 is the endpoint projection and ι is the inclusion. Applying the functor $\pi_0(-) = [S^0, -]$ to this sequence, we obtain the long exact sequence
$$\cdots \longrightarrow \pi_n(A) \longrightarrow \pi_n(X) \longrightarrow \pi_n(X, A) \xrightarrow{\partial} \pi_{n-1}(A) \longrightarrow \cdots \longrightarrow \pi_0(A) \longrightarrow \pi_0(X).$$
Define
$$J^n = \partial I^{n-1} \times I \cup I^{n-1} \times \{0\} \subset I^n,$$
with $J^1 = \{0\} \subset I$. We can write
$$\pi_n(X, A, *) = [(I^n, \partial I^n, J^n), (X, A, *)],$$
where the notation indicates the homotopy classes of maps of triples: maps and homotopies carry ∂I^n into A and J^n to the basepoint. Then
$$\partial : \pi_n(X, A) \longrightarrow \pi_{n-1}(A)$$

63

is obtained by restricting maps
$$(I^n, \partial I^n, J^n) \longrightarrow (X, A, *)$$
to maps
$$(I^{n-1} \times \{1\}, \partial I^{n-1} \times \{1\}) \longrightarrow (A, *),$$
while $\pi_n(A) \longrightarrow \pi_n(X)$ and $\pi_n(X) \longrightarrow \pi_n(X, A)$ are induced by the inclusions
$$(A, *) \subset (X, *) \quad \text{and} \quad (X, *, *) \subset (X, A, *).$$

3. Long exact sequences associated to fibrations

Let $p : E \longrightarrow B$ be a fibration, where B is path connected. Fix a basepoint $* \in B$, let $F = p^{-1}(*)$, and fix a basepoint $* \in F \subset E$. The inclusion $\phi : F \longrightarrow Fp$ is a homotopy equivalence, and, being pedantically careful to choose signs appropriately, we obtain the following diagram, in which two out of each three consecutive squares commute and the third commutes up to homotopy:

$$\begin{array}{ccccccccccccc}
\cdots & \longrightarrow & \Omega^2 E & \xrightarrow{-\Omega\iota} & \Omega Fi & \xrightarrow{-\Omega p_1} & \Omega F & \xrightarrow{-\Omega i} & \Omega E & \xrightarrow{\iota} & Fi & \xrightarrow{p_1} & F & \xrightarrow{i} & E \\
& & \downarrow \mathrm{id} & & \downarrow -\Omega p & & \downarrow \Omega\phi & & \downarrow \mathrm{id} & & \downarrow -p & & \downarrow \phi & & \downarrow \mathrm{id} \\
\cdots & \longrightarrow & \Omega^2 E & \xrightarrow{\Omega^2 p} & \Omega^2 B & \xrightarrow{-\Omega\iota} & \Omega Fp & \xrightarrow{-\Omega\pi} & \Omega E & \xrightarrow{-\Omega p} & \Omega B & \xrightarrow{\iota} & Fp & \xrightarrow{\pi} & E.
\end{array}$$

Here $Fi = P(E; *, F)$, $p(\xi) = p \circ \xi \in \Omega B$ for $\xi \in Fi$, and the next to last square commutes up to the homotopy $h : \iota \circ (-p) \simeq \phi \circ p_1$ specified by
$$h(\xi, t) = (\xi(t), p(\xi[1, t])),$$
where $\xi[1, t](s) = \xi(1 - s + st)$.

Passing to long exact sequences of homotopy groups and using the five lemma, together with a little extra argument in the case $n = 1$, we conclude that
$$p_* : \pi_n(E, F) \longrightarrow \pi_n(B)$$
is an isomorphism for $n \geq 1$. This can also be derived directly from the covering homotopy property.

Using ϕ_* to identify $\pi_* F$ with $\pi_*(Fp)$, we may rewrite the long exact sequence of the bottom row of the diagram as
$$\cdots \longrightarrow \pi_n(F) \longrightarrow \pi_n(E) \longrightarrow \pi_n(B) \xrightarrow{\partial} \pi_{n-1}(F) \longrightarrow \cdots \longrightarrow \pi_0(E) \longrightarrow \{*\}.$$
(At the end, a little path lifting argument shows that $\pi_0(F) \longrightarrow \pi_0(E)$ is a surjection.) This is one of the main tools for the computation of homotopy groups.

4. A few calculations

We observe some easily derived calculational facts about homotopy groups.

LEMMA. *If X is contractible, then $\pi_n(X) = 0$ for all $n \geq 0$.*

LEMMA. *If X is discrete, then $\pi_n(X) = 0$ for all $n > 0$.*

LEMMA. *If $p : E \longrightarrow B$ is a covering, then $p_* : \pi_n(E) \longrightarrow \pi_n(B)$ is an isomorphism for all $n \geq 2$.*

LEMMA. *$\pi_1(S^1) = \mathbb{Z}$ and $\pi_n(S^1) = 0$ if $n \neq 1$.*

LEMMA. *If $i \geq 2$, then $\pi_1(\mathbb{R}P^i) = \mathbb{Z}_2$ and $\pi_n(\mathbb{R}P^i) \cong \pi_n(S^i)$ for $n \neq 1$.*

LEMMA. *For all spaces X and Y and all n,*

$$\pi_n(X \times Y) \cong \pi_n(X) \times \pi_n(Y).$$

LEMMA. *If $i < n$, then $\pi_i(S^n) = 0$.*

PROOF. For any based map $f : S^i \longrightarrow S^n$, we can apply smooth (or simplicial) approximation to obtain a based homotopy from f to a map that misses a point p, and we can then deform f to the trivial map by contracting $S^n - \{p\}$ to the basepoint. □

There are three standard bundles, called the Hopf bundles, that can be used to obtain a bit more information about the homotopy groups of spheres. Recall that $\mathbb{C}P^1$ is the space of complex lines in $\mathbb{C}^2$. That is, $\mathbb{C}P^1 = (\mathbb{C} \times \mathbb{C} - \{0\})/(\sim)$, where $(z_1, z_2) \sim (\lambda z_1, \lambda z_2)$ for complex numbers λ, z_1, and z_2. Write $[z_1, z_2]$ for the equivalence class of (z_1, z_2). We obtain a homeomorphism $\mathbb{C}P^1 \longrightarrow S^2$ by identifying S^2 with the one-point compactification of $\mathbb{C}$ and mapping $[z_1, z_2]$ to z_2/z_1 if $z_1 \neq 0$ and to the point at ∞ if $z_1 = 0$. The Hopf map $\eta : S^3 \longrightarrow S^2$ is specified by $\eta(z_1, z_2) = [z_1, z_2]$, where S^3 is identified with the unit sphere in the complex plane $\mathbb{C}^2$. It is a worthwhile exercise to check that η is a bundle with fiber S^1. By use of the quaternions and Cayley numbers, we obtain analogous Hopf maps $\nu : S^7 \longrightarrow S^4$ and $\sigma : S^{15} \longrightarrow S^8$. Then ν is a bundle with fiber S^3 and σ is a bundle with fiber S^7. Since we have complete information on the homotopy groups of S^1, the long exact sequence of homotopy groups associated to η has the following direct consequence.

LEMMA. $\pi_2(S^2) \cong \mathbb{Z}$ *and* $\pi_n(S^3) \cong \pi_n(S^2)$ *for $n \geq 3$.*

We shall later prove the following more substantial result.

THEOREM. *For all $n \geq 1$, $\pi_n(S^n) \cong \mathbb{Z}$.*

It is left as an exercise to show that the long exact sequence associated to ν implies that $\pi_7(S^4)$ contains an element of infinite order, and σ can be used similarly to show the same for $\pi_{15}(S^8)$.

In fact, the homotopy groups $\pi_q(S^n)$ for $q > n > 1$ are all finite except for $\pi_{4n-1}(S^{2n})$, which is the direct sum of $\mathbb{Z}$ and a finite group.

The difficulty of computing homotopy groups is well illustrated by the fact that there is no non-contractible simply connected compact manifold (or finite CW complex) all of whose homotopy groups are known. We shall find many non-compact spaces whose homotopy groups we can determine completely. Such computations will rely on the following observation.

LEMMA. *If X is the colimit of a sequence of inclusions $X_i \longrightarrow X_{i+1}$ of based spaces, then the natural map*

$$\operatorname{colim}_i \pi_n(X_i) \longrightarrow \pi_n(X)$$

is an isomorphism for each n.

PROOF. This follows directly from the point-set topological fact that if K is a compact space, then a map $K \longrightarrow X$ has image in one of the X_i. □

5. Change of basepoint

We shall use our results on change of fibers to generalize our results on change of basepoint from the fundamental group to the higher absolute and relative homotopy groups. In the absolute case, we have the identification

$$\pi_n(X,x) = [(S^n, *), (X, x)],$$

where we assume that $n \geq 1$. Since the inclusion of the basepoint in S^n is a cofibration, evaluation at the basepoint gives a fibration $p : X^{S^n} \longrightarrow X$. We may identify $\pi_n(X,x)$ with $\pi_0(F_x)$ since a path in F_x is just a based homotopy $h : S^n \times I \longrightarrow X$ with respect to the basepoint x. Another way to see this is to observe that F_x is the nth loop space $\Omega^n X$, specified with respect to the basepoint x. A path class $[\xi] : I \longrightarrow X$ from x to x' induces a homotopy equivalence $\tau[\xi] : F_x \longrightarrow F_{x'}$, and we continue to write $\tau[\xi]$ for the induced bijection

$$\tau[\xi] : \pi_n(X,x) \longrightarrow \pi_n(X,x').$$

This bijection is an isomorphism of groups. One conceptual way to see this is to observe that addition is induced from the "pinch map" $S^n \longrightarrow S^n \vee S^n$ that is obtained by collapsing an equator to the basepoint. That is, the sum of maps $f, g : S^n \longrightarrow X$ is the composite

$$S^n \longrightarrow S^n \vee S^n \xrightarrow{f \vee g} X \vee X \xrightarrow{\nabla} X,$$

where ∇ is the folding map, which restricts to the identity map $X \longrightarrow X$ on each wedge summand. Evaluation at the basepoint of $S^n \vee S^n$ gives a fibration $X^{S^n \vee S^n} \longrightarrow X$, and the pinch map induces a map of fibrations

$$\begin{array}{ccc} X^{S^n \vee S^n} & \longrightarrow & X^{S^n} \\ \downarrow & & \downarrow \\ X & = & X. \end{array}$$

The fiber over x in the left-hand fibration is the product $F_x \times F_x$, where F_x is the fiber over x in the right-hand fibration. In fact, the induced map of fibers can be identified as the map $\Omega^n X \times \Omega^n X \longrightarrow \Omega^n X$ given by composition of loops (using the first loop coordinate say). By the naturality of translations of fibers with respect to maps of fibrations, we have a homotopy commutative diagram

$$\begin{array}{ccc} F_x \times F_x & \longrightarrow & F_x \\ \tau[\xi] \times \tau[\xi] \downarrow & & \downarrow \tau[\xi] \\ F_{x'} \times F_{x'} & \longrightarrow & F_{x'} \end{array}$$

in which the horizontal arrows induce addition on passage to π_0.

We can argue similarly in the relative case. The triple $(I^n, \partial I^n, J^n)$ is homotopy equivalent to the triple $(CS^{n-1}, S^{n-1}, *)$, as we see by quotienting out J^n. Therefore, for $a \in A$, we have the identification

$$\pi_n(X, A, a) \cong [(CS^{n-1}, S^{n-1}, *), (X, A, a)].$$

Using that the inclusions $\{*\} \longrightarrow S^{n-1}$ and $S^{n-1} \longrightarrow CS^{n-1}$ are both cofibrations, we can check that evaluation at $*$ specifies a fibration

$$p : (X, A)^{(CS^{n-1}, S^{n-1})} \longrightarrow A,$$

where the domain is the subspace of $X^{CS^{n-1}}$ consisting of the indicated maps of pairs. We may identify $\pi_n(X, A, a)$ with $\pi_0(F_a)$. A path class $[\alpha] : I \longrightarrow A$ from a to a' induces a homotopy equivalence $\tau[\alpha] : F_a \longrightarrow F_{a'}$, and we continue to write $\tau[\alpha]$ for the induced isomorphism

$$\tau[\alpha] : \pi_n(X, A, a) \longrightarrow \pi_n(X, A, a').$$

Our naturality results on change of fibers now directly imply the desired results on change of basepoint.

THEOREM. *If $f : (X, A) \longrightarrow (Y, B)$ is a map of pairs and $\alpha : I \longrightarrow A$ is a path from a to a', then the following diagram commutes:*

$$\begin{array}{ccc} \pi_n(X, A, a) & \xrightarrow{f_*} & \pi_n(Y, B, f(a)) \\ {\scriptstyle \tau[\alpha]} \downarrow & & \downarrow {\scriptstyle \tau[f \circ \alpha]} \\ \pi_n(X, A, a') & \xrightarrow{f_*} & \pi_n(Y, B, f(a')) \end{array}$$

If $h : f \simeq f'$ is a homotopy of maps of pairs and $h(a)(t) = h(a, t)$, then the following diagram commutes:

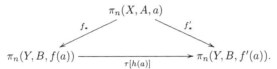

The analogous conclusions hold for the absolute homotopy groups.

Therefore, up to non-canonical isomorphism, the homotopy groups of (X, A) are independent of the choice of basepoint in a given path component of A.

COROLLARY. *A homotopy equivalence of spaces or of pairs of spaces induces an isomorphism on all homotopy groups.*

We shall soon show that the converse holds for a quite general class of spaces, namely the class of CW complexes, but we first need a few preliminaries.

6. n-Equivalences, weak equivalences, and a technical lemma

DEFINITION. A map $e : Y \longrightarrow Z$ is an n-equivalence if, for all $y \in Y$, the map

$$e_* : \pi_q(Y, y) \longrightarrow \pi_q(Z, e(y))$$

is an injection for $q < n$ and a surjection for $q \leq n$; e is said to be a weak equivalence if it is an n-equivalence for all n.

Thus any homotopy equivalence is a weak equivalence. The following technical lemma will be at the heart of our study of CW complexes, but it will take some getting used to. It gives a useful criterion for determining when a given map is an n-equivalence.

It is convenient to take CX to be the unreduced cone $X \times I/X \times \{1\}$ here. If $f, f' : (X, A) \longrightarrow (Y, B)$ are maps of pairs such that $f = f'$ on A, then we say that f and f' are homotopic relative to A if there is a homotopy $h : f \simeq f'$ such that h is constant on A, in the sense that $h(a, t) = f(a)$ for all $a \in A$ and $t \in I$; we write $h : f \simeq f'$ rel A. Observe that $\pi_{n+1}(X, x)$ can be viewed as the set of relative homotopy classes of maps $(CS^n, S^n) \longrightarrow (X, x)$.

LEMMA. *The following conditions on a map* $e: Y \longrightarrow Z$ *are equivalent.*

(i) *For any* $y \in Y$, $e_* : \pi_q(Y, y) \longrightarrow \pi_q(Z, e(y))$ *is an injection for* $q = n$ *and a surjection for* $q = n + 1$.

(ii) *Given maps* $f : CS^n \longrightarrow Z$, $g : S^n \longrightarrow Y$, *and* $h : S^n \times I \longrightarrow Z$ *such that* $f|S^n = h \circ i_0$ *and* $e \circ g = h \circ i_1$ *in the following diagram, there are maps* $\tilde{g}$ *and* $\tilde{h}$ *that make the entire diagram commute.*

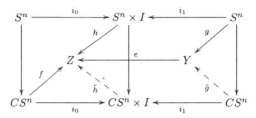

(iii) *The conclusion of (ii) holds when* $f|S^n = e \circ g$ *and* h *is the constant homotopy at this map.*

PROOF. Trivially (ii) implies (iii). We first show that (iii) implies (i). If $n = 0$, (iii) says (in part) that if $e(y)$ and $e(y')$ can be connected by a path in Z, then y and y' can be connected by a path in Y. If $n > 0$, then (iii) says (in part) that if $e \circ g$ is null homotopic, then g is null homotopic. Therefore $\pi_n(e)$ is injective. If we specialize (iii) by letting g be the constant map at a point $y \in Y$, then f is a map $(CS^n, S^n) \longrightarrow (Z, e(y))$, $\tilde{g}$ is a map $(CS^n, S^n) \longrightarrow (Y, y)$, and $\tilde{h} : f \simeq e \circ \tilde{g}$ rel S^n. Therefore $\pi_{n+1}(e)$ is surjective.

Thus assume (i). We must prove (ii), and we assume given f, g, and h making the solid arrow part of the diagram commute. The idea is to use (i) to show that the nth homotopy group of the fiber $F(e)$ is zero, to use the given part of the diagram to construct a map $S^n \longrightarrow F(e)$, and to use a null homotopy of that map to construct $\tilde{g}$ and $\tilde{h}$. However, since homotopy groups involve choices of basepoints and the diagram makes no reference to basepoints, the details require careful tracking of basepoints. Thus fix a basepoint $* \in S^n$, let $\bullet$ be the cone point of CS^n, and define

$$y_1 = g(*), \quad z_1 = e(y_1), \quad z_0 = f(*, 0), \quad \text{and} \quad z_{-1} = f(\bullet).$$

For $x \in S^n$, let $f_x : I \longrightarrow Z$ and $h_x : I \longrightarrow Z$ be the paths $f_x(s) = f(x, s)$ from $f(x, 0) = h(x, 0)$ to z_{-1} and $h_x(t) = h(x, t)$ from $h(x, 0)$ to $h(x, 1) = (e \circ g)(x)$. Consider the homotopy fiber

$$F(e; y_1) = \{(y, \zeta) | \zeta(0) = z_1 \text{ and } e(y) = \zeta(1)\} \subset Y \times Z^I.$$

This has basepoint $w_1 = (y_1, c_{z_1})$. By (i) and the exact sequence

$$\pi_{n+1}(Y, y_1) \xrightarrow{e_*} \pi_{n+1}(Z, z_1) \longrightarrow \pi_n(F(e; y_1), w_1) \longrightarrow \pi_n(Y, y_1) \xrightarrow{e_*} \pi_n(Z, z_1),$$

we see that $\pi_n(F(e; y_1), w_1) = 0$. Define $k_0 : S^n \longrightarrow F(e; y_1)$ by

$$k_0(x) = (g(x), h_x \cdot f_x^{-1} \cdot f_* \cdot h_*^{-1}).$$

While k_0 is not a based map, $k_0(*)$ is connected to the basepoint since $h_* \cdot f_*^{-1} \cdot f_* \cdot h_*^{-1}$ is equivalent to c_{z_1}. By HEP for the cofibration $\{*\} \longrightarrow S^n$, k_0 is homotopic to a based map. This based map is null homotopic in the based sense, hence k_0 is null

homotopic in the unbased sense. Let $k : S^n \times I \longrightarrow F(e; y_1)$ be a homotopy from k_0 to the trivial map at w_1. Write
$$k(x,t) = (\tilde{g}(x,t), \zeta(x,t)).$$
Then $\tilde{g}(x,1) = y_1$ for all $x \in S^n$, so that $\tilde{g}$ factors through a map $CS^n \longrightarrow Y$, and $\tilde{g} = g$ on S^n. We have a map $j : S^n \times I \times I$ given by $j(x,s,t) = \zeta(x,t)(s)$ that behaves as follows on the boundary of the square for each fixed $x \in S^n$, where $\tilde{g}_x(t) = \tilde{g}(x,t)$:

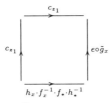

The desired homotopy $\tilde{h}$, written $\tilde{h}(x,s,t)$ where s is the cone coordinate and t is the interval coordinate, should behave as follows on the boundary of the square:

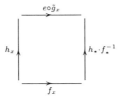

Thus we can obtain $\tilde{h}$ by composing j with a suitable reparametrization $I^2 \longrightarrow I^2$ of the square. □

PROBLEMS

1. Show that, if $n \geq 2$, then $\pi_n(X \vee Y)$ is isomorphic to
$$\pi_n(X) \oplus \pi_n(Y) \oplus \pi_{n+1}(X \times Y, X \vee Y).$$
2. Compute $\pi_n(\mathbb{R}P^n, \mathbb{R}P^{n-1})$ for $n \geq 2$. Deduce that the quotient map
$$(\mathbb{R}P^n, \mathbb{R}P^{n-1}) \to (\mathbb{R}P^n/\mathbb{R}P^{n-1}, *)$$
does not induce an isomorphism of homotopy groups.
3. Compute the homotopy groups of complex projective space $\mathbb{C}P^n$ in terms of the homotopy groups of spheres.
4. Verify that the "Hopf bundles" are in fact bundles.
5. Show that $\pi_7(S^4)$ contains an element of infinite order.
6. Compute all of the homotopy groups of $\mathbb{R}P^\infty$ and $\mathbb{C}P^\infty$.

CHAPTER 10

CW complexes

We introduce a large class of spaces, called CW complexes, between which a weak equivalence is necessarily a homotopy equivalence. Thus, for such spaces, the homotopy groups are, in a sense, a complete set of invariants. Moreover, we shall see that every space is weakly equivalent to a CW complex.

1. The definition and some examples of CW complexes

Let D^{n+1} be the unit disk $\{x \mid |x| \leq 1\} \subset \mathbb{R}^{n+1}$ with boundary S^n.

DEFINITION. (i) A CW complex X is a space X which is the union of an expanding sequence of subspaces X^n such that, inductively, X^0 is a discrete set of points (called vertices) and X^{n+1} is the pushout obtained from X^n by attaching disks D^{n+1} along "attaching maps" $j : S^n \longrightarrow X^n$. Thus X^{n+1} is the quotient space obtained from $X^n \cup (J_{n+1} \times D^{n+1})$ by identifying (j, x) with $j(x)$ for $x \in S^n$, where J_{n+1} is the discrete set of such attaching maps j. Each resulting map $D^{n+1} \longrightarrow X$ is called a "cell." The subspace X^n is called the n-skeleton of X.

(ii) More generally, given any space A, we define a relative CW complex (X, A) in the same fashion, but with X^0 replaced by the union of A and a (possibly empty) discrete set of points; we write $(X, A)^n$, or X^n when A is clear from the context, for the relative n-skeleton, and we say that (X, A) has dimension $\leq n$ if $X = X^n$.

(iii) A subcomplex A of a CW complex X is a subspace and a CW complex such that the composite of each cell $D^n \longrightarrow A$ of A and the inclusion of A in X is a cell of X. That is, A is the union of some of the cells of X. The pair (X, A) can then be viewed as a relative CW complex.

(iv) A map of pairs $f : (X, A) \longrightarrow (Y, B)$ between relative CW complexes is said to be "cellular" if $f(X^n) \subset Y^n$ for all n.

Of course, pushouts and unions are understood in the topological sense, with the compactly generated topologies. A subspace of X is closed if and only if its intersection with each X^n is closed.

EXAMPLES. (i) A graph is a one-dimensional CW complex.

(ii) Via a homeomorphism $I \times I \cong D^2$, the standard presentations of the torus $T = S^1 \times S^1$, the projective plane $\mathbb{R}P^2$, and the Klein bottle K as quotients of a square display these spaces as CW complexes with one or two vertices, two edges, and one 2-cell:

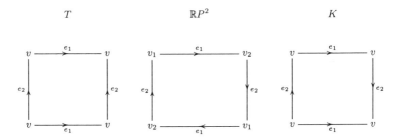

(iii) For $n \geq 1$, S^n is a CW complex with one vertex $\{*\}$ and one n-cell, the attaching map $S^{n-1} \longrightarrow \{*\}$ being the only possible map. Note that this entails a choice of homeomorphism $D^n/S^{n-1} \cong S^n$. If $m < n$, then the only cellular map $S^m \longrightarrow S^n$ is the trivial map. If $m \geq n$, then every based map $S^m \longrightarrow S^n$ is cellular.

(iv) $\mathbb{R}P^n$ is a CW complex with m-skeleton $\mathbb{R}P^m$ and with one m-cell for each $m \leq n$. The attaching map $j : S^{n-1} \longrightarrow \mathbb{R}P^{n-1}$ is the standard double cover. That is, $\mathbb{R}P^n$ is homeomorphic to $\mathbb{R}P^{n-1} \cup_j D^n$. Explicitly, write $\bar{x} = [x_1, \ldots, x_{n+1}]$, $\sum x_i^2 = 1$, for a typical point of $\mathbb{R}P^n$. Then $\bar{x}$ is in $\mathbb{R}P^{n-1}$ if and only if $x_{n+1} = 0$. The required homeomorphism is obtained by identifying D^n and its boundary sphere with the upper hemisphere

$$E_+^n = \{(x_1, \ldots, x_{n+1}) \mid \sum x_i^2 = 1 \quad \text{and} \quad x_{n+1} \geq 0\}$$

and its boundary sphere.

(v) $\mathbb{C}P^n$ is a CW complex whose $2m$-skeleton and $(2m+1)$-skeleton are both $\mathbb{C}P^m$ and which has one $2m$-cell for each $m \leq n$. The attaching map $S^{2n-1} \longrightarrow \mathbb{C}P^{n-1}$ is the standard bundle with fiber S^1, where S^{2n-1} is identified with the unit sphere in $\mathbb{C}^n$. We leave the specification of the required homeomorphism as an exercise.

2. Some constructions on CW complexes

We need to know that various constructions on spaces preserve CW complexes. We leave most of the proofs as exercises in the meaning of the definitions.

LEMMA. *If (X, A) is a relative CW complex, then the quotient space X/A is a CW complex with a vertex corresponding to A and one n-cell for each relative n-cell of (X, A).*

LEMMA. *For CW complexes X_i with basepoints that are vertices, the wedge of the X_i is a CW complex which contains each X_i as a subcomplex.*

LEMMA. *If A is a subcomplex of a CW complex X, Y is a CW complex, and $f : A \longrightarrow Y$ is a cellular map, then the pushout $Y \cup_f X$ is a CW complex that contains Y as a subcomplex and has one cell for each cell of X that is not in A. The quotient complex $(Y \cup_f X)/Y$ is isomorphic to X/A.*

LEMMA. *The colimit of a sequence of inclusions of subcomplexes $X_n \longrightarrow X_{n+1}$ in CW complexes is a CW complex that contains each of the X_i as a subcomplex.*

LEMMA. *The product $X \times Y$ of CW complexes X and Y is a CW complex with an n-cell for each pair consisting of a p-cell of X and q-cell of Y, where $p + q = n$.*

PROOF. For $p + q = n$, there are canonical homeomorphisms
$$(D^n, S^{n-1}) \cong (D^p \times D^q, D^p \times S^{q-1} \cup S^{p-1} \times D^q).$$
This allows us to define product cells. □

We shall look at the general case more closely later, but we point out one important special case for immediate use. Of course, the unit interval is a graph with two vertices and one edge.

LEMMA. *For a CW complex X, $X \times I$ is a CW complex that contains $X \times \partial I$ as a subcomplex and, in addition, has one $(n + 1)$-cell for each n-cell of X.*

A "cellular homotopy" $h : f \simeq f'$ between cellular maps $X \longrightarrow Y$ of CW complexes is a homotopy that is itself a cellular map $X \times I \longrightarrow Y$.

3. HELP and the Whitehead theorem

The following "homotopy extension and lifting property" is a powerful organizational principle for proofs of results about CW complexes. In the case
$$(X, A) = (D^n, S^{n-1}) \cong (CS^{n-1}, S^{n-1}),$$
it is the main point of the technical lemma proved at the end of the last chapter.

THEOREM (HELP). *Let (X, A) be a relative CW complex of dimension $\leq n$ and let $e : Y \longrightarrow Z$ be an n-equivalence. Then, given maps $f : X \longrightarrow Z$, $g : A \longrightarrow Y$, and $h : A \times I \longrightarrow Z$ such that $f|A = h \circ i_0$ and $e \circ g = h \circ i_1$ in the following diagram, there are maps $\tilde{g}$ and $\tilde{h}$ that make the entire diagram commute:*

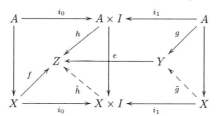

PROOF. Proceed by induction over skeleta, applying the case (D^n, S^{n-1}) one cell at a time to the n-cells of X not in A. □

In particular, if we take e to be the identity map of Y, we see that the inclusion $A \longrightarrow X$ is a cofibration. Observe that, by passage to colimits, we are free to take $n = \infty$ in the theorem.

We write $[X, Y]$ for homotopy classes of unbased maps in this chapter, and we have the following direct and important application of HELP.

THEOREM (Whitehead). *If X is a CW complex and $e : Y \longrightarrow Z$ is an n-equivalence, then $e_* : [X, Y] \longrightarrow [X, Z]$ is a bijection if $\dim X < n$ and a surjection if $\dim X = n$.*

PROOF. Apply HELP to the pair $(X, \emptyset)$ to see the surjectivity. Apply HELP to the pair $(X \times I, X \times \partial I)$, taking h to be a constant homotopy, to see the injectivity. □

THEOREM (Whitehead). *An n-equivalence between CW complexes of dimension less than n is a homotopy equivalence. A weak equivalence between CW complexes is a homotopy equivalence.*

PROOF. Let $e : Y \longrightarrow Z$ satisfy either hypothesis. Since $e_* : [Z, Y] \longrightarrow [Z, Z]$ is a bijection, there is a map $f : Z \longrightarrow Y$ such that $e \circ f \simeq \text{id}$. Then $e \circ f \circ e \simeq e$, and, since $e_* : [Y, Y] \longrightarrow [Y, Z]$ is also a bijection, this implies that $f \circ e \simeq \text{id}$. □

If X is a finite CW complex, in the sense that it has finitely many cells, and if $\dim X > 1$ and X is not contractible, then it is known that X has infinitely many non-zero homotopy groups. The Whitehead theorem is thus surprisingly strong: in its first statement, if low dimensional homotopy groups are mapped isomorphically, then so are all higher homotopy groups.

4. The cellular approximation theorem

Cellular maps are under much better algebraic control than general maps, as will become both clear and important later. Fortunately, any map between CW complexes is homotopic to a cellular map. We need a lemma.

DEFINITION. A space X is said to be *n-connected* if $\pi_q(X, x) = 0$ for $0 \leq q \leq n$ and all x. A pair (X, A) is said to be *n-connected* if $\pi_0(A) \longrightarrow \pi_0(X)$ is surjective and $\pi_q(X, A, a) = 0$ for $1 \leq q \leq n$ and all a. It is equivalent that the inclusion $A \longrightarrow X$ be an *n*-equivalence.

LEMMA. *A relative CW complex (X, A) with no m-cells for $m \leq n$ is n-connected. In particular, (X, X^n) is n-connected for any CW complex X.*

PROOF. Consider $f : (I^q, \partial I^q, J^q) \longrightarrow (X, A, a)$, where $q \leq n$. Since the image of f is compact, we may assume that (X, A) has finitely many cells. By induction on the number of cells, we may assume that $X = A \cup_j D^r$, where $r > n$. By smooth (or simplicial) approximation, there is a map $f' : I^q \longrightarrow X$ such that $f' = f$ on ∂I^q, $f' \simeq f$ rel ∂I^q and f' misses a point p in the interior of D^r. Clearly we can deform $X - \{p\}$ onto A and so deform f' to a map into A. □

THEOREM (Cellular approximation). *Any map $f : (X, A) \longrightarrow (Y, B)$ between relative CW complexes is homotopic relative to A to a cellular map.*

PROOF. We proceed by induction over skeleta. To start the induction, note that any point of Y is connected by a path to a point in Y^0 and apply this to the images of points of $X^0 - A$ to obtain a homotopy of $f|X^0$ to a map into Y^0. Assume given $g_n : X^n \longrightarrow Y^n$ and $h_n : X^n \times I \longrightarrow Y$ such that $h_n : f|X^n \simeq \iota_n \circ g_n$, where $\iota_n : Y^n \longrightarrow Y$ is the inclusion. For an attaching map $j : S^n \longrightarrow X^n$ of a cell $\tilde{j} : D^{n+1} \longrightarrow X$, we apply HELP to the following diagram:

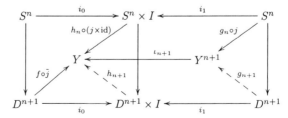

where $g_n \circ j : S^n \longrightarrow Y^n$ is composed with the inclusion $Y^n \longrightarrow Y^{n+1}$; HELP applies since ι_{n+1} is an $(n+1)$-equivalence. □

COROLLARY. *For CW complexes X and Y, any map $X \longrightarrow Y$ is homotopic to a cellular map, and any two homotopic cellular maps are cellularly homotopic.*

5. Approximation of spaces by CW complexes

The following result says that there is a functor $\Gamma : h\mathcal{U} \longrightarrow h\mathcal{U}$ and a natural transformation $\gamma : \Gamma \longrightarrow \mathrm{Id}$ that assign a CW complex ΓX and a weak equivalence $\gamma : \Gamma X \longrightarrow X$ to a space X.

THEOREM (Approximation by CW complexes). *For any space X, there is a CW complex ΓX and a weak equivalence $\gamma : \Gamma X \longrightarrow X$. For a map $f : X \longrightarrow Y$ and another such CW approximation $\gamma : \Gamma Y \longrightarrow Y$, there is a map $\Gamma f : \Gamma X \longrightarrow \Gamma Y$, unique up to homotopy, such that the following diagram is homotopy commutative:*

$$\begin{array}{ccc} \Gamma X & \xrightarrow{\Gamma f} & \Gamma Y \\ \gamma \downarrow & & \downarrow \gamma \\ X & \xrightarrow{f} & Y. \end{array}$$

If X is n-connected, $n \geq 1$, then ΓX can be chosen to have a unique vertex and no q-cells for $1 \leq q \leq n$.

PROOF. The existence and uniqueness up to homotopy of Γf will be immediate since the Whitehead theorem will give a bijection

$$\gamma_* : [\Gamma X, \Gamma Y] \longrightarrow [\Gamma X, Y].$$

Proceeding one path component at a time, we may as well assume that X is path connected, and we may then work with based spaces and based maps. We construct ΓX as the colimit of a sequence of cellular inclusions

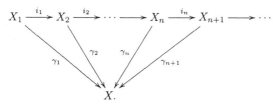

Let X_1 be a wedge of spheres S^q, $q \geq 1$, one for each pair (q,j), where $j : S^q \longrightarrow X$ represents a generator of the group $\pi_q(X)$. On the (q,j)th wedge summand, the map γ_1 is the given map j. Clearly $\gamma_1 : X_1 \longrightarrow X$ induces an epimorphism on all homotopy groups. We give X_1 the CW structure induced by the standard CW structures on the spheres S^q. Inductively, suppose that we have constructed CW complexes X_m, cellular inclusions i_{m-1}, and maps γ_m for $m \leq n$ such that $\gamma_m \circ i_{m-1} = \gamma_{m-1}$ and $(\gamma_m)_* : \pi_q(X_m) \longrightarrow \pi_q(X)$ is a surjection for all q and a bijection for $q < m$. We construct

$$X_{n+1} = X_n \cup (\bigvee_{(f,g)} (S^n \wedge I_+)),$$

where the wedge is taken over cellular representatives $f, g : S^n \longrightarrow X_n$ in each pair of homotopy classes $[f], [g] \in \pi_n(X_n)$ such that $[f] \neq [g]$ but $[\gamma_n \circ f] = [\gamma_n \circ g]$. We attach the (f,g)th reduced cylinder $S^n \wedge I_+$ to X_n by identifying $(s, 0)$ with $f(s)$ and $(s, 1)$ with $g(s)$ for $s \in S^n$. Let $i_n : X_n \longrightarrow X_{n+1}$ be the inclusion and observe that $(i_n)_*[f] = (i_n)_*[g]$. Define $\gamma_{n+1} : X_{n+1} \longrightarrow X$ by means of γ_n on X_n and a chosen homotopy $h : S^n \wedge I_+ \longrightarrow X$ from $\gamma_n \circ f$ to $\gamma_n \circ g$ on the (f, g)th cylinder. Then $(\gamma_{n+1})_* : \pi_q(X_{n+1}) \longrightarrow \pi_q(X)$ is a surjection for all q, because $(\gamma_n)_*$ is so, and a bijection for $q \leq n$ by construction. We have not changed the homotopy groups in dimensions less than n since we have not changed the n-skeleton. Since f and g are cellular and since, as is easily verified, $S^n \wedge I_+$ admits a CW structure with $S^n \wedge (\partial I)_+$ as a subcomplex, we conclude from the pushout property of CW complexes that X_{n+1} is a CW complex that contains X_n as a subcomplex. Then the colimit ΓX of the X_n is a CW complex that contains all of the X_i as subcomplexes, and the induced map $\gamma : \Gamma X \longrightarrow X$ induces an isomorphism on all homotopy groups since the homotopy groups of ΓX are the colimits of the homotopy groups of the X_n. If X is n-connected, then we have used no q-cells for $q \leq n$ in the construction. $\square$

6. Approximation of pairs by CW pairs

We will need a relative generalization of the previous result, but the reader should not dwell on the details: there are no new ideas.

THEOREM. *For any pair of spaces (X, A) and any CW approximation $\gamma : \Gamma A \longrightarrow A$, there is a CW approximation $\gamma : \Gamma X \longrightarrow X$ such that ΓA is a subcomplex of ΓX and γ restricts to the given γ on ΓA. If $f : (X, A) \longrightarrow (Y, B)$ is a map of pairs and $\gamma : (\Gamma Y, \Gamma B) \longrightarrow (Y, B)$ is another such CW approximation of pairs, there is a map $\Gamma f : (\Gamma X, \Gamma A) \longrightarrow (\Gamma Y, \Gamma B)$, unique up to homotopy, such that the following diagram of pairs is homotopy commutative:*

$$\begin{array}{ccc} (\Gamma X, \Gamma A) & \xrightarrow{\Gamma f} & (\Gamma Y, \Gamma B) \\ \gamma \downarrow & & \downarrow \gamma \\ (X, A) & \xrightarrow{f} & (Y, B). \end{array}$$

If (X, A) is n-connected, then $(\Gamma X, \Gamma A)$ can be chosen to have no relative q-cells for $q \leq n$.

PROOF. We proceed exactly as above. We may assume that X has a basepoint in A and that X, but not necessarily A, is path connected. We then start the construction above with

$$X_1 = \Gamma A \vee (\bigvee_{(q,j)} S^n),$$

letting γ_1 restrict to the given γ on Γ. If (X, A) is n-connected, then $\pi_q(A) \longrightarrow \pi_q(X)$ is bijective for $q < n$ and surjective for $q = n$, hence we need only use spheres S^q with $q > n$ to arrange the surjectivity of $\pi_*(X_1) \longrightarrow \pi_*(X)$. To construct Γf, we first construct it on A and then extend to all of X by use of HELP:

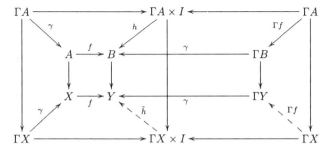

The uniqueness up to homotopy of Γf is proved similarly. □

7. Approximation of excisive triads by CW triads

We will need another, and considerably more subtle, relative approximation theorem. A triad $(X; A, B)$ is a space X together with subspaces A and B. This must not be confused with a triple (X, A, B), which would require $B \subset A \subset X$. A triad $(X; A, B)$ is said to be excisive if X is the union of the interiors of A and B. Such triads play a fundamental role in homology and cohomology theory, and some version of the arguments to follow must play a role in any treatment. We prefer to use these arguments to prove a strong homotopical result, rather than its pale homological reflection that is seen in standard treatments of the subject.

A CW triad $(X; A, B)$ is a CW complex X with subcomplexes A and B such that $X = A \cup B$.

THEOREM. Let $(X; A, B)$ be an excisive triad and let $C = A \cap B$. Then there is a CW triad $(\Gamma X; \Gamma A, \Gamma B)$ and a map of triads

$$\gamma : (\Gamma X; \Gamma A, \Gamma B) \longrightarrow (X; A, B)$$

such that, with $\Gamma C = \Gamma A \cap \Gamma B$, the maps

$$\gamma : \Gamma C \longrightarrow C, \quad \gamma : \Gamma A \longrightarrow A, \quad \gamma : \Gamma B \longrightarrow B, \quad \text{and} \quad \gamma : \Gamma X \longrightarrow X$$

are all weak equivalences. If (A, C) is n-connected, then $(\Gamma A, \Gamma C)$ can be chosen to have no q-cells for $q \leq n$, and similarly for (B, C). Up to homotopy, CW approximation of excisive triads is functorial in such a way that γ is natural.

PROOF. Choose a CW approximation $\gamma : \Gamma C \longrightarrow C$ and use the previous result to extend it to CW approximations

$$\gamma : (\Gamma A, \Gamma C) \longrightarrow (A, C) \quad \text{and} \quad \gamma : (\Gamma B, \Gamma C) \longrightarrow (B, C).$$

We then define ΓX to be the pushout $\Gamma A \cup_{\Gamma C} \Gamma B$ and let $\gamma : \Gamma X \longrightarrow X$ be given by the universal property of pushouts. Certainly $\Gamma C = \Gamma A \cap \Gamma B$. All of the conclusions except for the assertion that $\gamma : \Gamma X \longrightarrow X$ is a weak equivalence follow immediately from the result for pairs, and the lemma and theorem below will complete the proof. □

A CW triad $(X; A, B)$ is not excisive, since A and B are closed in X, but it is equivalent to an excisive triad. To see this, we describe a simple but important general construction. Suppose that maps $i : C \longrightarrow A$ and $j : C \longrightarrow B$ are given. Define the double mapping cylinder

$$M(i,j) = A \cup (C \times I) \cup B$$

to be the space obtained from $C \times I$ by gluing A to $C \times \{0\}$ along i and gluing B to $C \times \{1\}$ along j. Let $A \cup_C B$ denote the pushout of i and j and observe that we obtain a natural quotient map $q : M(i,j) \longrightarrow A \cup_C B$ by collapsing the cylinder, sending (c,t) to the image of c in the pushout.

LEMMA. *For a cofibration $i : C \longrightarrow A$ and any map $j : C \longrightarrow B$, the quotient map $q : M(i,j) \longrightarrow A \cup_C B$ is a homotopy equivalence.*

PROOF. Because i is a cofibration, the retraction $r : Mi \longrightarrow A$ is a cofiber homotopy equivalence. That is, there is a homotopy inverse map and a pair of homotopies under C. These maps and homotopies induce maps of the pushouts that are obtained by gluing B to Mi and to C, and q is induced by r. □

When i is a cofibration and j is an inclusion, with $X = A \cup B$ and $C = A \cap B$, we can think of q as giving a map of triads

$$q : (M(i,j); A \cup (C \times [0, 2/3)), (C \times (1/3, 1]) \cup B) \longrightarrow (A \cup_C B; A, B).$$

The domain triad is excisive, and q restricts to homotopy equivalences from the domain subspaces and their intersection to the target subspaces A, B, and C. This applies when $(X; A, B)$ is a CW triad with $C = A \cap B$. Now our theorem on the approximation of excisive triads is a consequence of the following result.

THEOREM. *If $e : (X; A, B) \longrightarrow (X'; A', B')$ is a map of excisive triads such that the maps*

$$e : C \longrightarrow C', \quad e : A \longrightarrow A', \quad \text{and} \quad e : B \longrightarrow B'$$

are weak equivalences, where $C = A \cap B$ and $C' = A' \cap B'$, then $e : X \longrightarrow X'$ is a weak equivalence.

PROOF. By our technical lemma giving equivalent conditions for a map e to be a weak equivalence, it suffices to show that if $f|S^n = e \circ g$ in the following diagram, then there exists a map $\tilde{g}$ such that $\tilde{g}|S^n = g$ and $f \simeq e \circ \tilde{g}$ rel S^n:

$$\begin{array}{ccc} X & \xrightarrow{e} & X' \\ {\scriptstyle g}\uparrow & \nwarrow {\scriptstyle \tilde{g}} & \uparrow {\scriptstyle f} \\ S^n & \longrightarrow & D^{n+1} \end{array}$$

We may assume without loss of generality that $S^n \subset U \subset D^{n+1}$, where U is open in D^{n+1} and g is the restriction of a map $\hat{g} : U \longrightarrow X$ such that $f|U = e \circ \hat{g}$. To see this, define a deformation $d : D^{n+1} \times I \longrightarrow D^{n+1}$ by

$$d(x,t) = \begin{cases} 2x/(2-t) & \text{if } |x| \leq (2-t)/2 \\ x/|x| & \text{if } |x| \geq (2-t)/2. \end{cases}$$

Then $d(x,0) = x$, $d(x,t) = x$ if $x \in S^n$, and d_1 maps the boundary collar $\{x \mid |x| \geq 1/2\}$ onto S^n. Let U be the open boundary collar $\{x \mid |x| > 1/2\}$. Define $\hat{g} = g \circ d_1 : U \longrightarrow X$ and define $f' = f \circ d_1 : D^{n+1} \longrightarrow X'$. Then $\hat{g}|S^n = g$, $e \circ \hat{g} = f'|U$, and $f' \simeq f$ rel S^n. Thus the conclusion will hold for f if it holds with f replaced by f'.

With this assumption on g and f, we claim first that the closed sets
$$C_A = g^{-1}(X - \text{int } A) \cup \overline{f^{-1}(X' - A')}$$
and
$$C_B = g^{-1}(X - \text{int } B) \cup \overline{f^{-1}(X' - B')},$$
have empty intersection. Indeed, these sets are contained in the sets $\hat{C}_A$ and $\hat{C}_B$ that are obtained by replacing g by $\hat{g}$ in the definitions of C_A and C_B, and we claim that $\hat{C}_A \cap \hat{C}_B = \emptyset$. Certainly
$$\hat{g}^{-1}(X - \text{int } A) \cap \hat{g}^{-1}(X - \text{int } B) = \emptyset$$
since $(X - \text{int } A) \cap (X - \text{int } B) = \emptyset$. Similarly,
$$f^{-1}(X' - \text{int } A') \cap f^{-1}(X' - \text{int } B') = \emptyset.$$
Since $\overline{f^{-1}(X' - A')} \subset f^{-1}(X' - \text{int } A')$ and similarly for B, this implies that
$$\overline{f^{-1}(X' - A')} \cap \overline{f^{-1}(X' - B')} = \emptyset.$$
Now suppose that $v \in \hat{C}_A \cap \hat{C}_B$. In view of the possibilities that we have ruled out, we may assume that
$$v \in \hat{g}^{-1}(X - \text{int } A) \cap \overline{f^{-1}(X' - B')} \subset \hat{g}^{-1}(\text{int } B) \cap \overline{f^{-1}(X' - B')}.$$
Since $\hat{g}^{-1}(\text{int } B)$ is an open subset of D^n, there must be a point
$$u \in \hat{g}^{-1}(\text{int } B) \cap f^{-1}(X' - B').$$
Then $\hat{g}(u) \in \text{int } B \subset B$ but $f(u) \notin B'$. This contradicts $f|U = e \circ \hat{g}$.

We can subdivide D^{n+1} sufficiently finely (as a simplicial or CW complex) that no cell intersects both C_A and C_B. Let K_A be the union of those cells σ such that
$$g(\sigma \cap S^n) \subset \text{int } A \quad \text{and} \quad f(\sigma) \subset \text{int } A'$$
and define K_B similarly. If σ does not intersect C_A, then $\sigma \subset K_A$, and if σ does not intersect C_B, then $\sigma \subset K_B$. Therefore $D^{n+1} = K_A \cup K_B$. By HELP, we can obtain a map $\bar{g}$ such that the lower triangle in the diagram

$$\begin{array}{ccc} A \cap B & \xrightarrow{e} & A' \cap B' \\ {\scriptstyle g}\uparrow & \underset{\bar{g}}{\nwarrow} & \uparrow {\scriptstyle f} \\ S^n \cap (K_A \cap K_B) & \longrightarrow & K_A \cap K_B \end{array}$$

commutes, together with a homotopy $\bar{h} : (K_A \cap K_B) \times I \longrightarrow A' \cap B'$ such that
$$\bar{h} : f \simeq e \circ \bar{g} \text{ rel } S^n \cap (K_A \cap K_B).$$
Define $\bar{g}_A : K_A \cap (S^n \cup K_B) \longrightarrow A$ to be g on $K_A \cap S^n$ and $\bar{g}$ on $K_A \cap K_B$. Since $f = e \circ g$ on $K_A \cap S^n$ and $\bar{h} : f \simeq e \circ \bar{g}$ on $K_A \cap K_B$, $\bar{h}$ induces a homotopy
$$\bar{h}_A : f|K_A \cap (S^n \cup B) \simeq e \circ g_A \text{ rel } S^n \cap K_A.$$

Applying HELP again, we can obtain maps $\tilde{g}_A$ and $\tilde{h}_A$ such that the following diagram commutes:

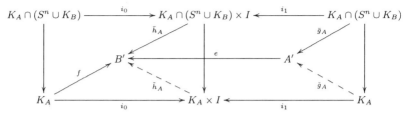

We have a symmetric diagram with the roles of K_A and K_B reversed. The maps $\tilde{g}_A$ and $\tilde{g}_B$ agree on $K_A \cap K_B$ and together define the desired map $\tilde{g} : D^{n+1} \longrightarrow X$. The homotopies $\tilde{h}_A$ and $\tilde{h}_B$ agree on $(K_A \cap K_B) \times I$ and together define the desired homotopy $\tilde{h}_A : f \simeq e \circ \tilde{g}$ rel S^n. □

PROBLEMS

1. Show that complex projective space $\mathbb{C}P^n$ is a CW complex with one $2q$-cell for each q, $0 \leq q \leq n$.
2. Let $X = \{x | x = 0 \text{ or } x = 1/n \text{ for a positive integer } n\} \subset \mathbb{R}$. Show that X does not have the homotopy type of a CW complex.
3. Assume given maps $f : X \longrightarrow Y$ and $g : Y \longrightarrow X$ such that $g \circ f$ is homotopic to the identity. (We say that Y "dominates" X.) Suppose that Y is a CW complex. Prove that X has the homotopy type of a CW complex.

Define the Euler characteristic $\chi(X)$ of a finite CW complex X to be the alternating sum $\sum (-1)^n \gamma_n(X)$, where $\gamma_n(X)$ is the number of n-cells of X. Let A be a subcomplex of a CW complex X, let Y be a CW complex, let $f : A \longrightarrow Y$ be a cellular map, and let $Y \cup_f X$ be the pushout of f and the inclusion $A \longrightarrow X$.

4. Show that $Y \cup_f X$ is a CW complex with Y as a subcomplex and X/A as a quotient complex. Formulate and prove a formula relating the Euler characteristics $\chi(A)$, $\chi(X)$, $\chi(Y)$, and $\chi(Y \cup_f X)$ when X and Y are finite.
5. * Think about proving from what we have done so far that $\chi(X)$ depends only on the homotopy type of X, not on its decomposition as a finite CW complex.

CHAPTER 11

The homotopy excision and suspension theorems

The fundamental obstruction to the calculation of homotopy groups is the failure of excision: for an excisive triad $(X; A, B)$, the inclusion $(A, A \cap B) \longrightarrow (X, B)$ fails to induce an isomorphism of homotopy groups in general. It is this that distinguishes homotopy groups from the far more computable homology groups. However, we do have such an isomorphism in a range of dimensions. This implies the Freudenthal suspension theorem, which gives that $\pi_{n+q}(\Sigma^n X)$ is independent of n if q is small relative to n. We shall rely on the consequence $\pi_n(S^n) \cong \mathbb{Z}$ in our construction of homology groups.

1. Statement of the homotopy excision theorem

We shall prove the following theorem later in this chapter, but we first explain its consequences.

DEFINITION. A map $f : (A, C) \longrightarrow (X, B)$ of pairs is an n-equivalence, $n \geq 1$, if
$$(f_*)^{-1}(\mathrm{im}(\pi_0(B) \longrightarrow \pi_0(X))) = \mathrm{im}(\pi_0(C) \longrightarrow \pi_0(A))$$
(which holds automatically when A and X are path connected) and, for all choices of basepoint in C,
$$f_* : \pi_q(A, C) \longrightarrow \pi_q(X, B)$$
is a bijection for $q < n$ and a surjection for $q = n$.

Recall that a pair (A, C) is n-connected, $n \geq 0$, if $\pi_0(C) \longrightarrow \pi_0(A)$ is surjective and $\pi_q(A, C) = 0$ for $q \leq n$.

THEOREM (Homotopy excision). Let $(X; A, B)$ be an excisive triad such that $C = A \cap B$ is non-empty. Assume that (A, C) is $(m-1)$-connected and (B, C) is $(n-1)$-connected, where $m \geq 2$ and $n \geq 1$. Then the inclusion $(A, C) \longrightarrow (X, B)$ is an $(m+n-2)$-equivalence.

This specializes to give a relationship between the homotopy groups of pairs (X, A) and of quotients X/A and to prove the Freudenthal suspension theorem.

THEOREM. Let $f : X \longrightarrow Y$ be an $(n-1)$-equivalence between $(n-2)$-connected spaces, where $n \geq 2$; thus $\pi_{n-1}(f)$ is an epimorphism. Then the quotient map $\pi : (Mf, X) \longrightarrow (Cf, *)$ is a $(2n-2)$-equivalence. In particular, Cf is $(n-1)$-connected. If X and Y are $(n-1)$-connected, then $\pi : (Mf, X) \longrightarrow (Cf, *)$ is a $(2n-1)$-equivalence.

PROOF. We are writing Cf for the unreduced cofiber Mf/X. We have the excisive triad $(Cf; A, B)$, where
$$A = Y \cup (X \times [0, 2/3]) \quad \text{and} \quad B = (X \times [1/3, 1])/(X \times \{1\}).$$

Thus $C \equiv A \cap B = X \times [1/3, 2/3]$. It is easy to check that π is homotopic to a composite

$$(Mf, X) \xrightarrow{\simeq} (A, C) \longrightarrow (Cf, B) \xrightarrow{\simeq} (Cf, *),$$

the first and last arrows of which are homotopy equivalences of pairs. The hypothesis on f and the long exact sequence of the pair (Mf, X) imply that (Mf, X) and therefore also (A, C) are $(n-1)$-connected. In view of the connecting isomorphism $\partial : \pi_{q+1}(CX, X) \longrightarrow \pi_q(X)$ and the evident homotopy equivalence of pairs $(B, C) \simeq (CX, X)$, (B, C) is also $(n-1)$-connected, and it is n-connected if X is $(n-1)$-connected. The homotopy excision theorem gives the conclusions. □

We shall later use the following bit of the result to prove the Hurewicz theorem relating homotopy groups to homology groups.

COROLLARY. *Let $f : X \longrightarrow Y$ be a based map between $(n-1)$-connected nondegenerately based spaces, where $n \geq 2$. Then Cf is $(n-1)$-connected and*

$$\pi_n(Mf, X) \longrightarrow \pi_n(Cf, *)$$

is an isomorphism. Moreover, the canonical map $\eta : Ff \longrightarrow \Omega Cf$ induces an isomorphism

$$\pi_{n-1}(Ff) \longrightarrow \pi_n(Cf).$$

PROOF. Here in the based context, we are thinking of the reduced mapping cylinder and cofiber, but the maps to them from the unreduced constructions are homotopy equivalences since our basepoints are nondegenerate. Thus the first statement is immediate from the theorem. For the second, if $j : X \longrightarrow Mf$ is the inclusion, then we have a map

$$Fr : Fj = P(Mf; *, X) \longrightarrow Ff$$

induced by the retraction $r : Mf \longrightarrow Y$. By a comparison of long exact sequences, $(Fr)_* : \pi_q(Mf, X) \longrightarrow \pi_{q-1}(Ff)$ is an isomorphism for all q. Moreover, η factors through a map $Fj \longrightarrow \Omega Cf$, as we noted at the end of Chapter 8 §7. Thus the second statement follows from the first. □

Specializing f to be a cofibration and changing notation, we obtain the following version of the previous theorem.

THEOREM. *Let $i : A \longrightarrow X$ be a cofibration and an $(n-1)$-equivalence between $(n-2)$-connected spaces, where $n \geq 2$. Then the quotient map $(X, A) \longrightarrow (X/A, *)$ is a $(2n-2)$-equivalence, and it is a $(2n-1)$-equivalence if A and X are $(n-1)$-connected.*

PROOF. The vertical arrows are homotopy equivalences of pairs in the commutative diagram

$$\begin{array}{ccc} (Mi, A) & \xrightarrow{\pi} & (Ci, *) \\ {\scriptstyle r}\downarrow & & \downarrow{\scriptstyle \psi} \\ (X, A) & \longrightarrow & (X/A, *). \end{array}$$

□

2. The Freudenthal suspension theorem

A specialization of the last result gives the Freudenthal suspension theorem. For a based space X, define the suspension homomorphism

$$\Sigma : \pi_q(X) \longrightarrow \pi_{q+1}(\Sigma X)$$

by letting

$$\Sigma f = f \wedge \mathrm{id} : S^{q+1} \cong S^q \wedge S^1 \longrightarrow X \wedge S^1 = \Sigma X.$$

THEOREM (Freudenthal suspension). *Assume that X is nondegenerately based and $(n-1)$-connected, where $n \geq 1$. Then Σ is a bijection if $q < 2n - 1$ and a surjection if $q = 2n - 1$.*

PROOF. We give a different description of Σ. Consider the "reversed" cone $C'X = X \wedge I$, where I is given the basepoint 0 rather than 1. Thus

$$C'X = X \times I / X \times \{0\} \cup \{*\} \times I.$$

For a map $f : (I^q, \partial I^q) \longrightarrow (X, *)$, the product $f \times \mathrm{id} : I^{q+1} \longrightarrow X \times I$ passes to quotients to give a map of triples

$$(I^{q+1}, \partial I^{q+1}, J^q) \longrightarrow (C'X, X, *)$$

whose restriction to $I^q \times \{1\}$ is f and which induces Σf when we quotient out $X \times \{1\}$. That is, the following diagram commutes, where $\rho : C'X \longrightarrow \Sigma X$ is the quotient map:

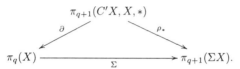

Since $C'X$ is contractible, ∂ is an isomorphism. Since the inclusion $X \longrightarrow C'X$ is a cofibration and an n-equivalence between $(n-1)$-connected spaces, ρ is a $2n$-equivalence by the last theorem of the previous section. The conclusion follows. ⊔

This implies the promised calculation of $\pi_n(S^n)$.

THEOREM. *For all $n \geq 1$, $\pi_n(S^n) = \mathbb{Z}$ and $\Sigma : \pi_n(S^n) \longrightarrow \pi_{n+1}(S^{n+1})$ is an isomorphism.*

PROOF. We saw by use of the Hopf bundle $S^3 \longrightarrow S^2$ that $\pi_2(S^2) = \mathbb{Z}$, and the suspension theorem applies to give the conclusion for $n \geq 2$. A little extra argument is needed to check that Σ is an isomorphism for $n = 1$; one can inspect the connecting homomorphism of the Hopf bundle or refer ahead to the observation that the Hurewicz homomorphism commutes with the corresponding suspension isomorphism in homology. □

The dimensional range of the suspension theorem is sharp. We saw before that $\pi_3(S^2) = \pi_3(S^3)$, which is $\mathbb{Z}$. The suspension theorem applies to show that

$$\Sigma : \pi_3(S^2) \longrightarrow \pi_4(S^3)$$

is an epimorphism, and it is known that $\pi_4(S^3) = \mathbb{Z}_2$.

Applying suspension repeatedly, we can form a colimit

$$\pi_q^s(X) = \mathrm{colim}\ \pi_{q+n}(\Sigma^n X).$$

This group is called the qth stable homotopy group of X. For $q < n - 1$, the maps of the colimit system are isomorphisms and therefore

$$\pi_q^s(X) = \pi_{q+n}(\Sigma^n X) \text{ if } q < n - 1.$$

The calculation of the stable homotopy groups of spheres, $\pi_q^s(S^0)$, is one of the deepest and most studied problems in algebraic topology. Important problems of geometric topology, such as the enumeration of the distinct differential structures on S^q for $q \geq 5$, have been reduced to the determination of these groups.

3. Proof of the homotopy excision theorem

This is a deep result, and it is remarkable that a direct homotopical proof, in principle an elementary one, is possible. Most standard texts, if they treat this topic at all, give a far more sophisticated proof of a significantly weaker result. However, the reader may prefer to skip this argument on a first reading. The idea is clear enough. We are trying to show that a certain map of pairs induces an isomorphism in a range of dimensions. We capture the relevant map as part of a long exact sequence, and we prove that the third term in the long exact sequence vanishes in the required range.

However, we start with an auxiliary long exact sequence that we shall also need. Recall that a triple (X, A, B) consists of spaces $B \subset A \subset X$ and must not be confused with a triad.

PROPOSITION. *For a triple (X, A, B) and any basepoint in B, the following sequence is exact:*

$$\cdots \longrightarrow \pi_q(A, B) \xrightarrow{i_*} \pi_q(X, B) \xrightarrow{j_*} \pi_q(X, A) \xrightarrow{k_* \circ \partial} \pi_{q-1}(A, B) \longrightarrow \cdots.$$

Here $i : (A, B) \longrightarrow (X, B)$, $j : (X, B) \longrightarrow (X, A)$, and $k : (A, *) \longrightarrow (A, B)$ are the inclusions.

PROOF. The proof is a purely algebraic deduction from the long exact sequences of the various pairs in sight and is left as an exercise for the reader. □

We now define the "triad homotopy groups" that are needed to implement the idea of the proof sketched above.

DEFINITION. For a triad $(X; A, B)$ with basepoint $* \in C = A \cap B$, define

$$\pi_q(X; A, B) = \pi_{q-1}(P(X; *, B), P(A; *, C)),$$

where $q \geq 2$. More explicitly, $\pi_q(X; A, B)$ is the set of homotopy classes of maps of tetrads

$$(I^q; I^{q-2} \times \{1\} \times I, I^{q-1} \times \{1\}, J^{q-2} \times I \cup I^{q-1} \times \{0\})$$
$$\downarrow$$
$$(X; A, B, *),$$

where $J^{q-2} = \partial I^{q-2} \times I \cup I^{q-2} \times \{0\} \subset I^{q-1}$. The long exact sequence of the pair in the first form of the definition is

$$\cdots \longrightarrow \pi_{q+1}(X; A, B) \longrightarrow \pi_q(A, C) \longrightarrow \pi_q(X, B) \longrightarrow \pi_q(X; A, B) \longrightarrow \cdots.$$

3. PROOF OF THE HOMOTOPY EXCISION THEOREM

Now we return to the homotopy excision theorem. Its conditions $m \geq 1$ and $n \geq 1$ merely give that $\pi_0(C) \longrightarrow \pi_0(A)$ and $\pi_0(C) \longrightarrow \pi_0(B)$ are surjective, and any extraneous components of A or B would not affect the relevant homotopy groups. The condition $m \geq 2$ implies that (X, B) is 1-connected. By the long exact sequence just given, the theorem is equivalent to the following one.

THEOREM. *Under the hypotheses of the homotopy excision theorem,*
$$\pi_q(X; A, B) = 0 \text{ for } 2 \leq q \leq m + n - 2$$
and all choices of basepoint $* \in C$.

In this form, the conclusion is symmetric in A and B and vacuous if $m+n \leq 3$. Thus our hypotheses $m \geq 2$ and $n \geq 1$ are the minimal ones under which our strategy can apply.

In order to have some hope of tackling the problem in direct terms, we first reduce it to the case when A and B are each obtained from C by attaching a single cell. We may approximate our given excisive triad by a weakly equivalent CW triad. This does not change the triad homotopy groups. More precisely, by our connectivity hypotheses, we may assume that X is a CW complex that is the union of subcomplexes A and B with intersection C, where (A, C) has no relative q-cells for $q < m$ and (B, C) has no relative q-cells for $q < n$. Since any map $I^q \longrightarrow X$ has image contained in a finite subcomplex, we may assume that X has finitely many cells. We may also assume that (A, C) and (B, C) each have at least one cell since otherwise the result holds trivially.

We claim first that, inductively, it suffices to prove the result when (A, C) has exactly one cell. Indeed, suppose that $C \subset A' \subset A$, where A is obtained from A' by attaching a single cell and (A', C) has one less cell than (A, C). Let $X' = A' \cup_C B$. If the result holds for the triads $(X'; A', B)$ and $(X; A, X')$, then the result holds for the triad $(X; A, B)$ by application of the five lemma to the following diagram:

$$\begin{array}{ccccccccc} \pi_{q+1}(A, A') & \longrightarrow & \pi_q(A', C) & \longrightarrow & \pi_q(A, C) & \longrightarrow & \pi_q(A, A') & \longrightarrow & \pi_{q-1}(A', C) \\ \downarrow & & \downarrow & & \downarrow & & \downarrow & & \downarrow \\ \pi_{q+1}(X, X') & \longrightarrow & \pi_q(X', B) & \longrightarrow & \pi_q(X, B) & \longrightarrow & \pi_q(X, X') & \longrightarrow & \pi_{q-1}(X', B). \end{array}$$

The rows are the exact sequences of the triples (A, A', C) and (X, X', B). Note for the case $q = 1$ that all pairs in the diagram are 1-connected.

We claim next that, inductively, it suffices to prove the result when (B, C) also has exactly one cell. Indeed, suppose that $C \subset B' \subset B$, where B is obtained from B' by attaching a single cell and (B', C) has one less cell than (B, C) and let $X' = A \cup_C B'$. If the result holds for the triads $(X'; A, B')$ and $(X; X', B)$, then the result holds for the triad $(X; A, B)$ since the inclusion $(A, C) \longrightarrow (X, B)$ factors as the composite
$$(A, C) \longrightarrow (X', B') \longrightarrow (X, B).$$

Thus we may assume that $A = C \cup D^m$ and $B = C \cup D^n$, where $m \geq 2$ and $n \geq 1$, and we fix a basepoint $* \in C$. Assume given a map of tetrads
$$(I^q; I^{q-2} \times \{1\} \times I, I^{q-1} \times \{1\}, J^{q-2} \times I \cup I^{q-1} \times \{0\})$$
$$\downarrow f$$
$$(X; A, B, *),$$

where $2 \leq q \leq m + n - 2$. We must prove that f is null homotopic as a map of tetrads. For interior points $x \in D^m$ and $y \in D^n$, we have inclusions of based triads

$$(A; A, A - x) \subset (X - \{y\}; A, X - \{x, y\}) \subset (X; A, X - \{x\}) \supset (X; A, B).$$

The first and third of these induce isomorphisms on triad homotopy groups in view of the radial deformation away from y of $X - \{y\}$ onto A and the radial deformation away from x of $X - \{x\}$ onto B. It is trivial to check that $\pi_*(A; A, A') = 0$ for any $A' \subset A$. We shall show that, for well chosen points x and y, f regarded as a map of based triads into $(X; A, X - \{x\})$ is homotopic to a map f' that has image in $(X - \{y\}; A, X - \{x, y\})$. This will imply that f is null homotopic.

Let $D^m_{1/2} \subset D^m$ and $D^n_{1/2} \subset D^n$ be the subdisks of radius $1/2$. We can cubically subdivide I^q into subcubes I^q_α such that $f(I^q_\alpha)$ is contained in the interior of D^m if it intersects $D^m_{1/2}$ and $f(I^q_\alpha)$ is contained in the interior of D^n if it intersects $D^n_{1/2}$. By simplicial approximation, f is homotopic as a map of tetrads to a map g whose restriction to the $(n-1)$-skeleton of I^q with its subdivided cell structure does not cover $D^n_{1/2}$ and whose restriction to the $(m-1)$-skeleton of I_q does not cover $D^m_{1/2}$. Moreover, we can arrange that the dimension of $g^{-1}(y)$ is at most $q - n$ for a point $y \in D^n_{1/2}$ that is not in the image under g of the $(n-1)$-skeleton of I^q. This is the main point of the proof, and to be completely rigorous about it we would have to digress to introduce a bit of dimension theory. Alternatively, we could use smooth approximation to arrive at g and y with appropriate properties. Since the intuition should be clear, we shall content ourselves with showing how the conclusion of the theorem follows.

Let $\pi : I^q \longrightarrow I^{q-1}$ be the projection on the first $q - 1$ coordinates and let K be the prism $\pi^{-1}(\pi(g^{-1}(y)))$. Then K can have dimension at most one more than the dimension of $g^{-1}(y)$, so that

$$\dim K \leq q - n + 1 \leq m - 1.$$

Therefore $g(K)$ cannot cover $D^m_{1/2}$. Choose a point $x \in D^m_{1/2}$ such that $x \notin g(K)$. Since $g(\partial I^{q-1} \times I) \subset A$, we see that $\pi(g^{-1}(x)) \cup \partial I^{q-1}$ and $\pi(g^{-1}(y))$ are disjoint closed subsets of I^{q-1}. By Uryssohn's lemma, we may choose a map $v : I^{q-1} \longrightarrow I$ such that

$$v(\pi(g^{-1}(x)) \cup \partial I^{q-1}) = 0 \quad \text{and} \quad v(\pi(g^{-1}(y))) = 1.$$

Define $h : I^{q+1} \longrightarrow I^q$ by

$$h(r, s, t) = (r, s - stv(r)) \text{ for } r \in I^{q-1} \quad \text{and} \quad s, t \in I.$$

Then let $f' = g \circ h_1$, where $h_1(r, s) = h(r, s, 1)$. We claim that f' is as desired. Observe that

$$h(r, s, 0) = (r, s), \quad h(r, 0, t) = (r, 0), \quad \text{and} \quad h(r, s, t) = (r, s) \text{ if } r \in \partial I^{q-1}.$$

Moreover,

$$h(r, s, t) = (r, s) \text{ if } h(r, s, t) \in g^{-1}(x)$$

since $r \in \pi(g^{-1}(x))$ implies $v(r) = 0$ and

$$h(r, s, t) = (r, s - st) \text{ if } h(r, s, t) \in g^{-1}(y)$$

since $r \in \pi(g^{-1}(y))$ implies $v(r) = 1$. Then $g \circ h$ is a homotopy of maps of tetrads

$$(I^q; I^{q-2} \times \{1\} \times I, I^{q-1} \times \{1\}, J^{q-2} \times I \cup I^{q-1} \times \{0\})$$
$$\downarrow$$
$$(X; A, X - \{x\}, *)$$

from g to f', and f' has image in $(X - \{y\}; A, X - \{x, y\})$, as required.

CHAPTER 12

A little homological algebra

Let R be a commutative ring. The main example will be $R = \mathbb{Z}$. We develop some rudimentary homological algebra in the category of R-modules. We shall say more later. For now, we give the minimum that will be needed to develop cellular and singular homology theory.

1. Chain complexes

A chain complex over R is a sequence of maps of R-modules

$$\cdots \longrightarrow X_{i+1} \xrightarrow{d_{i+1}} X_i \xrightarrow{d_i} X_{i-1} \longrightarrow \cdots$$

such that $d_i \circ d_{i+1} = 0$ for all i. We generally abbreviate $d = d_i$. A cochain complex over R is an analogous sequence

$$\cdots \longrightarrow Y^{i-1} \xrightarrow{d^{i-1}} Y^i \xrightarrow{d^i} Y^{i+1} \longrightarrow \cdots$$

with $d^i \circ d^{i-1} = 0$. In practice, we usually require chain complexes to satisfy $X_i = 0$ for $i < 0$ and cochain complexes to satisfy $Y^i = 0$ for $i < 0$. Without these restrictions, the notions are equivalent since a chain complex $\{X_i, d_i\}$ can be rewritten as a cochain complex $\{X^{-i}, d^{-i}\}$, and vice versa.

An element of the kernel of d_i is called a cycle and an element of the image of d_{i+1} is called a boundary. We say that two cycles are "homologous" if their difference is a boundary. We write $B_i(X) \subset Z_i(X) \subset X_i$ for the submodules of boundaries and cycles, respectively, and we define the ith homology group $H_i(X)$ to be the quotient module $Z_i(X)/B_i(X)$. We write $H_*(X)$ for the sequence of R-modules $H_i(X)$. We understand "graded R-modules" to be sequences of R-modules such as this (and we never take the sum of elements in different gradings).

2. Maps and homotopies of maps of chain complexes

A map $f : X \longrightarrow X'$ of chain complexes is a sequence of maps of R-modules $f_i : X_i \longrightarrow X'_i$ such that $d'_i \circ f_i = f_{i-1} \circ d_i$ for all i. That is, the following diagram commutes for each i:

$$\begin{array}{ccc} X_i & \xrightarrow{f_i} & X'_i \\ d_i \downarrow & & \downarrow d'_i \\ X_{i-1} & \xrightarrow{f_{i-1}} & X'_{i-1}. \end{array}$$

It follows that $f_i(B_i(X)) \subset B_i(X')$ and $f_i(Z_i(X)) \subset Z_i(X')$. Therefore f induces a map of R-modules $f_* = H_i(f) : H_i(X) \longrightarrow H_i(X')$.

89

A chain homotopy $s : f \simeq g$ between chain maps $f, g : X \longrightarrow X'$ is a sequence of homomorphisms $s_i : X_i \longrightarrow X'_{i+1}$ such that

$$d'_{i+1} \circ s_i + s_{i-1} \circ d_i = f_i - g_i$$

for all i. Chain homotopy is an equivalence relation since if $t : g \simeq h$, then $s + t = \{s_i + t_i\}$ is a chain homotopy $f \simeq h$.

LEMMA. *Chain homotopic maps induce the same homomorphism of homology groups.*

PROOF. Let $s : f \simeq g$, $f, g : X \longrightarrow X'$. If $x \in Z_i(X)$, then

$$f_i(x) - g_i(x) = d'_{i+1} s_i(x),$$

so that $f_i(x)$ and $g_i(x)$ are homologous. □

3. Tensor products of chain complexes

The tensor product (over R) of chain complexes X and Y is specified by letting

$$(X \otimes Y)_n = \sum_{i+j=n} X_i \otimes Y_j.$$

When X_i and Y_i are zero for $i < 0$, the sum is finite, but we don't need to assume this. The differential is specified by

$$d(x \otimes y) = d(x) \otimes y + (-1)^i x \otimes d(y)$$

for $x \in X_i$ and $y \in Y_j$. The sign ensures that $d \circ d = 0$. We may write this as

$$d = d \otimes \mathrm{id} + \mathrm{id} \otimes d.$$

The sign is dictated by the general rule that whenever two entities to which degrees m and n can be assigned are permuted, the sign $(-1)^{mn}$ should be inserted. In the present instance, when calculating $(\mathrm{id} \otimes d)(x \otimes y)$, we must permute the map d of degree -1 with the element x of degree i.

We regard R-modules M as chain complexes concentrated in degree zero, and thus with zero differential. For a chain complex X, there results a chain complex $X \otimes M$; $H_*(X \otimes M)$ is called the homology of X with coefficients in M.

Define a chain complex $\mathscr{I}$ by letting $\mathscr{I}_0$ be the free Abelian group with two generators $[0]$ and $[1]$, letting $\mathscr{I}_1$ be the free Abelian group with one generator $[I]$ such that $d([I]) = [0] - [1]$, and letting $\mathscr{I}_i = 0$ for all other i.

LEMMA. *A chain homotopy $s : f \simeq g$ between chain maps $f, g : X \longrightarrow X'$ determines and is determined by a chain map $h : X \otimes \mathscr{I} \longrightarrow X'$ such that $h(x, [0]) = f(x)$ and $h(x, [1]) = g(x)$.*

PROOF. Let s correspond to h via $(-1)^i s(x) = h(x \otimes [I])$ for $x \in X_i$. The relation

$$d'_{i+1}(s_i(x)) = f_i(x) - g_i(x) - s_{i-1}(d_i(x))$$

corresponds to the relation $d'h = hd$ by the definition of our differential on $\mathscr{I}$. The sign in the correspondence would disappear if we replaced by $X \otimes \mathscr{I}$ by $\mathscr{I} \otimes X$. □

4. Short and long exact sequences

A sequence $M' \xrightarrow{f} M \xrightarrow{g} M''$ of modules is exact if $\operatorname{im} f = \ker g$. If $M' = 0$, this means that g is a monomorphism; if $M'' = 0$, it means that f is an epimorphism. A longer sequence is exact if it is exact at each position. A short exact sequence of chain complexes is a sequence

$$0 \longrightarrow X' \xrightarrow{f} X \xrightarrow{g} X'' \longrightarrow 0$$

that is exact in each degree. Here 0 denotes the chain complex that is the zero module in each degree.

PROPOSITION. *A short exact sequence of chain complexes naturally gives rise to a long exact sequence of R-modules*

$$\cdots \longrightarrow H_q(X') \xrightarrow{f_*} H_q(X) \xrightarrow{g_*} H_q(X'') \xrightarrow{\partial} H_{q-1}(X') \longrightarrow \cdots.$$

PROOF. Write $[x]$ for the homology class of a cycle x. We define the "connecting homomorphism" $\partial : H_q(X'') \longrightarrow H_{q-1}(X')$ by $\partial[x''] = [x']$, where $f(x') = d(x)$ for some x such that $g(x) = x''$. There is such an x since g is an epimorphism, and there is such an x' since $gd(x) = dg(x) = 0$. It is a standard exercise in "diagram chasing" to verify that ∂ is well defined and the sequence is exact. Naturality means that a commutative diagram of short exact sequences of chain complexes gives rise to a commutative diagram of long exact sequences of R-modules. The essential point is the naturality of the connecting homomorphism, which is easily checked. □

PROBLEMS

For a graded vector space $V = \{V_n\}$ with $V_n = 0$ for all but finitely many n and with all V_n finite dimensional, define the Euler characteristic $\chi(V)$ to be $\sum (-1)^n \dim V_n$.

1. Let V', V, and V'' be such graded vector spaces and suppose there is a long exact sequence

   $$\cdots \longrightarrow V'_n \longrightarrow V_n \longrightarrow V''_n \longrightarrow V'_{n-1} \longrightarrow \cdots.$$

   Prove that $\chi(V) = \chi(V') + \chi(V'')$.

2. If $\{V_n, d_n\}$ is a chain complex, show that $\chi(V) = \chi(H_*(V))$.

3. Let $0 \longrightarrow \pi \xrightarrow{f} \rho \xrightarrow{g} \sigma \longrightarrow 0$ be an exact sequence of Abelian groups and let C be a chain complex of flat (= torsion free) Abelian groups. Write $H_*(C; \pi) = H_*(C \otimes \pi)$. Construct a natural long exact sequence

   $$\cdots \longrightarrow H_q(C; \pi) \xrightarrow{f_*} H_q(C; \rho) \xrightarrow{g_*} H_q(C; \sigma) \xrightarrow{\beta} H_{q-1}(C; \pi) \longrightarrow \cdots.$$

   The connecting homomorphism β is called a Bockstein operation.

CHAPTER 13

Axiomatic and cellular homology theory

Homology groups are the basic computable invariants of spaces. Unlike homotopy groups, these are stable invariants, the same for a space and its suspension, and it is this that makes them computable. In this and the following two chapters, we first give both an axiomatic and a cellular description of homology, next revert to an axiomatic development of the properties of homology, and then prove the Hurewicz theorem and use it to prove the uniqueness of homology.

1. Axioms for homology

Fix an Abelian group π and consider pairs of spaces (X, A). We shall see that π determines a "homology theory on pairs (X, A)." We say that a map $(X, A) \longrightarrow (Y, B)$ of pairs is a weak equivalence if its maps $A \longrightarrow B$ and $X \longrightarrow Y$ are weak equivalences.

THEOREM. *For integers q, there exist functors $H_q(X, A; \pi)$ from the homotopy category of pairs of spaces to the category of Abelian groups together with natural transformations $\partial : H_q(X, A; \pi) \longrightarrow H_{q-1}(A; \pi)$, where $H_q(X; \pi)$ is defined to be $H_q(X, \emptyset; \pi)$. These functors and natural transformations satisfy and are characterized by the following axioms.*

- *DIMENSION If X is a point, then $H_0(X; \pi) = \pi$ and $H_q(X; \pi) = 0$ for all other integers.*
- *EXACTNESS The following sequence is exact, where the unlabeled arrows are induced by the inclusions $A \longrightarrow X$ and $(X, \emptyset) \longrightarrow (X, A)$:*

$$\cdots \longrightarrow H_q(A; \pi) \longrightarrow H_q(X; \pi) \longrightarrow H_q(X, A; \pi) \xrightarrow{\partial} H_{q-1}(A; \pi) \longrightarrow \cdots .$$

- *EXCISION If $(X; A, B)$ is an excisive triad, so that X is the union of the interiors of A and B, then the inclusion $(A, A \cap B) \longrightarrow (X, B)$ induces an isomorphism*

$$H_*(A, A \cap B; \pi) \longrightarrow H_*(X, B; \pi).$$

- *ADDITIVITY If (X, A) is the disjoint union of a set of pairs (X_i, A_i), then the inclusions $(X_i, A_i) \longrightarrow (X, A)$ induce an isomorphism*

$$\sum_i H_*(X_i, A_i; \pi) \longrightarrow H_*(X, A; \pi).$$

- *WEAK EQUIVALENCE If $f : (X, A) \longrightarrow (Y, B)$ is a weak equivalence, then*

$$f_* : H_*(X, A; \pi) \longrightarrow H_*(Y, B; \pi)$$

is an isomorphism.

Here, by a standard abuse, we write f_* instead of $H_*(f)$ or $H_q(f)$. Our approximation theorems for spaces, pairs, maps, homotopies, and excisive triads imply that such a theory determines and is determined by an appropriate theory defined on CW pairs, as spelled out in the following CW version of the theorem.

THEOREM. *For integers q, there exist functors $H_q(X, A; \pi)$ from the homotopy category of pairs of CW complexes to the category of Abelian groups together with natural transformations $\partial : H_q(X, A) \longrightarrow H_{q-1}(A; \pi)$, where $H_q(X; \pi)$ is defined to be $H_q(X, \emptyset; \pi)$. These functors and natural transformations satisfy and are characterized by the following axioms.*

- *DIMENSION If X is a point, then $H_0(X; \pi) = \pi$ and $H_q(X; \pi) = 0$ for all other integers.*
- *EXACTNESS The following sequence is exact, where the unlabeled arrows are induced by the inclusions $A \longrightarrow X$ and $(X, \emptyset) \longrightarrow (X, A)$:*

$$\cdots \longrightarrow H_q(A; \pi) \longrightarrow H_q(X; \pi) \longrightarrow H_q(X, A; \pi) \xrightarrow{\partial} H_{q-1}(A; \pi) \longrightarrow \cdots .$$

- *EXCISION If X is the union of subcomplexes A and B, then the inclusion $(A, A \cap B) \longrightarrow (X, B)$ induces an isomorphism*

$$H_*(A, A \cap B; \pi) \longrightarrow H_*(X, B; \pi).$$

- *ADDITIVITY If (X, A) is the disjoint union of a set of pairs (X_i, A_i), then the inclusions $(X_i, A_i) \longrightarrow (X, A)$ induce an isomorphism*

$$\sum_i H_*(X_i, A_i; \pi) \longrightarrow H_*(X, A; \pi).$$

Such a theory determines and is determined by a theory as in the previous theorem.

PROOF. We prove the last statement and return to the rest later. Since a CW triad (which, we recall, was required to be the union of its given subcomplexes) is homotopy equivalent to an excisive triad, it is immediate that the restriction to CW pairs of a theory on pairs of spaces gives a theory on pairs of CW complexes. Conversely, given a theory on CW pairs, we may define a theory on pairs of spaces by turning the weak equivalence axiom into a definition. That is, we fix a CW approximation functor Γ from the homotopy category of pairs of spaces to the homotopy category of CW pairs and we define

$$H_*(X, A; \pi) = H_*(\Gamma X, \Gamma A; \pi).$$

Similarly, we define ∂ for (X, A) to be ∂ for $(\Gamma X, \Gamma A)$. For a map $f : (X, A) \longrightarrow (Y, B)$ of pairs, we define $f_* = (\Gamma f)_*$. It is clear from our earlier results that this does give a well defined homology theory on pairs of spaces. □

Clearly, up to canonical isomorphism, this construction of a homology theory on pairs of spaces is independent of the choice of our CW approximation functor Γ. The reader may have seen singular homology before. As we shall explain later, the classical construction of singular homology amounts to a choice of a particularly nice CW approximation functor, one that is actually functorial on the point-set level, before passage to homotopy categories.

2. Cellular homology

We must still construct $H_*(X, A; \pi)$ on CW pairs. We shall give a seemingly ad hoc construction, but we shall later see that precisely this construction is in fact forced upon us by the axioms. We concentrate on the case $\pi = \mathbb{Z}$, and we abbreviate notation by setting $H_*(X, A) = H_*(X, A; \mathbb{Z})$.

2. CELLULAR HOMOLOGY

Let X be a CW complex. We shall define the cellular chain complex $C_*(X)$. We let $C_n(X)$ be the free Abelian group with one generator $[j]$ for each n-cell j. We must define a differential $d_n : C_n(X) \longrightarrow C_{n-1}(X)$. We shall first give a direct definition in terms of the cell structure and then give a more conceptual description in terms of cofiber sequences. It will be convenient to work with unreduced cones, cofibers, and suspensions in this section; that is, we do not choose basepoints and so we do not collapse out lines through basepoints. (We shall discuss this difference more formally in the next chapter.) We still have the basic result that if $i : A \longrightarrow X$ is a cofibration, then collapsing the cone on A to a point gives a homotopy equivalence $\psi : Ci \longrightarrow X/A$. We shall use the notation ψ^{-1} for any chosen homotopy inverse to such a homotopy equivalence. We again obtain $\pi : Ci \longrightarrow \Sigma A$ by collapsing the base X of the cofiber to a point.

Our first definition of d_n involves the calculation of the degrees of maps between spheres. A map $f : S^n \longrightarrow S^n$ induces a homomorphism $f_* : \pi_n(S^n) \longrightarrow \pi_n(S^n)$, which is given by multiplication by an integer called the degree of f. As in our discussion earlier for π_1, f_* is defined using a change of basepoint isomorphism, but $\deg(f)$ is independent of the choice of the path connecting $*$ to $f(*)$. Of course, this only makes sense for $n \geq 1$.

To define and calculate degrees, the domain and target of f must both be S^n. However, there are three models of S^n that are needed in our discussion: the standard sphere $S^n \subset D^{n+1}$, the quotient D^n/S^{n-1}, and the (unreduced) suspension ΣS^{n-1}. We must fix suitably compatible homeomorphisms relating these "n-spheres." We define a homeomorphism

$$\nu_n : D^n/S^{n-1} \longrightarrow S^n$$

by

$$\nu_n(tx_1,\ldots,tx_n) = (ux_1,\ldots,ux_n, 2t-1)$$

for $0 \leq t \leq 1$ and $(x_1,\ldots,x_n) \in S^{n-1}$, where $u = (1-(2t-1)^2)^{1/2}$. Thus ν_n sends the ray from 0 to $(x_1,\ldots,x_n)$ to the longitude that runs from the south pole $(0,\ldots,0,-1)$ through the equatorial point $(x_1,\ldots,x_n,0)$ to the north pole $(0,\ldots,0,1)$. We define a homeomorphism

$$\iota_n : S^n \longrightarrow \Sigma S^{n-1}$$

by

$$\iota_n(x_1,\ldots,x_{n+1}) = (vx_1,\ldots,vx_n) \wedge (x_{n+1}+1)/2,$$

where $v = 1/(\sum_{i=1}^n x_i^2)^{1/2}$. This makes sense since if $x_i = 0$ for $1 \leq i \leq n$, then $x_{n+1} = \pm 1$, so that $(x_{n+1}+1)/2 = 0$ or 1 and $\iota_n(x_1,\ldots,x_{n+1})$ is a cone point. In effect, ι_n makes the last coordinate the suspension coordinate. We define a homeomorphism of pairs

$$\xi_n : (D^n, S^{n-1}) \longrightarrow (CS^{n-1}, S^{n-1})$$

by

$$\xi_n(tx_1,\ldots,tx_n) = (x_1,\ldots,x_n) \wedge (1-t),$$

and we continue to write ξ_n for the induced homeomorphism

$$D^n/S^{n-1} \cong CS^{n-1}/S^{n-1} = \Sigma S^{n-1}.$$

Observe that

$$\iota_n \circ \nu_n = -\xi_n : D^n/S^{n-1} \longrightarrow \Sigma S^{n-1},$$

where the minus is interpreted as the sign map $y \wedge t \longrightarrow y \wedge (1-t)$ on ΣS^{n-1}. We saw in our treatment of cofiber sequences that, up to homotopy, the maps

$$CS^{n-1} \cup_{S^{n-1}} CS^{n-1} \longrightarrow \Sigma S^{n-1}$$

obtained by collapsing out the first and second cone also differ by this sign map. By an easy diagram chase, these observations imply the following compatibility statement. It will be used to show that the two definitions of d_n that we shall give are in fact the same.

LEMMA. *The following diagram is homotopy commutative:*

$$\begin{array}{ccc} D^n \cup_{S^{n-1}} CS^{n-1} & \xrightarrow{\pi} & \Sigma S^{n-1} \\ \psi \downarrow & & \uparrow \iota_n \\ D^n/S^{n-1} & \xrightarrow{\nu_n} & S^n. \end{array}$$

Returning to our CW complex X, we think of an n-cell j as a map of pairs

$$j : (D^n, S^{n-1}) \longrightarrow (X^n, X^{n-1}).$$

There results a homeomorphism

$$\alpha : \bigvee_j D^n/S^{n-1} \longrightarrow X^n/X^{n-1}$$

whose restriction to the jth wedge summand is induced by j. Define

$$\pi_j : X^n/X^{n-1} \longrightarrow S^n$$

to be the composite of α^{-1} with the map given by ν_n on the jth wedge summand and the constant map at the basepoint on all other wedge summands. If $n = 0$, we interpret D^0 to be a point and interpret S^{-1} and X^{-1} to be empty. With our convention that $X/\emptyset = X_+$, we see that X^0/X^{-1} can be identified with the wedge of one copy of S^0 for each vertex j, and π_j is still defined. Here we take $S^0 = \{\pm 1\}$, with basepoint 1.

For an n-cell j and an $(n-1)$-cell i, where $n \geq 1$, we have a composite

$$S^{n-1} \xrightarrow{j} X^{n-1} \xrightarrow{\rho} X^{n-1}/X^{n-2} \xrightarrow{\pi_i} S^{n-1}.$$

When $n = 1$, we interpret ρ to be the inclusion $X^0 \longrightarrow X^0_+$. When $n \geq 2$, let $a_{i,j}$ be the degree of this composite and define

$$d_n[j] = \sum_i a_{i,j}[i].$$

When $n = 1$, specify coefficents $a_{i,j}$ implicitly by defining

$$d_1[j] = [j(1)] - [j(-1)].$$

We claim that $d_{n-1} \circ d_n = 0$, and we define

$$H_*(X) = H_*(C_*(X)).$$

To see that $d_{n-1} \circ d_n = 0$, we use the theory of cofiber sequences to obtain a more conceptual description of d_n. We define the "topological boundary map"

$$\partial_n : X^n/X^{n-1} \longrightarrow \Sigma(X^{n-1}/X^{n-2})$$

to be the composite

$$X^n/X^{n-1} \xrightarrow{\psi^{-1}} Ci \xrightarrow{\pi} \Sigma X^{n-1} \xrightarrow{\Sigma \rho} \Sigma(X^{n-1}/X^{n-2}),$$

where $i : X^{n-1} \longrightarrow X^n$ is the inclusion.

2. CELLULAR HOMOLOGY

We claim that ∂_n induces d_n upon application of a suitable functor, and we need some preliminaries to show this. For certain based spaces X, we adopt the following provisional definition of the "reduced nth homology group" of X.

DEFINITION. Let X be a based $(n-1)$-connected space. Define $\tilde{H}'_n(X)$ as follows.

$n = 0$: The free Abelian group generated by the set $\pi_0(X) - \{*\}$ of non-basepoint components of X.
$n = 1$: The Abelianization $\pi_1(X)/[\pi_1(X), \pi_1(X)]$ of the fundamental group of X.
$n \geq 2$: The nth homotopy group of X.

Up to canonical isomorphism, $\tilde{H}'_n(X)$ is independent of the choice of basepoint in its path component. In fact, we can define $\tilde{H}'_n(X)$ in terms of unbased homotopy classes of maps $S^n \longrightarrow X$. We also need a suspension homomorphism, and we adopt another provisional definition.

DEFINITION. Let X be a based $(n-1)$-connected space. Define
$$\Sigma : \tilde{H}'_n(X) \longrightarrow \tilde{H}'_{n+1}(\Sigma X)$$
by letting $\Sigma[f] = [\Sigma f \circ \iota_{n+1}]$ for $f : S^n \longrightarrow X$; that is, $\Sigma[f]$ is represented by
$$S^{n+1} \xrightarrow{\iota_{n+1}} \Sigma S^n \xrightarrow{\Sigma f} \Sigma X.$$
This only makes sense for $n \geq 1$. For $n = 0$, and a point $x \in X$ that is not in the component of the basepoint, we define $\Sigma[x] = [f_*^{-1} \cdot f_x]$, where $* \in X$ is the basepoint and f_x is the path $t \longrightarrow x \wedge t$ from one cone point to the other in the unreduced suspension ΣX.

LEMMA. If X is a wedge of n-spheres, then
$$\Sigma : \tilde{H}'_n(X) \longrightarrow \tilde{H}'_{n+1}(\Sigma X)$$
is an isomorphism.

PROOF. We claim first that $\tilde{H}'_n(X)$ is the free Abelian group with generators given by the inclusions of the wedge summands. Since maps and homotopies of maps $S^n \longrightarrow X$ have images in compact subspaces, it suffices to check this on finite wedges, and, when $n \geq 2$, we can give the pair $(\times_i S^n, \vee_i S^n)$ the structure of a CW pair with no relative q-cells for $q < 2n - 1$. The claim follows by cellular approximation of maps and homotopies. Now the conclusion of the lemma follows from the case of a single sphere in view of the canonical direct sum decompositions. □

Returning to our CW complex X, we take the homotopy classes $[j \circ \nu_n^{-1}]$ of the composites
$$S^n \xrightarrow{\nu_n^{-1}} D^n/S^{n-1} \xrightarrow{j} X^n/X^{n-1}$$
as canonical basis elements of $\tilde{H}'_n(X^n/X^{n-1})$.

LEMMA. The differential $d_n : C_n(X) \longrightarrow C_{n-1}(X)$ can be identified with the composite
$$d'_n : \tilde{H}'_n(X^n/X^{n-1}) \xrightarrow{(\partial_n)_*} \tilde{H}'_n(\Sigma(X^{n-1}/X^{n-2})) \xrightarrow{\Sigma^{-1}} \tilde{H}'_{n-1}(X^{n-1}/X^{n-2}).$$

PROOF. The identification of the groups is clear: we let the basis element $[j]$ of $C_n(X)$ correspond to the basis element $[j \circ \nu_n^{-1}]$ of $\tilde{H}'_n(X^n/X^{n-1})$. For an n-cell j and an $(n-1)$-cell i, the following diagram is homotopy commutative by the naturality of ψ and π, the definition of $a_{i,j}$, and our lemma relating different models of the n-sphere:

$$\begin{array}{ccccccc}
S^n & \xrightarrow{a_{i,j}} & & & & & S^n \\
{\scriptstyle \nu_n^{-1}}\downarrow & {\scriptstyle \iota_n}\searrow & & & & {\scriptstyle \iota_n}\nearrow & \\
D^n/S^{n-1} & \xrightarrow{\psi^{-1}} & D^n \cup CS^{n-1} & \xrightarrow{\pi} & \Sigma S^{n-1} & \xrightarrow{a_{i,j}} & \Sigma S^{n-1} \\
{\scriptstyle j}\downarrow & & \downarrow{\scriptstyle j \cup Cj} & & \downarrow{\scriptstyle \Sigma j} & & \downarrow{\scriptstyle \Sigma \pi_i} \\
X^n/X^{n-1} & \xrightarrow{\psi^{-1}} & X^n \cup CX^{n-1} & \xrightarrow{\pi} & \Sigma X^{n-1} & \xrightarrow{\Sigma \rho} & \Sigma(X^{n-1}/X^{n-2}).
\end{array}$$

An inspection of circles shows that this diagram homotopy commutes even when $n = 1$. The bottom composite is the topological boundary map. Write d'_n in matrix form,

$$d'_n[j \circ \nu_n^{-1}] = \Sigma^{-1}(\partial_n)_*[j \circ \nu_n^{-1}] = \sum_i a'_{i,j}[i \circ \nu_{n-1}^{-1}].$$

Then, since the composite

$$\pi_i \circ (i \circ \nu_{n-1}^{-1}) : S^{n-1} \longrightarrow S^{n-1}$$

is the identity map for each i, $a'_{i,j}$ is the degree of the map $S^n \longrightarrow S^n$ that we obtain by traversing the diagram counterclockwise from the top left to the top right. The diagram implies that $a'_{i,j} = a_{i,j}$. □

LEMMA. $d_{n-1} \circ d_n = 0$.

PROOF. The composite $\Sigma \partial_{n-1} \circ \partial_n$ is homotopic to the trivial map since the following diagram is homotopy commutative and the composite $\Sigma \pi \circ \Sigma i$ is the trivial map:

$$\begin{array}{ccccccc}
X^n \cup CX^{n-1} & \xrightarrow{\pi} & \Sigma X^{n-1} & \xrightarrow{\Sigma i} & \Sigma(X^{n-1} \cup CX^{n-2}) & \xrightarrow{\Sigma \pi} & \Sigma^2 X^{n-2} \\
{\scriptstyle \psi}\downarrow & & \downarrow{\scriptstyle \Sigma \rho} & & \downarrow{\scriptstyle \Sigma \psi} & & \downarrow{\scriptstyle \Sigma^2 \rho} \\
X^n/X^{n-1} & \xrightarrow{\partial_n} & \Sigma(X^{n-1}/X^{n-2}) & = & \Sigma(X^{n-1}/X^{n-2}) & \xrightarrow{\Sigma \partial_{n-1}} & \Sigma^2 X^{n-2}/X^{n-3}.
\end{array}$$

The conclusion follows from our identification of d_n and the naturality of Σ. □

3. Verification of the axioms

For a CW complex X with base vertex $*$, define $\tilde{C}_*(X) = C_*(X)/C_*(*)$ and define $\tilde{H}_*(X) = H_*(\tilde{C}_*(X))$. This is the reduced homology of X. For a subcomplex A of X, define

$$C_*(X, A) = C_*(X)/C_*(A) \cong \tilde{C}_*(X/A)$$

and define

$$H_*(X, A) = H_*(C_*(X, A)) \cong \tilde{H}_*(X/A).$$

The long exact homology sequence of the exact sequence of chain complexes

$$0 \longrightarrow C_*(A) \longrightarrow C_*(X) \longrightarrow C_*(X, A) \longrightarrow 0$$

gives the connecting homorphisms $\partial : H_q(X, A) \longrightarrow H_{q-1}(A)$ and the long exact sequence called for in the exactness axiom.

If $f : X \longrightarrow Y$ is a cellular map, it induces maps $X^n/X^{n-1} \longrightarrow Y^n/Y^{n-1}$ that commute up to homotopy with the topological boundary maps and so induce homomorphisms $f_n : C_n(X) \longrightarrow C_n(Y)$ that commute with the differentials. That is, f_* is a chain map, and it induces a homomorphism $H_*(X) \longrightarrow H_*(Y)$.

For any CW complex X, $X \times I$ is a CW complex whose cellular chains are isomorphic to $C_*(X) \otimes C_*(I)$, as we shall verify in the next section. Here $C_*(I) \cong \mathscr{I}$ has basis elements $[0]$ and $[1]$ of degree zero and $[I]$ of degree one such that $d([I]) = [0] - [1]$. We have observed that, for chain complexes C and D, a chain map $C \otimes \mathscr{I} \longrightarrow D$ can be identified with a chain homotopy between its restrictions to $C \otimes \mathbb{Z}[0]$ and $C \otimes \mathbb{Z}[1]$. A cellular homotopy $h : X \times I \longrightarrow Y$ induces just such a chain map, hence cellularly homotopic maps induce the same homomorphism on homology. The analogous conclusion for pairs follows by consideration of quotient complexes.

The dimension and additivity axioms are obvious. If X is a point, then $C_*(X) = \mathbb{Z}$, concentrated in degree zero. The cellular chain complex of $\amalg X_i$ is the direct sum of the chain complexes $C_*(X_i)$, and similarly for pairs. Excision is also obvious. If $X = A \cup B$, then the inclusion $A/A \cap B \longrightarrow X/B$ is an isomorphism of CW complexes.

We have dealt so far with the case of integral homology. For more general coefficient groups π, we define

$$C_*(X, A; \pi) = C_*(X, A) \otimes \pi$$

and proceed in exactly the same fashion to define homology groups and verify the axioms. Observe that a homomorphism of groups $\pi \longrightarrow \rho$ induces a natural transformation

$$H_q(X, A; \pi) \longrightarrow H_q(X, A; \rho)$$

that commutes with the connecting homomorphisms.

4. The cellular chains of products

A nice fact about cellular homology is that the definition leads directly to an algebraic procedure for the calculation of the homology of Cartesian products. We explain the topological point here and return to the algebra later, where we discuss the Künneth theorem for the computation of the homology of tensor products of chain complexes.

THEOREM. *If X and Y are CW complexes, then $X \times Y$ is a CW complex such that*

$$C_*(X \times Y) \cong C_*(X) \otimes C_*(Y).$$

PROOF. We have already seen that $X \times Y$ is a CW complex. Its n-skeleton is

$$(X \times Y)^n = \bigcup_{p+q=n} X^p \times Y^q,$$

and it has one n-cell, denoted $i \times j$, for each p-cell i and q-cell j. We define an isomorphism of graded Abelian groups

$$\kappa : C_*(X) \otimes C_*(Y) \longrightarrow C_*(X \times Y)$$

by setting

$$\kappa([i] \otimes [j]) = (-1)^{pq}[i \times j].$$

It is clear from the definition of the product cell structure that κ commutes up to sign with the differentials, and the insertion of the coefficient $(-1)^{pq}$ ensures that the signs work out. As we shall see, the coefficient appears because we write suspension coordinates on the right rather than on the left. To be precise about this verification, we fix homeomorphisms

$$(D^n, S^{n-1}) \cong (I^n, \partial I^n)$$

by radial contraction to the unit cube centered at 0 followed by translation to the unit cube centered at $1/2$. This fixes homeomorphisms

$$\iota_{p,q} : (D^n, S^{n-1}) \cong (I^n, \partial I^n) = (I^p \times I^q, I^p \times \partial I^q \cup \partial I^p \times I^q)$$
$$\cong (D^p \times D^q, D^p \times S^{q-1} \cup S^{p-1} \times D^q)$$

and thus fixes the product cells $i \times j$. For each p and q, the following diagrams are homotopy commutative, where $n = p + q$ and where

$$t : (\Sigma S^{p-1}) \wedge S^q = S^{p-1} \wedge S^1 \wedge S^q \longrightarrow S^{p-1} \wedge S^q \wedge S^1 = \Sigma(S^{p-1} \wedge S^q)$$

is the transposition map. Note that, by a quick check when $q = 1$ and induction, t has degree $(-1)^q$.

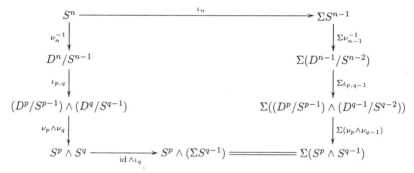

and

$$\begin{array}{ccc}
S^n & \xrightarrow{\iota_n} & \Sigma S^{n-1} \\
\downarrow \nu_n^{-1} & & \downarrow \Sigma \nu_{n-1}^{-1} \\
D^n/S^{n-1} & & \Sigma(D^{n-1}/S^{n-2}) \\
\downarrow \iota_{p,q} & & \downarrow \Sigma \iota_{p-1,q} \\
(D^p/S^{p-1}) \wedge (D^q/S^{q-1}) & & \Sigma((D^{p-1}/S^{p-2}) \wedge (D^q/S^{q-1})) \\
\downarrow \nu_p \wedge \nu_q & & \downarrow \Sigma(\nu_{p-1} \wedge \nu_q) \\
S^p \wedge S^q & \xrightarrow{\iota_p \wedge \mathrm{id}} (\Sigma S^{p-1}) \wedge S^q \xrightarrow{(-1)^q t} & \Sigma(S^{p-1} \wedge S^q).
\end{array}$$

The homotopy commutativity would be clear if we worked only with cubes and replaced the maps ι_n and $\iota_{p,q}$ with the evident identifications. The only point at issue then would be which copy of $I/\partial I$ is to be interpreted as the suspension coordinate. The homotopy commutativity of the diagrams as written follows directly.

Now comparison of our description of the cellular differential in terms of the topological boundary map and the algebraic description of the differential on tensor products shows that κ is an isomorphism of chain complexes. □

5. Some examples: T, K, and $\mathbb{R}P^n$

Cellular chains make some computations quite trivial. For example, since S^n is a CW complex with one vertex and one n-cell, we see immediately that

$$\tilde{H}_n(S^n; \pi) \cong \pi \quad \text{and} \quad \tilde{H}_q(S^n; \pi) = 0 \quad \text{for} \quad q \neq n.$$

A little less obviously, if we look back at the CW decompositions of the torus T, the projective plane $\mathbb{R}P^2$, and the Klein bottle K and if we let j denote the unique 2-cell in each case, then we find the following descriptions of the cellular chains and integral homologies by quick direct inspections. We agree to write $\mathbb{Z}_n$ for the cyclic group $\mathbb{Z}/n\mathbb{Z}$.

EXAMPLES. (i) The cell complex $C_*(T)$ has one basis element $[v]$ in degree zero, two basis elements $[e_1]$ and $[e_2]$ in degree one, and one basis element $[j]$ in degree two. All basis elements are cycles, hence $H_*(T; \mathbb{Z}) = C_*(T)$.

(ii) The cell complex $C_*(\mathbb{R}P^2)$ has two basis elements $[v_1]$ and $[v_2]$ in degree zero, two basis elements $[e_1]$ and $[e_2]$ in degree one, and one basis element $[j]$ in degree two. The differentials are given by

$$d([e_1]) = [v_1] - [v_2], \quad d([e_2]) = [v_2] - [v_1], \quad \text{and} \quad d([j]) = 2[e_1] + 2[e_2].$$

Therefore $H_0(\mathbb{R}P^2; \mathbb{Z}) = \mathbb{Z}$ with basis element the homology class of $[v_1]$ (or $[v_2]$), $H_1(\mathbb{R}P^2; \mathbb{Z}) = \mathbb{Z}_2$ with non-zero element the homology class of $e_1 + e_2$, and $H_q(\mathbb{R}P^2; \mathbb{Z}) = 0$ for $q \geq 2$.

(iii) The cell complex $C_*(K)$ has one basis element $[v]$ in degree zero, two basis elements $[e_1]$ and $[e_2]$ in degree one, and one basis element $[j]$ in degree two. The only non-zero differential is $d([j]) = 2[e_2]$. Therefore $H_0(K; \mathbb{Z}) = \mathbb{Z}$ with basis element the homology class of $[v]$, $H_1(K; \mathbb{Z}) = \mathbb{Z} \oplus \mathbb{Z}_2$ with $\mathbb{Z}$ generated by the class of $[e_1]$ and $\mathbb{Z}_2$ generated by the class of $[e_2]$, and $H_q(K; \mathbb{Z}) = 0$ for $q \geq 2$.

However, these examples are misleading. While homology groups are far easier to compute than homotopy groups, direct chain level calculation is seldom the method of choice. Rather, one uses chains as a tool for developing more sophisticated algebraic techniques, notably spectral sequences. We give an illustration that both shows that chain level calculations are sometimes practicable even when there are many non-zero differentials to determine and indicates why one might not wish to attempt such calculations for really complicated spaces.

We shall use cellular chains to compute the homology of $\mathbb{R}P^n$. We think of $\mathbb{R}P^n$ as the quotient of S^n obtained by identifying antipodal points, and we need to know the degree of the antipodal map.

LEMMA. *The degree of the antipodal map* $a_n : S^n \longrightarrow S^n$ *is* $(-1)^{n+1}$.

PROOF. Since $a_1 \simeq \text{id}$ via an obvious rotation, the result is clear for $n = 1$. The homeomorphism $\iota_n : S^n \longrightarrow \Sigma S^{n-1}$ satisfies

$$\iota_n(-x_1, \ldots, -x_{n+1}) = (-vx_1, \ldots, -vx_n) \wedge (1 - x_{n+1})/2,$$

where $v = 1/(\sum x_i^2)^{1/2}$. That is, $\iota_n \circ a_n = -(\Sigma a_{n-1}) \circ \iota_n$. The conclusion follows by induction on n. □

We shall give $\mathbb{R}P^n$ a CW structure with one q-cell for $0 \leq q \leq n$ by passage to quotients from a CW structure on S^n with two q-cells for $0 \leq q \leq n$. (Note that this cell structure on $\mathbb{R}P^2$ will be more economical than the one used in the calculation above.) The q-skeleton of S^n will be S^q, which we identify with the subspace of S^n whose points have last $n - q$ coordinates zero. We denote the two q-cells of S^n by $j_\pm^q$. The two vertices are the points ± 1 of S^0. Let $E_\pm^q$ be the upper and lower hemispheres in S^q, so that

$$S^q = E_+^q \cup E_-^q \quad \text{and} \quad S^{q-1} = E_+^q \cap E_-^q.$$

We shall write

$$\pi_\pm : S^q \longrightarrow E_\pm^q / S^{q-1}$$

for the quotient maps that identify the lower or upper hemispheres to the basepoint. Of course, these are homotopy equivalences. We define homeomorphisms

$$j_\pm^q : D^q \longrightarrow E_\pm^q \subset S^q$$

by

$$j_\pm^q(x_1, \ldots, x_q) = (\pm x_1, \ldots, \pm x_q, \pm(1 - \sum x_i^2)^{1/2}).$$

This decomposes S^q as the the union of the images of two q-cells. The intersection of these images is S^{q-1} since

$$(x_1, \ldots, x_q, (1 - \sum x_i^2)^{1/2}) = (-y_1, \ldots, -y_q, -(1 - \sum y_i^2)^{1/2})$$

if and only if $x_i = -y_i$ for each i and $\sum x_i^2 = 1$. Clearly

$$j_+^q \mid S^{q-1} = \text{id} \quad \text{and} \quad j_-^q \mid S^{q-1} = a_{q-1}.$$

Inspection of definitions shows that the following diagram commutes:

$$\begin{array}{ccccc}
S^{q-1} & \xrightarrow{\pi_+} & E_+^{q-1}/S^{q-2} & \xleftarrow{j_+^{q-1}} & D^{q-1}/S^{q-2} \\
{\scriptstyle a_{q-1}} \downarrow & & {\scriptstyle a_{q-1}} \downarrow & & \| \\
S^{q-1} & \xrightarrow{\pi_-} & E_-^{q-1}/S^{q-2} & \xleftarrow{j_-^{q-1}} & D^{q-1}/S^{q-2}.
\end{array}$$

Since $a_{q-1}^2 = \text{id}$, it follows that we also obtain a commutative diagram if we interchange $+$ and $-$. If we invert the homeomorphisms $j_\pm^{q-1}$ and compose on the right with the homeomorphism $i_{q-1} : D^{q-1}/S^{q-2} \longrightarrow S^{q-1}$, then the degrees of the four resulting composite homotopy equivalences give the coefficients of the differential d_q. By composing i_{q-1} with a homeomorphism of degree -1 if necessary, we can arrange that the degree of $i_{q-1} \circ (j_+^{q-1})^{-1} \circ \pi_+$ is 1. We then deduce from the lemma and the definition of the differential on the cellular chains that

$$d_q[j_+^q] = (-1)^q d_q[j_-^q] = [j_+^{q-1}] + (-1)^q[j_-^{q-1}]$$

for all $q \geq 1$.

Now, identifying antipodal points, we obtain the promised CW decomposition of $\mathbb{R}P^n$. If $p : S^n \longrightarrow \mathbb{R}P^n$ is the quotient map, then

$$p \circ j_+^q = p \circ j_-^q : (D^q, S^{q-1}) \longrightarrow (\mathbb{R}P^q, \mathbb{R}P^{q-1}).$$

We call this map j^q and see that these maps give $\mathbb{R}P^n$ a CW structure. Therefore $C_q(\mathbb{R}P^n) = \mathbb{Z}$ with basis element $[j^q]$ for $q \geq 0$. Moreover, it is immediate from the calculation just given that
$$d[j^q] = (1 + (-1)^q)[j^{q-1}]$$
for all $q \geq 1$. This is zero if q is odd and multiplication by 2 if q is even, and we read off that
$$H_q(\mathbb{R}P^n;\mathbb{Z}) = \begin{cases} \mathbb{Z} & \text{if } q = 0 \\ \mathbb{Z}_2 & \text{if } 0 < q < n \text{ and } q \text{ is odd} \\ \mathbb{Z} & \text{if } q = n \text{ is odd} \\ 0 & \text{otherwise.} \end{cases}$$
If we work mod 2, taking $\mathbb{Z}_2$ as coefficient group, then the answer takes a nicer form, namely
$$H_q(\mathbb{R}P^n;\mathbb{Z}_2) = \begin{cases} \mathbb{Z}_2 & \text{if } 0 \leq q \leq n \\ 0 & \text{if } q > n. \end{cases}$$

This calculation well illustrates general facts about the homology of compact connected closed n-manifolds M that we shall prove later. The nth integral homology group of such a manifold M is $\mathbb{Z}$ if M is orientable and zero if M is not orientable. The nth mod 2 homology group of M is $\mathbb{Z}_2$ whether or not M is orientable.

PROBLEMS

1. If X is a finite CW complex, show that $\chi(X) = \chi(H_*(X;k))$ for any field k.
2. Let A be a subcomplex of a CW complex X, let Y be a CW complex, and let $f: A \longrightarrow Y$ be a cellular map. What is the relationship between $H_*(X,A)$ and $H_*(Y \cup_f X, Y)$? Is there a similar relationship between $\pi_*(X,A)$ and $\pi_*(Y \cup_f X, Y)$? If not, give a counterexample.
3. Fill in the details of the computation of the differentials on the cellular chains in the examples in §5.
4. Compute $H_*(S^m \times S^n)$ for $m \geq 1$ and $n \geq 1$. Convince yourself that you can do this by use of CW structures, by direct deduction from the axioms, and by the Künneth theorem (for which see Chapter 17).
5. Let p be an odd prime number. Regard the cyclic group π of order p as the group of pth roots of unity contained in S^1. Regard S^{2n-1} as the unit sphere in $\mathbb{C}^n$, $n \geq 2$. Then $\pi \subset S^1$ acts freely on S^{2n-1} via
$$\zeta(z_1,\ldots,z_n) = (\zeta z_1,\ldots,\zeta z_n).$$
Let $L^n = S^{2n-1}/\pi$ be the orbit space; it is called a lens space and is an odd primary analogue of $\mathbb{R}P^n$. The obvious quotient map $S^{2n-1} \longrightarrow L^n$ is a universal covering.
   (a) Compute the integral homology of L^n, $n \geq 2$, by mimicking the calculation of $H_*(\mathbb{R}P^n)$.
   (b) Compute $H_*(L^n;\mathbb{Z}_p)$, where $\mathbb{Z}_p = \mathbb{Z}/p\mathbb{Z}$.

CHAPTER 14

Derivations of properties from the axioms

Returning to the axiomatic approach to homology, we assume given a theory on pairs of spaces and make some deductions from the axioms. We abbreviate notations by setting $E_q(X,A) = H_q(X,A;\pi)$. However, the arguments in this chapter make no use whatever of the dimension axiom. A "generalized homology theory" E_* is defined to be a system of functors $E_q(X,A)$ and natural transformations $\partial : E_q(X,A) \longrightarrow E_{q-1}(A)$ that satisfy all of our axioms except for the dimension axiom. Similarly, we have the notion of a generalized homology theory on CW pairs, and the results of the first section of the previous chapter generalize directly to give the following result.

THEOREM. *A homology theory E_* on pairs of spaces determines and is determined by its restriction to a homology theory E_* on pairs of CW complexes.*

The study of such generalized homology theories pervades modern algebraic topology, and we shall describe some examples later on. The brave reader may be willing to think of E_* in such generality in this chapter. The timorous reader may well prefer to think of $E_*(X,A)$ concretely, following our proposal that $E_*(X,A)$ be taken as an alternative notation for $H_*(X,A;\pi)$.

1. Reduced homology; based versus unbased spaces

One of the themes of this chapter is the relationship between homology theories on pairs of spaces and reduced homology theories on based spaces. The latter are more convenient in most advanced work in algebraic topology. For a based space X, we define the reduced homology of X to be

$$\tilde{E}_q(X) = E_q(X,*).$$

Since the basepoint is a retract of X, there results a direct sum decomposition

$$E_*(X) \cong \tilde{E}_*(X) \oplus E_*(*)$$

that is natural with respect to based maps. For $* \in A \subset X$, the summand $E_*(*)$ maps isomorphically under the map $E_*(A) \longrightarrow E_*(X)$, and the exactness axiom implies that there is a reduced long exact sequence

$$\cdots \longrightarrow \tilde{E}_q(A) \longrightarrow \tilde{E}_q(X) \longrightarrow E_q(X,A) \overset{\partial}{\longrightarrow} \tilde{E}_{q-1}(A) \longrightarrow \cdots.$$

We can obtain the unreduced homology groups as special cases of the reduced ones. For an unbased space X, we define a based space X_+ by adjoining a disjoint basepoint to X. By the additivity axiom, we see immediately that

$$E_*(X) = \tilde{E}_*(X_+).$$

Similarly, a map $f : X \longrightarrow Y$ of unbased spaces induces a map $f_+ : X_+ \longrightarrow Y_+$ of based spaces, and the map f_* on unreduced homology coincides with the map $(f_+)_*$ on reduced homology.

We shall make considerable use of cofiber sequences in this chapter. To be consistent about this, we should always work with reduced cones and cofibers. However, it is more convenient to make the convention that we work with unreduced cones and cofibers when we apply unreduced homology theories, and we work with reduced cones and cofibers when we apply reduced homology theories. In fact, the unreduced cone on a space Y coincides with the reduced cone on Y_+: the line through the disjoint basepoint is identified to the cone point when constructing the reduced cone on Y_+. Therefore the unreduced cofiber of an unbased map f coincides with the reduced cofiber of the based map f_+. Our convention really means that we are always working with reduced cofibers, but when we are studying unreduced homology theories we are implicitly applying the functor $(-)_+$ to put ourselves in the based context before constructing cones and cofibers.

The observant reader will have noticed that the unreduced suspension of X is *not* the reduced suspension on X_+. Rather, under either interpretation of suspension, $\Sigma(X_+)$ is homotopy equivalent to the wedge of $\Sigma(X)$ and a circle.

2. Cofibrations and the homology of pairs

We use cofibrations to show that the homology of pairs of spaces is in principle a special case of the reduced homology of spaces.

THEOREM. *For any cofibration* $i : A \longrightarrow X$, *the quotient map* $q : (X, A) \longrightarrow (X/A, *)$ *induces an isomorphism*

$$E_*(X, A) \longrightarrow E_*(X/A, *) = \tilde{E}_*(X/A).$$

PROOF. Consider the (unreduced) cofiber

$$Ci = X \cup_i CA = X \cup_i A \times I/A \times \{1\}.$$

We have an excisive triad

$$(Ci; X \cup_i A \times [0, 2/3], A \times [1/3, 1]/A \times \{1\}).$$

The excision axiom gives that the top inclusion in the following commutative diagram induces an isomorphism on passage to homology:

$$\begin{CD}
(X \cup_i A \times [0, 2/3], A \times [1/3, 2/3]) @>>> (Ci, A \times [1/3, 1]/A \times \{1\}) \\
@VrVV @VV\psi V \\
(X, A) @>>q> (X/A, *)
\end{CD}$$

The map r is obtained by restriction of the retraction $Mi \longrightarrow X$ and is a homotopy equivalence of pairs. The map ψ collapses CA to a point and is also a homotopy equivalence of pairs. The conclusion follows. □

As in our construction of cellular homology, we choose a homotopy inverse $\psi^{-1} : X/A \longrightarrow Ci$ and consider the composite

$$X/A \xrightarrow{\psi^{-1}} Ci \xrightarrow{\pi} \Sigma A$$

to be a topological boundary map

$$\partial : X/A \longrightarrow \Sigma A.$$

Observe that we may replace any inclusion $i : A \longrightarrow X$ by the canonical cofibration $A \longrightarrow Mi$ and then apply the result just given to obtain

$$E_*(X, A) \cong \tilde{E}_*(Ci).$$

3. Suspension and the long exact sequence of pairs

We have a fundamentally important consequence of the results of the previous section, which should be contrasted with what happened with homotopy groups. Recall that a basepoint $* \in X$ is nondegenerate if the inclusion $\{*\} \longrightarrow X$ is a cofibration. This ensures that the inclusion of the line through the basepoint in the unreduced suspension of X is a cofibration, so that the map from the unreduced suspension to the suspension that collapses out the line through the basepoint is a homotopy equivalence. We apply reduced homology here, so we use reduced cones and suspensions.

THEOREM. *For a nondegenerately based space X, there is a natural isomorphism*

$$\Sigma : \tilde{E}_q(X) \cong \tilde{E}_{q+1}(\Sigma X).$$

PROOF. Since CX is contractible, its reduced homology is identically zero. By the reduced long exact sequence, there results an isomorphism

$$\tilde{E}_{q+1}(\Sigma X) \cong \tilde{E}_{q+1}(CX/X) \xrightarrow{\partial} \tilde{E}_q(X). \quad \square$$

An easy diagram chase gives the following consequence, which describes the axiomatically given connecting homomorphism of the pair (X, A) in terms of the topological boundary map $\partial : X/A \longrightarrow \Sigma A$ and the suspension isomorphism.

COROLLARY. *Let $* \in A \subset X$, where $i : A \longrightarrow X$ is a cofibration between nondegenerately based spaces. In the long exact sequence*

$$\cdots \longrightarrow \tilde{E}_q(A) \longrightarrow \tilde{E}_q(X) \longrightarrow \tilde{E}_q(X/A) \xrightarrow{\partial} \tilde{E}_{q-1}(A) \longrightarrow \cdots$$

of the pair (X, A), the connecting homomorphism ∂ is the composite

$$\tilde{E}_q(X/A) \xrightarrow{\partial_*} \tilde{E}_q(\Sigma A) \xrightarrow{\Sigma^{-1}} \tilde{E}_{q-1}(A).$$

Since S^0 consists of two points, $\tilde{E}_*(S^0) = E_*(*)$. Since S^n is the suspension of S^{n-1}, we have the following special case of the suspension isomorphism.

COROLLARY. *For any n and q,*

$$\tilde{E}_q(S^n) \cong E_{q-n}(*).$$

Of course, for the theory $H_*(X; \pi)$, this was immediate from our construction in terms of cellular chains.

4. Axioms for reduced homology

In the study of generalized homology theories, it is most convenient to restrict attention to reduced homology theories defined on nondegenerately based spaces. The results of the previous sections imply that we can do so without loss of generality. Again the reader has the choice of bravery or timorousness in interpreting E_*, but we opt for bravery:

DEFINITION. A reduced homology theory $\tilde{E}_*$ consists of functors $\tilde{E}_q$ from the homotopy category of nondegenerately based spaces to the category of Abelian groups that satisfy the following axioms.

- EXACTNESS If $i : A \longrightarrow X$ is a cofibration, then the sequence
$$\tilde{E}_q(A) \longrightarrow \tilde{E}_q(X) \longrightarrow \tilde{E}_q(X/A)$$
is exact.
- SUSPENSION For each integer q, there is a natural isomorphism
$$\Sigma : \tilde{E}_q(X) \cong \tilde{E}_{q+1}(\Sigma X).$$
- ADDITIVITY If X is the wedge of a set of nondegenerately based spaces X_i, then the inclusions $X_i \longrightarrow X$ induce an isomorphism
$$\sum_i \tilde{E}_*(X_i) \longrightarrow \tilde{E}_*(X).$$
- WEAK EQUIVALENCE If $f : X \longrightarrow Y$ is a weak equivalence, then
$$f_* : \tilde{E}_*(X) \longrightarrow \tilde{E}_*(Y)$$
is an isomorphism.

The reduced form of the dimension axiom would read
$$\tilde{H}_0(S^0) = \pi \quad \text{and} \quad \tilde{H}_q(S^0) = 0 \text{ for } q \neq 0.$$

THEOREM. *A homology theory E_* on pairs of spaces determines and is determined by a reduced homology theory $\tilde{E}_*$ on nondegenerately based spaces.*

PROOF. Given a theory on pairs, we define $\tilde{E}_*(X) = E_*(X, *)$ and deduce the new axioms. For additivity, the specified wedge is the quotient $(\amalg X_i)/(\amalg \{*_i\})$, where $*_i$ is the basepoint of X_i, and our result on quotients of cofibrations applies to compute its homology. Conversely, assume given a reduced homology theory $\tilde{E}_*$, and define
$$E_*(X) = \tilde{E}_*(X_+) \quad \text{and} \quad E_*(X, A) = \tilde{E}_*(C(i_+)),$$
where $C(i_+)$ is the cofiber of the based inclusion $i_+ : A_+ \longrightarrow X_+$. Equivalently, $C(i_+)$ is the unreduced cofiber of $i : A \longrightarrow X$ with its cone point as basepoint. We must show that the suspension axiom and our restricted exactness axiom imply the original, seemingly much stronger, exactness and excision axioms. We have the long exact cofiber sequence associated to the based inclusion $i_+ : A_+ \longrightarrow X_+$, in which each consecutive pair of maps is equivalent to a cofibration and the associated quotient map. Noting that $X_+/A_+ = X/A$, we define the connecting homomorphism $\partial_q : E_q(X, A) \longrightarrow E_{q-1}(A)$ to be the composite
$$\tilde{E}_q(X_+/A_+) \xrightarrow{\partial_*} \tilde{E}_q(\Sigma A_+) \xrightarrow{\Sigma^{-1}} \tilde{E}_{q-1}(A_+)$$
and find that the exactness and suspension axioms for $\tilde{E}_*$ imply the exactness axiom for E_*. For excision, we could carry out a similarly direct homotopical argument,

but it is simpler to observe that this follows from the equivalence of theories on pairs of spaces with theories on pairs of CW complexes together with the next two theorems. For the additivity axiom, we note that the cofiber of a disjoint union of maps is the wedge of the cofibers of the given maps. □

COROLLARY. *For nondegenerately based spaces X, $E_*(X)$ is naturally isomorphic to $\tilde{E}_*(X) \oplus E_*(*)$.*

PROOF. The long exact sequence in E_* of the pair $(X, *)$ is naturally split in each degree by means of the homomorphism induced by the projection $X \longrightarrow \{*\}$. □

We require of based CW complexes that the basepoint be a vertex. It is certainly a nondegenerate basepoint. We give the circle its standard CW structure and so deduce a CW structure on the suspension of a based CW complex.

DEFINITION. A reduced homology theory $\tilde{E}_*$ on based CW complexes consists of functors $\tilde{E}_q$ from the homotopy category of based CW complexes to the category of Abelian groups that satisfy the following axioms.

- EXACTNESS If A is a subcomplex of X, then the sequence

$$\tilde{E}_q(A) \longrightarrow \tilde{E}_q(X) \longrightarrow \tilde{E}_q(X/A)$$

is exact.
- SUSPENSION For each integer q, there is a natural isomorphism

$$\Sigma : \tilde{E}_q(X) \cong \tilde{E}_{q+1}(\Sigma X).$$

- ADDITIVITY If X is the wedge of a set of based CW complexes X_i, then the inclusions $X_i \longrightarrow X$ induce an isomorphism

$$\sum_i \tilde{E}_*(X_i) \longrightarrow \tilde{E}_*(X).$$

THEOREM. *A reduced homology theory $\tilde{E}_*$ on nondegenerately based spaces determines and is determined by its restriction to a reduced homology theory on based CW complexes.*

PROOF. This is immediate by CW approximation of based spaces. □

THEOREM. *A homology theory E_* on CW pairs determines and is determined by a reduced homology theory $\tilde{E}_*$ on based CW complexes.*

PROOF. Given a theory on pairs, we define $\tilde{E}_*(X) = E_*(X, *)$ and deduce the new axioms directly. Conversely, given a reduced theory on based CW complexes, we define

$$E_*(X) = \tilde{E}_*(X_+) \quad \text{and} \quad E_*(X, A) = \tilde{E}_*(X/A).$$

Of course X/A is homotopy equivalent to $C(i_+)$, where $i_+ : A_+ \longrightarrow X_+$ is the inclusion. The arguments for exactness and additivity are the same as those given in the analogous result for nondegenerately based spaces, but now excision is obvious since if $(X; A, B)$ is a CW triad, then the inclusion $A/A \cap B \longrightarrow X/B$ is an isomorphism of based CW complexes. □

110   DERIVATIONS OF PROPERTIES FROM THE AXIOMS

5. Mayer-Vietoris sequences

The Mayer-Vietoris sequences are long exact sequences associated to excisive triads that will play a fundamental role in our later proof of the Poincaré duality theorem. We need two preliminaries, both of independent interest. The first is the long exact sequence of a triple (X, A, B) of spaces $B \subset A \subset X$, which is just like its analogue for homotopy groups.

PROPOSITION. *For a triple (X, A, B), the following sequence is exact:*

$$\cdots \longrightarrow E_q(A, B) \xrightarrow{i_*} E_q(X, B) \xrightarrow{j_*} E_q(X, A) \xrightarrow{\partial} E_{q-1}(A, B) \longrightarrow \cdots.$$

Here $i : (A, B) \longrightarrow (X, B)$ and $j : (X, B) \longrightarrow (X, A)$ are inclusions and ∂ is the composite

$$E_q(X, A) \xrightarrow{\partial} E_{q-1}(A) \longrightarrow E_{q-1}(A, B).$$

PROOF. There are two easy arguments. One can either use diagram chasing from the various long exact sequences of pairs or one can apply CW approximation to replace (X, A, B) by a triple of CW complexes. After the replacement, we have that $X/A \cong (X/B)/(A/B)$ as a CW complex, and the desired sequence is the reduced exact sequence of the pair $(X/B, A/B)$. □

LEMMA. *Let $(X; A, B)$ be an excisive triad and set $C = A \cap B$. The map*

$$E_*(A, C) \oplus E_*(B, C) \longrightarrow E_*(X, C)$$

induced by the inclusions of (A, C) and (B, C) in (X, C) is an isomorphism.

PROOF. Again, there are two easy proofs. One can either pass to homology from the diagram

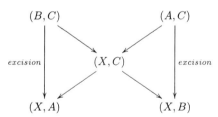

and use algebra or one can approximate $(X; A, B)$ by a CW triad, for which

$$X/C \cong A/C \vee B/C$$

as a CW complex. □

THEOREM (Mayer-Vietoris sequence). *Let $(X; A, B)$ be an excisive triad and set $C = A \cap B$. The following sequence is exact:*

$$\cdots \longrightarrow E_q(C) \xrightarrow{\psi} E_q(A) \oplus E_q(B) \xrightarrow{\phi} E_q(X) \xrightarrow{\Delta} E_{q-1}(C) \longrightarrow \cdots.$$

Here, if $i : C \longrightarrow A$, $j : C \longrightarrow B$, $k : A \longrightarrow X$, and $\ell : B \longrightarrow X$ are the inclusions, then

$$\psi(c) = (i_*(c), j_*(c)), \quad \phi(a, b) = k_*(a) - \ell_*(b),$$

and Δ is the composite

$$E_q(X) \longrightarrow E_q(X, B) \cong E_q(A, C) \xrightarrow{\partial} E_{q-1}(C).$$

5. MAYER-VIETORIS SEQUENCES

PROOF. Note that the definition of ϕ requires a sign in order to make $\phi \circ \psi = 0$. The proof of exactness is algebraic diagram chasing and is left as an exercise. The following diagram may help:

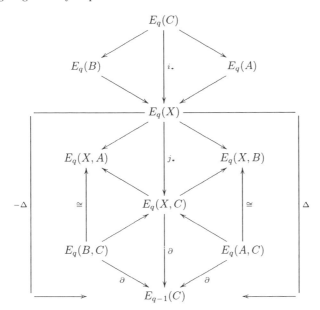

Here the arrow labeled "$-\Delta$" is in fact $-\Delta$ by an algebraic argument from the direct sum decomposition of $E_q(X,C)$. Alternatively, one can use CW approximation. For a CW triad, there is a short exact sequence

$$0 \longrightarrow C_*(C) \longrightarrow C_*(A) \oplus C_*(B) \longrightarrow C_*(X) \longrightarrow 0$$

whose associated long exact sequence is the Mayer-Vietoris sequence. □

We shall also need a relative analogue, but the reader may wish to ignore this for now. It will become important when we study manifolds with boundary.

THEOREM (Relative Mayer-Vietoris sequence). *Let $(X; A, B)$ be an excisive triad and set $C = A \cap B$. Assume that X is contained in some ambient space Y. The following sequence is exact:*

$$\cdots \longrightarrow E_q(Y,C) \xrightarrow{\psi} E_q(Y,A) \oplus E_q(Y,B) \xrightarrow{\phi} E_q(Y,X) \xrightarrow{\Delta} E_{q-1}(Y,C) \longrightarrow \cdots.$$

Here, if $i : (Y,C) \longrightarrow (Y,A)$, $j : (Y,C) \longrightarrow (Y,B)$, $k : (Y,A) \longrightarrow (Y,X)$, and $\ell : (Y,B) \longrightarrow (Y,X)$ are the inclusions, then

$$\psi(c) = (i_*(c), j_*(c)), \quad \phi(a,b) = k_*(a) - \ell_*(b),$$

and Δ is the composite

$$E_q(Y,X) \xrightarrow{\partial} E_{q-1}(X,B) \cong E_{q-1}(A,C) \longrightarrow E_{q-1}(Y,C).$$

PROOF. This too is left as an exercise, but it is formally the same exercise. The relevant diagram is the following one:

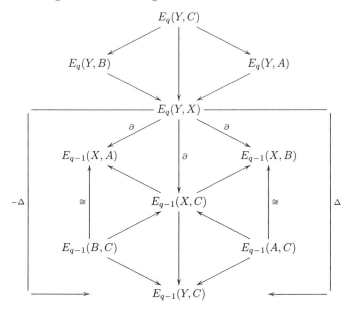

Alternatively, one can use CW approximation. For a CW triad $(X; A, B)$, with X a subcomplex of a CW complex Y, there is a short exact sequence

$$0 \longrightarrow C_*(Y/C) \longrightarrow C_*(Y/A) \oplus C_*(Y/B) \longrightarrow C_*(Y/X) \longrightarrow 0$$

whose associated long exact sequence is the relative Mayer-Vietoris sequence. □

A comparison of definitions gives a relationship between these sequences.

COROLLARY. *The absolute and relative Mayer-Vietoris sequences are related by the following commutative diagram:*

$$\begin{array}{ccccccc}
E_q(Y,C) & \xrightarrow{\psi} & E_q(Y,A) \oplus E_q(Y,B) & \xrightarrow{\phi} & E_q(Y,X) & \xrightarrow{\Delta} & E_{q-1}(Y,C) \\
\downarrow \partial & & \downarrow \partial+\partial & & \downarrow \partial & & \downarrow \partial \\
E_{q-1}(C) & \xrightarrow{\psi} & E_{q-1}(A) \oplus E_{q-1}(B) & \xrightarrow{\phi} & E_{q-1}(X) & \xrightarrow{\Delta} & E_{q-2}(C).
\end{array}$$

6. The homology of colimits

In this section, we let X be the union of an expanding sequence of subspaces X_i, $i \geq 0$. We have seen that the compactness of spheres S^n and cylinders $S^n \times I$ implies that, for any choice of basepoint in X_0, the natural map

$$\text{colim } \pi_*(X_i) \longrightarrow \pi_*(X)$$

is an isomorphism. We shall use the additivity and weak equivalence axioms and the Mayer-Vietoris sequence to prove the analogue for homology.

6. THE HOMOLOGY OF COLIMITS

THEOREM. *The natural map*
$$\operatorname{colim} E_*(X_i) \longrightarrow E_*(X)$$
is an isomorphism.

We record an algebraic description of the colimit of a sequence for use in the proof.

LEMMA. *Let* $f_i : A_i \longrightarrow A_{i+1}$ *be a sequence of homomorphisms of Abelian groups. Then there is a short exact sequence*
$$0 \longrightarrow \sum_i A_i \xrightarrow{\alpha} \sum_i A_i \xrightarrow{\beta} \operatorname{colim} A_i \longrightarrow 0,$$
where $\alpha(a_i) = a_i - f_i(a_i)$ *for* $a_i \in A_i$ *and the restriction of* β *to* A_i *is the canonical map given by the definition of a colimit.*

By the additivity axiom, we may as well assume that X and the X_i are path connected. The proof makes use of a useful general construction called the "telescope" of the X_i, denoted tel X_i. Let $j_i : X_i \longrightarrow X_{i+1}$ be the given inclusions and consider the mapping cylinders
$$M_{i+1} = (X_i \times [i, i+1]) \cup X_{i+1}$$
that are obtained by identifying $(x, i+1)$ with $j_i(x)$ for $x \in X_i$. Inductively, let $Y_0 = X_0 \times \{0\}$ and suppose that we have constructed $Y_i \supset X_i \times \{i\}$. Define Y_{i+1} to be the double mapping cylinder $Y_i \cup M_{i+1}$ obtained by identifying $(x, i) \in Y_i$ with $(x, i) \in M_{i+1}$ for $x \in X_i$. Define tel X_i to be the union of the Y_i, with the colimit topology. Thus
$$\operatorname{tel} X_i = \bigcup_i X_i \times [i, i+1],$$
with the evident identifications at the ends of the cylinders.

Using the retractions of the mapping cylinders, we obtain composite retractions $r_i : Y_i \longrightarrow X_i$ such that the following diagrams commute

$$\begin{array}{ccc} Y_i & \xrightarrow{\subset} & Y_{i+1} \\ {\scriptstyle r_i}\downarrow & & \downarrow{\scriptstyle r_{i+1}} \\ X_i & \xrightarrow{j_i} & X_{i+1} \end{array}$$

Since the r_i are homotopy equivalences and since homotopy groups commute with colimits, it follows that we obtain a weak equivalence
$$r : \operatorname{tel} X_i \longrightarrow X$$
on passage to colimits. By the weak equivalence axiom, r induces an isomorphism on homology. It therefore suffices to prove that the natural map
$$\operatorname{colim} E_*(X_i) \cong \operatorname{colim} E_*(Y_i) \longrightarrow E_*(\operatorname{tel} X_i)$$
is an isomorphism. We define subspaces A and B of tel X_i by choosing $\varepsilon < 1$ and letting
$$A = X_0 \times [0, 1] \coprod \coprod_{i \geq 1} X_{2i-1} \times [2i - \varepsilon, 2i] \cup X_{2i} \times [2i, 2i + 1]$$
and
$$B = \coprod_{i \geq 0} X_{2i} \times [2i + 1 - \varepsilon, 2i + 1] \cup X_{2i+1} \times [2i + 1, 2i + 2].$$

We let $C = A \cap B$ and find that
$$C = \coprod_{i \geq 0} X_i \times [i+1-\varepsilon, i+1].$$
This gives an excisive triad, and a quick inspection shows that we have canonical homotopy equivalences
$$A \simeq \coprod_{i \geq 0} X_{2i}, \quad B \simeq \coprod_{i \geq 0} X_{2i+1}, \quad \text{and} \quad C \simeq \coprod_{i \geq 0} X_i.$$
Moreover, under these equivalences the inclusion $C \longrightarrow A$ has restrictions
$$\text{id}: X_{2i} \longrightarrow X_{2i} \quad \text{and} \quad j_{2i+1}: X_{2i+1} \longrightarrow X_{2i+2},$$
while the inclusion $C \longrightarrow B$ has restrictions
$$j_{2i}: X_{2i} \longrightarrow X_{2i+1} \quad \text{and} \quad \text{id}: X_{2i+1} \longrightarrow X_{2i+1}.$$
By the additivity axiom,
$$E_*(A) = \sum_i E_*(X_{2i}), \quad E_*(B) = \sum_i E_*(X_{2i+1}), \quad \text{and} \quad E_*(C) = \sum_i E_*(X_i).$$
We construct the following commutative diagram, whose top row is the Mayer-Vietoris sequence of the triad $(\text{tel } X_i; A, B)$ and whose bottom row is a short exact sequence as displayed in our algebraic description of colimits:

$$\begin{array}{ccccccccc}
\cdots & \longrightarrow & E_q(C) & \longrightarrow & E_q(A) \oplus E_q(B) & \longrightarrow & E_q(\text{tel } X_i) & \longrightarrow & \cdots \\
& & \cong \downarrow & & \cong \downarrow & & \cong \downarrow & & \\
\cdots & \longrightarrow & \sum_i E_q(X_i) & \stackrel{\alpha'}{\longrightarrow} & \sum_i E_q(X_i) & \stackrel{\beta'}{\longrightarrow} & E_q(X) & \longrightarrow & \cdots \\
& & \Sigma(-1)^i \downarrow & & \Sigma_i(-1)^i \downarrow & & \xi \downarrow & & \\
0 & \longrightarrow & \sum_i E_q(X_i) & \stackrel{\alpha}{\longrightarrow} & \sum_i E_q(X_i) & \stackrel{\beta}{\longrightarrow} & \text{colim } E_q(X_i) & \longrightarrow & 0.
\end{array}$$

By the definition of the maps in the Mayer-Vietoris sequence, $\alpha'(x_i) = x_i + (j_i)_*(x_i)$ and $\beta'_i(x_i) = (-1)^i(k_i)_*(x_i)$ for $x_i \in E_q(X_i)$, where $k_i: X_i \longrightarrow X$ is the inclusion. The commutativity of the lower left square is just the relation
$$(\sum_i (-1)^i) \alpha'(x_i) = (-1)^i(x_i - (j_i)_*(x_i)).$$
The diagram implies the required isomorphism ξ.

REMARK. There is a general theory of "homotopy colimits," which are up to homotopy versions of colimits. The telescope is the homotopy colimit of a sequence. The double mapping cylinder that we used in approximating excisive triads by CW triads is the homotopy pushout of a diagram of the shape $\bullet \longleftarrow \bullet \longrightarrow \bullet$. We implicitly used homotopy coequalizers in constructing CW approximations of spaces.

PROBLEM

1. Complete the proof that the Mayer-Vietoris sequence is exact.

CHAPTER 15

The Hurewicz and uniqueness theorems

We now return to the context of "ordinary homology theories," namely those that satisfy the dimension axiom. We prove a fundamental relationship, called the Hurewicz theorem, between homotopy groups and homology groups. We then use it to prove the uniqueness of ordinary homology with coefficients in π.

1. The Hurewicz theorem

Although the reader may prefer to think in terms of the cellular homology theory already constructed, the proof of the Hurewicz theorem depends only on the axioms. It is this fact that will allow us to use the result to prove the uniqueness of homology theories in the next section. We take $\pi = \mathbb{Z}$ and delete it from the notation. The dimension axiom implicitly fixes a generator i_0 of $\tilde{H}_0(S^0)$, and we choose generators i_n of $\tilde{H}_n(S^n)$ inductively by setting $\Sigma i_n = i_{n+1}$.

DEFINITION. For based spaces X, define the Hurewicz homomorphism

$$h : \pi_n(X) \longrightarrow \tilde{H}_n(X)$$

by

$$h([f]) = f_*(i_n).$$

LEMMA. If $n \geq 1$, then h is a homomorphism for all X.

PROOF. For maps $f, g : S^n \longrightarrow X$, $[f + g]$ is represented by the composite

$$S^n \xrightarrow{p} S^n \vee S^n \xrightarrow{f \vee g} X \vee X \xrightarrow{\nabla} X,$$

where p is the pinch map and ∇ is the codiagonal map; that is, ∇ restricts to the identity on each wedge summand. Since $p_*(i_n) = i_n + i_n$ and ∇ induces addition on $\tilde{H}_*(X)$, the conclusion follows. □

LEMMA. The Hurewicz homomorphism is natural and the following diagram commutes for $n \geq 0$:

$$\begin{array}{ccc} \pi_n(X) & \longrightarrow & \tilde{H}_n(X) \\ \Sigma \downarrow & & \downarrow \Sigma \\ \pi_{n+1}(\Sigma X) & \xrightarrow{h} & \tilde{H}_{n+1}(\Sigma X). \end{array}$$

PROOF. The naturality of h is clear, and the naturality of Σ on homology implies the commutativity of the diagram:

$$(h \circ \Sigma)([f]) = (\Sigma f)_*(\Sigma i_n) = \Sigma(f_*(i_n)) = \Sigma(h([f])). \quad \square$$

LEMMA. *Let X be a wedge of n-spheres. Then*
$$h : \pi_n(X) \longrightarrow \tilde{H}_n(X)$$
is the Abelianization homomorphism if $n = 1$ and is an isomorphism if $n > 1$.

PROOF. When X is a single sphere, $h[\mathrm{id}] = i_n$ and the conclusion is obvious. In general, $\pi_n(X)$ is the free group if $n = 1$ or the free Abelian group if $n \geq 2$ with generators given by the inclusions of the wedge summands. Since h maps these generators to the canonical generators of the free Abelian group $\tilde{H}_n(X)$, the conclusion follows. □

That is all that we shall need in the next section, but we can generalize the lemma to arbitrary $(n-1)$-connected based spaces X.

THEOREM (Hurewicz). *Let X be any $(n-1)$-connected based space. Then*
$$h : \pi_n(X) \longrightarrow \tilde{H}_n(X)$$
is the Abelianization homomorphism if $n = 1$ and is an isomorphism if $n > 1$.

PROOF. We can assume without loss of generality that X is a CW complex with a single vertex, based attaching maps, and no q-cells for $1 \leq q < n$. The inclusion of the $(n+1)$-skeleton in X induces an isomorphism on π_n by the cellular approximation theorem and induces an isomorphism on $\tilde{H}_n$ by our cellular construction of homology or by a deduction from the axioms that will be given in the next section. Thus we may assume without loss of generality that $X = X^{n+1}$. Then X is the cofiber of a map $f : K \longrightarrow L$, where K and L are both wedges of n-spheres. We have the following commutative diagram:

$$\begin{array}{ccccccc}
\pi_n(K) & \longrightarrow & \pi_n(L) & \longrightarrow & \pi_n(X) & \longrightarrow & 0 \\
\downarrow & & \downarrow & & \downarrow & & \\
\tilde{H}_n(K) & \longrightarrow & \tilde{H}_n(L) & \longrightarrow & \tilde{H}_n(X) & \longrightarrow & 0.
\end{array}$$

The lemma gives the conclusion for the two left vertical arrows. Since X/L is a wedge of $(n+1)$-spheres, the bottom row is exact by our long exact homology sequences and the known homology of wedges of spheres. When $n = 1$, a corollary of the van Kampen theorem gives that $\pi_1(X)$ is the quotient of $\pi_1(L)$ by the normal subgroup generated by the image of $\pi_1(K)$. An easy algebraic exercise shows that the sequence obtained from the top row by passage to Abelianizations is therefore exact. If $n > 1$, the homotopy excision theorem implies that the top row is exact. To see this, factor f as the composite of the inclusion $K \longrightarrow Mf$ and the deformation retraction $r : Mf \longrightarrow L$. Since $X = Cf$, we have the following commutative diagram, in which the top row is exact:

$$\begin{array}{ccccccc}
\pi_n(K) & \longrightarrow & \pi_n(Mf) & \longrightarrow & \pi_n(Mf, K) & \longrightarrow & 0 \\
\| & & \downarrow r_* & & \downarrow & & \\
\pi_n(K) & \longrightarrow & \pi_n(L) & \longrightarrow & \pi_n(X) & \longrightarrow & 0.
\end{array}$$

Since K and L are $(n-1)$-connected and $n > 1$, a corollary of the homotopy excision theorem gives that X is $(n-1)$-connected and $\pi_n(Mf, K) \longrightarrow \pi_n(X)$ is an isomorphism. □

2. The uniqueness of the homology of CW complexes

We assume given an ordinary homology theory on CW pairs and describe how it must be computed. We focus on integral homology, taking $\pi = \mathbb{Z}$ and deleting it from the notation. With a moment's reflection on the case $n = 0$, we see that the Hurewicz theorem gives a natural isomorphism

$$\tilde{H}'_n(X) \longrightarrow \tilde{H}_n(X)$$

for $(n-1)$-connected based spaces X. Here the groups on the left are defined in terms of homotopy groups and were used in our construction of cellular chains, while the groups on the right are those of our given homology theory. We use the groups on the right to construct cellular chains in our given theory, and we find that the isomorphism is compatible with differentials. From here, to prove uniqueness, we only need to check from the axioms that our given theory is computable from the homology groups of these cellular chain complexes.

Thus let X be a CW complex. For each integer n, define

$$C_n(X) = H_n(X^n, X^{n-1}) \cong \tilde{H}_n(X^n/X^{n-1}).$$

Define

$$d : C_n(X) \longrightarrow C_{n-1}(X)$$

to be the composite

$$H_n(X^n, X^{n-1}) \xrightarrow{\partial} H_{n-1}(X^{n-1}) \longrightarrow H_{n-1}(X^{n-1}, X^{n-2}).$$

It is not hard to check that $d \circ d = 0$.

THEOREM. $C_*(X)$ is isomorphic to the cellular chain complex of X.

PROOF. Since X^n/X^{n-1} is the wedge of an n-sphere for each n-cell of X, we see by the additivity axiom that $C_n(X)$ is the free Abelian group with one generator $[j]$ for each n-cell j. We must compare the differential with the one that we defined earlier. Let $i : X^{n-1} \longrightarrow X^n$ be the inclusion. We see from our proof of the suspension isomorphism that d coincides with the composite

$$\tilde{H}_n(X^n/X^{n-1}) \cong \tilde{H}_n(Ci) \to \tilde{H}_n(\Sigma X^{n-1}) \xrightarrow{\Sigma^{-1}} \tilde{H}_{n-1}(X^{n-1}) \to \tilde{H}_{n-1}(X^{n-1}/X^{n-2}).$$

By the naturality of the Hurewicz homomorphism and its commutation with suspension, this coincides with the differential that we defined originally. □

Similarly, if we start with a homology theory $H_*(-; \pi)$, we can use the axioms to construct a chain complex $C_*(X; \pi)$, and a comparison of definitions then gives an isomorphism of chain complexes

$$C_*(X; \pi) \cong C_*(X) \otimes \pi.$$

We have identified our axiomatically derived chain complex of X with the cellular chain complex of X, and we again adopt the notation $C_*(X, A) = \tilde{C}_*(X/A)$.

THEOREM. There is a natural isomorphism

$$H_*(X, A) \cong H_*(C_*(X, A))$$

under which the natural transformation ∂ agrees with the natural transformation induced by the connecting homomorphisms associated to the short exact sequences

$$0 \longrightarrow C_*(A) \longrightarrow C_*(X) \longrightarrow C_*(X, A) \longrightarrow 0.$$

PROOF. In view of our comparison of theories on pairs of spaces and theories on pairs of CW complexes and our comparison of theories on pairs with reduced theories, it suffices to obtain a natural isomorphism of reduced theories on based CW complexes X. By the additivity axiom, we may as well assume that X is connected. More precisely, we must obtain a system of natural isomorphisms

$$\tilde{H}_n(X) \cong H_*(\tilde{C}_n(X))$$

that commute with the suspension isomorphisms.

By the dimension and additivity axioms, we know the homology of wedges of spheres. Since X^n/X^{n-1} is a wedge of n-spheres, the long exact homology sequence associated to the cofiber sequence

$$X^{n-1} \longrightarrow X^n \longrightarrow X^n/X^{n-1}$$

and an induction on n imply that

$$\tilde{H}_q(X^{n-1}) \longrightarrow \tilde{H}_q(X^n)$$

is an isomorphism for $q < n - 1$ and

$$\tilde{H}_q(X^n) = 0$$

for $q > n$. Of course, the analogues for cellular homology are obvious. Note in particular that $\tilde{H}_n(X^{n+1}) \cong \tilde{H}_n(X^{n+i})$ for all $i > 1$. Since homology commutes with colimits on sequences of inclusions, this implies that the inclusion $X^{n+1} \longrightarrow X$ induces an isomorphism

$$\tilde{H}_n(X^{n+1}) \longrightarrow \tilde{H}_n(X).$$

Using these facts, we easily check from the exactness axiom that the rows and columns are exact in the following commutative diagram:

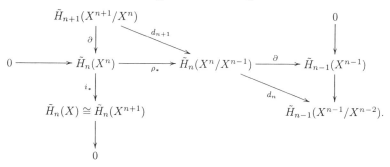

Define $\alpha : \tilde{H}_n(X) \longrightarrow H_n(\tilde{C}_*(X))$ by letting $\alpha(x)$ be the homology class of $\rho_*(y)$ for any y such that $i_*(y) = x$. It is an exercise in diagram chasing and the definition of the homology of a chain complex to check that α is a well defined isomorphism.

The reduced chain complex of ΣX can be identified with the suspension of the reduced chain complex of X. That is,

$$\tilde{C}_{n+1}(\Sigma X) \cong \tilde{C}_n(X),$$

compatibly with the differential. All maps in the diagram commute with suspension, and this implies that the isomorphisms α commute with the suspension isomorphisms. □

PROBLEMS

1. Let π be any group. Construct a connected CW complex $K(\pi,1)$ such that $\pi_1(K(\pi,1)) = \pi$ and $\pi_q(K(\pi,1)) = 0$ for $q \neq 1$.
2. * In Problem 1, it is rarely the case that $K(\pi,1)$ can be constructed as a compact manifold. What is a necessary condition on π for this to happen?
3. Let $n \geq 1$ and let π be an Abelian group. Construct a connected CW complex $M(\pi,n)$ such that $\tilde{H}_n(X;\mathbb{Z}) = \pi$ and $\tilde{H}_q(X;\mathbb{Z}) = 0$ for $q \neq n$. (Hint: construct $M(\pi,n)$ as the cofiber of a map between wedges of spheres.) The spaces $M(\pi,n)$ are called Moore spaces.
4. Let $n \geq 1$ and let π be an Abelian group. Construct a connected CW complex $K(\pi,n)$ such that $\pi_n(X) = \pi$ and $\pi_q(X) = 0$ for $q \neq n$. (Hint: start with $M(\pi,n)$, using the Hurewicz theorem, and kill its higher homotopy groups.) The spaces $K(\pi,n)$ are called Eilenberg-Mac Lane spaces.
5. There are familiar spaces that give $K(\mathbb{Z},1)$, $K(\mathbb{Z}_2,1)$, and $K(\mathbb{Z},2)$. Name them.
6. Let X be any connected CW complex whose only non-vanishing homotopy group is $\pi_n(X) \cong \pi$. Construct a homotopy equivalence $K(\pi,n) \longrightarrow X$, where $K(\pi,n)$ is the Eilenberg-Mac Lane space that you have constructed.
7. * For groups π and ρ, compute $[K(\pi,n), K(\rho,n)]$; here $[-,-]$ means based homotopy classes of based maps.

CHAPTER 16

Singular homology theory

We explain, without giving full details, how the standard approach to singular homology theory fits into our framework. We also introduce simplicial sets and spaces and their geometric realization. These notions play a fundamental role in modern algebraic topology.

1. The singular chain complex

The standard topological n-simplex is the subspace
$$\Delta_n = \{(t_0,\ldots,t_n) | 0 \le t_i \le 1,\ \sum t_i = 1\}$$
of $\mathbb{R}^{n+1}$. There are "face maps"
$$\delta_i : \Delta_{n-1} \longrightarrow \Delta_n,\ \ 0 \le i \le n,$$
specified by
$$\delta_i(t_0,\ldots,t_{n-1}) = (t_0,\ldots,t_{i-1},0,t_i,\ldots,t_{n-1})$$
and "degeneracy maps"
$$\sigma_i : \Delta_{n+1} \longrightarrow \Delta_n,\ \ 0 \le i \le n,$$
specified by
$$\sigma_i(t_0,\ldots,t_{n+1}) = (t_0,\ldots,t_{i-1},t_i+t_{i+1},\ldots,t_{n+1}).$$

For a space X, define $S_n X$ to be the set of continuous maps $f : \Delta_n \longrightarrow X$. In particular, regarding a point of X as the map that sends 1 to x, we may identify the underlying set of X with $S_0 X$. Define the ith face operator
$$d_i : S_n X \longrightarrow S_{n-1} X,\ \ 0 \le i \le n,$$
by
$$d_i(f)(u) = f(\delta_i(u)),$$
where $u \in \Delta_{n-1}$, and define the ith degeneracy operator
$$s_i : S_n X \longrightarrow S_{n+1} X,\ \ 0 \le i \le n,$$
by
$$s_i(f)(v) = f(\sigma_i(v)),$$
where $v \in \Delta_{n+1}$. The following identities are easily checked:
$$d_i \circ d_j = d_{j-1} \circ d_i \ \ \text{if}\ \ i < j$$
$$d_i \circ s_j = \begin{cases} s_{j-1} \circ d_i & \text{if}\ \ i < j \\ \text{id} & \text{if}\ \ i = j\ \ \text{or}\ \ i = j+1 \\ s_j \circ d_{i-1} & \text{if}\ \ i > j+1. \end{cases}$$
$$s_i \circ s_j = s_{j+1} \circ s_i\ \ \text{if}\ \ i \le j.$$

A map $f : \Delta_n \longrightarrow X$ is called a singular n-simplex. It is said to be nondegenerate if it is not of the form $s_i(g)$ for any i and g. Let $C_n(X)$ be the free Abelian group generated by the nondegenerate n-simplexes, and think of $C_n(X)$ as the quotient of the free Abelian group generated by all singular n-simplexes by the subgroup generated by the degenerate n-simplexes. Define

$$d = \sum_{i=0}^{n}(-1)^i d_i : C_n(X) \longrightarrow C_{n-1}(X).$$

The identities ensure that $C_*(X)$ is then a well defined chain complex. In fact,

$$d \circ d = \sum_{i=0}^{n-1}\sum_{j=0}^{n}(-1)^{i+j} d_i \circ d_j,$$

and, for $i < j$, the (i,j)th and $(j-1,i)$th summands add to zero. This gives that $d \circ d = 0$ before quotienting out the degenerate simplexes, and the degenerate simplexes span a subcomplex.

The singular homology of X is usually defined in terms of this chain complex:

$$H_*(X; \pi) = H_*(C_*(X) \otimes \pi).$$

2. Geometric realization

One can prove the compatibility of this definition with our definition by checking the axioms and quoting the uniqueness of homology. We instead describe how the new definition fits into our original definition in terms of CW approximation and cellular chain complexes. We define a space ΓX, called the "geometric realization of the total singular complex of X," as follows. As a set

$$\Gamma X = \coprod_{n \geq 0}(S_n X \times \Delta_n)/(\sim),$$

where the equivalence relation $\sim$ is generated by

$$(f, \delta_i u) \sim (d_i(f), u) \quad \text{for} \quad f : \Delta_n \longrightarrow X \quad \text{and} \quad u \in \Delta_{n-1}$$

and

$$(f, \sigma_i v) \sim (s_i(f), v) \quad \text{for} \quad f : \Delta_n \longrightarrow X \quad \text{and} \quad v \in \Delta_{n+1}.$$

Topologize ΓX by giving

$$\coprod_{0 \leq n \leq q}(S_n X \times \Delta_n)/(\sim)$$

the quotient topology and then giving ΓX the topology of the union. Define $\gamma : \Gamma X \longrightarrow X$ by

$$\gamma|f, u| = f(u) \quad \text{for} \quad f : \Delta_n \longrightarrow X \quad \text{and} \quad u \in \Delta_n,$$

where $|f, u|$ denotes the equivalence class of (f, u). Now the following two theorems imply that that this construction provides a canonical way of realizing our original construction of homology.

THEOREM. *For any space X, ΓX is a CW complex with one n-cell for each nondegenerate singular n-simplex, and the cellular chain complex $C_*(\Gamma X)$ is naturally isomorphic to the singular chain complex $C_*(X)$.*

THEOREM. *For any space X, the map $\gamma : \Gamma X \longrightarrow X$ is a weak equivalence.*

Thus the singular chain complex of X is the cellular chain complex of a functorial CW approximation of X, and this shows that our original construction of homology coincides with the classical construction in terms of singular chains. Each approach has its mathematical and pedagogical advantages.

3. Proofs of the theorems

We give a detailed outline of how the required CW decomposition of ΓX is obtained and sketch the proof that γ is a weak equivalence.

Let $\bar{X} = \coprod_{n \geq 0} S_n X \times \Delta_n$. Define functions
$$\lambda : \bar{X} \longrightarrow \bar{X} \quad \text{and} \quad \rho : \bar{X} \longrightarrow \bar{X}$$
by
$$\lambda(f, u) = (g, \sigma_{j_1} \cdots \sigma_{j_p} u)$$
if $f = s_{j_p} \cdots s_{j_1} g$ where g is nondegenerate and $0 \leq j_1 < \cdots < j_p$ and
$$\rho(f, u) = (d_{i_1} \cdots d_{i_q} f, v)$$
if $u = \delta_{i_q} \cdots \delta_{i_1} v$ where v is an interior point and $0 \leq i_1 < \cdots < i_q$. Note that the unique point of Δ_0 is interior. Say that a point (f, u) is nondegenerate if f is nondegenerate and u is interior. A combinatorial argument from the definitions gives the following observation.

LEMMA. *The composite $\lambda \circ \rho$ carries each point of $\bar{X}$ to the unique nondegenerate point to which it is equivalent.*

Let $(\Gamma X)^n$ be the image in ΓX of $\coprod_{m \leq n} S_m X \times \Delta_m$. Then
$$(\Gamma X)^n - (\Gamma X)^{n-1} = \{\text{nondegenerate } n\text{-simplexes}\} \times \{\Delta_n - \partial \Delta_n\}.$$
This implies that ΓX is a CW complex whose n-cells are the maps
$$\tilde{f} : (\Delta_n, \partial \Delta_n) \longrightarrow ((\Gamma X)^n, (\Gamma X)^{n-1})$$
specified by $\tilde{f}(u) = |f, u|$ for a nondegenerate n-simplex f. Here we think of $(\Delta_n, \partial \Delta_n)$ as the domains of cells via oriented homeomorphisms with (D^n, S^{n-1}).

To compute d on the cellular chains $C_*(\Gamma X)$, we must compute the degrees of the composites
$$S^{n-1} \cong \partial \Delta_n \xrightarrow{\tilde{f}} (\Gamma X)^{n-1}/(\Gamma X)^{n-2} \xrightarrow{\pi_{\tilde{g}}} \Delta_{n-1}/\partial \Delta_{n-1} \cong S^{n-1}$$
for nondegenerate n-simplexes f and $(n-1)$-simplexes g. The only relevant g are the $d_i f$ since f traverses these g on the various faces of $\partial \Delta_n$. Let $\bar{\delta}_i : \partial \Delta_n \longrightarrow \Delta_{n-1}/\partial \Delta_{n-1}$ be the map that collapses all faces of $\partial \Delta_n$ other than $\delta_i \Delta_{n-1}$ to the basepoint and is δ_i^{-1} on $\delta_i \Delta_{n-1}$. Then, with $g = d_i f$, the composite above reduces to the map
$$S^{n-1} \cong \partial \Delta_n \xrightarrow{\bar{\delta}_i} \Delta_{n-1}/\partial \Delta_{n-1} \cong S^{n-1}.$$
It is not hard to check that the degree of this map is $(-1)^i$ (provided that we choose our homeomorphisms sensibly). If $n = 2$, the three maps δ_i are given by
$$\delta_0(1 - t, t) = (0, 1 - t, t)$$
$$\delta_1(1 - t, t) = (1 - t, 0, t)$$
$$\delta_2(1 - t, t) = (1 - t, t, 0)$$

and we can visualize the maps $\bar{\delta}_i$ as follows:

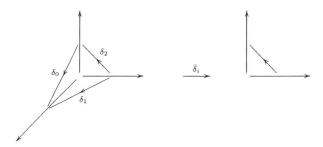

The alternation of orientations and thus of signs should be clear. This shows that $C_*(\Gamma X) = C_*(X)$, as claimed.

We must still explain why $\gamma : \Gamma X \longrightarrow X$ is a weak equivalence. In fact, it is tautologically obvious that γ induces an epimorphism on all homotopy groups: a map of pairs
$$f : (\Delta_n, \partial\Delta_n) \longrightarrow (X, x)$$
determines the map of pairs
$$\tilde{f} : (\Delta_n, \partial\Delta_n) \longrightarrow (\Gamma X, |x, 1|)$$
specified by $\tilde{f}(u) = |f, u|$, and $\gamma \circ \tilde{f} = f$. Injectivity is more delicate, and we shall only give a sketch. Given a map $g : (\Delta_n, \partial\Delta_n) \longrightarrow (\Gamma X, |x, 1|)$, we may first apply cellular approximation to obtain a homotopy of g with a cellular map and we may then subdivide the domain and apply a further homotopy so as to obtain a map $g' \simeq g$ such that g' is simplicial, in the sense that $g' \circ e$ is a cell of ΓX for every cell e of the subdivision of Δ_n. Suppose that $\gamma \circ g$ and thus $\gamma \circ g'$ is homotopic to the constant map c_x at the point x. We may view a homotopy $h : \gamma \circ g' \simeq c_x$ as a map
$$h : (\Delta_n \times I, \partial\Delta_n \times I \cup \Delta_n \times \{1\}) \longrightarrow (X, x).$$
We can simplicially subdivide $\Delta_n \times I$ so finely that our subdivided $\Delta_n = \Delta_n \times \{0\}$ is a subcomplex. We can then lift h simplex by simplex to a simplicial map
$$\tilde{h} : (\Delta_n \times I, \partial\Delta_n \times I \cup \Delta_n \times \{1\}) \longrightarrow (\Gamma X, |x, 1|)$$
such that $\tilde{h}$ restricts to $\tilde{g}'$ on $\Delta_n \times \{0\}$ and $\gamma \circ \tilde{h} = h$.

4. Simplicial objects in algebraic topology

A simplicial set K_* is a sequence of sets K_n, $n \geq 0$, connected by face and degeneracy operators $d_i : K_n \longrightarrow K_{n-1}$ and $s_i : K_n \longrightarrow K_{n+1}$, $0 \leq i \leq n$, that satisfy the commutation relations that we displayed for the total singular complex $S_*X = \{S_n X\}$ of a space X. Thus S_* is a functor from spaces to simplicial sets.

We may define the geometric realization $|K_*|$ of general simplicial sets exactly as we defined the geometric realization $\Gamma X = |S_*X|$ of the total singular complex of a topological space. In fact, the total singular complex and geometric realization functors are adjoint. If $\mathscr{SS}$ is the category of simplicial sets and $\mathscr{U}$ the category of spaces, then
$$\mathscr{U}(|K_*|, X) \cong \mathscr{SS}(K_*, S_*X).$$

The identity map of S_*X on the right corresponds to $\gamma : |S_*X| \longrightarrow X$ on the left. In general, for a map $f_* : K_* \longrightarrow S_*X$ of simplicial sets, the corresponding map of spaces is the composite

$$|K_*| \xrightarrow{|f_*|} |S_*X| \xrightarrow{\gamma} X.$$

In fact, one can develop homotopy theory and homology in the category of simplicial sets in a fashion parallel to and, in a suitable sense, equivalent to the development that we have here given for topological spaces. For example, we have the chain complex $C_*(K_*)$ defined exactly as we defined the singular chain complex, using the alternating sum of the face maps, and there result homology groups

$$H_*(K_*; \pi) = H_*(C_*(K_*) \otimes \pi).$$

Exactly as in the case of S_*X, $|K_*|$ is a CW complex and $C_*(K_*)$ is naturally isomorphic to the cellular chain complex $C_*(|K_*|)$. Singular homology is the special case obtained by taking $K_* = S_*X$ for spaces X. The passage back and forth between simplicial sets and topological spaces plays a major role in many applications.

The ideas generalize. One can define a simplicial object in any category $\mathscr{C}$ as a sequence of objects K_n of $\mathscr{C}$ connected by face and degeneracy maps in $\mathscr{C}$ that satisfy the commutation relations that we have displayed. Thus we have simplicial groups, simplicial Abelian groups, simplicial spaces, and so forth. We can think of simplicial sets as discrete simplicial spaces, and we then see that geometric realization generalizes directly to a functor $|-|$ from the category $\mathscr{S}\mathscr{U}$ of simplicial spaces to the category $\mathscr{U}$ of spaces. This provides a very useful way of constructing spaces with desirable properties.

We note one of the principal features of geometric realization. Define the product $X_* \times Y_*$ of simplicial spaces X_* and Y_* to be the simplicial space whose space of n-simplexes is $X_n \times Y_n$, with faces and degeneracies $d_i \times d_i$ and $s_i \times s_i$. The projections induce maps of simplicial spaces from $X_* \times Y_*$ to X_* and Y_*. On passage to geometric realization, these give the coordinates of a map

$$|X_* \times Y_*| \longrightarrow |X_*| \times |Y_*|.$$

It turns out that this map is always a homeomorphism.

Now restrict attention to simplicial sets K_* and L_*. Then the homeomorphism just specified is a map between CW complexes. However, it is not a cellular map; rather, it takes the n-skeleton of $|K_* \times L_*|$ to the $2n$-skeleton of $|K_*| \times |L_*|$. It is homotopic to a cellular map, no longer a homeomorphism, and there results a chain homotopy equivalence

$$C_*(|K_* \times L_*|) \longrightarrow C_*(|K_*|) \otimes C_*(|L_*|).$$

In particular, for spaces X and Y, there is a natural chain homotopy equivalence from the singular chain complex $C_*(X \times Y)$ to the tensor product $C_*(X) \otimes C_*(Y)$. One can be explicit about this but, pedagogically, one technical advantage of approaching homology via CW complexes is that it leaves us free to work directly with the natural cell structures on Cartesian products of CW complexes and to postpone the introduction of chain homotopy equivalences such as these to a later stage of the development.

5. Classifying spaces and $K(\pi, n)$s

We illustrate these ideas by defining the "classifying spaces" and "universal bundles" associated to topological groups G and describing how this leads to a beautiful conceptual construction of the Eilenberg-MacLane spaces $K(\pi, n)$ associated to discrete Abelian groups π. Recall that these are spaces such that

$$\pi_q(K(\pi, n)) = \begin{cases} \pi & \text{if } q = n \\ 0 & \text{if } q \neq n. \end{cases}$$

We define a map $p_* : E_*(G) \longrightarrow B_*(G)$ of simplicial topological spaces. Let $E_n(G) = G^{n+1}$ and $B_n(G) = G^n$, and let $p_n : G^{n+1} \longrightarrow G^n$ be the projection on the first n coordinates. The faces and degeneracies are defined on $E_n(G)$ by

$$d_i(g_1, \ldots, g_{n+1}) = \begin{cases} (g_2, \ldots, g_{n+1}) & \text{if } i = 0 \\ (g_1, \ldots, g_{i-1}, g_i g_{i+1}, g_{i+2}, \ldots, g_{n+1}) & \text{if } 1 \leq i \leq n \end{cases}$$

and

$$s_i(g_1, \ldots, g_{n+1}) = (g_1, \ldots, g_{i-1}, e, g_i, \ldots, g_{n+1}) \quad \text{if } 0 \leq i \leq n.$$

The faces and degeneracies on $B_n(G)$ are defined in the same way, except that the last coordinate g_{n+1} is omitted and the last face operation d_n takes the form

$$d_n(g_1, \ldots, g_n) = (g_1, \ldots, g_{n-1}).$$

Certainly p_* is a map of simplicial spaces. If we let G act from the right on $E_n(G)$ by multiplication on the last coordinate,

$$(g_1, \ldots, g_n, g_{n+1})g = (g_1, \ldots, g_n, g_{n+1}g),$$

then $E_*(G)$ is a simplicial G-space. That is, the action of G commutes with the face and degeneracy maps. We may view $B_n(G)$ as the orbit space $E_n(G)/G$. We define

$$E(G) = |E_*(G)|, \quad B(G) = |B_*(G)|, \quad \text{and} \quad p = |p_*(G)| : E(G) \longrightarrow B(G).$$

Then $E(G)$ inherits a free right action by G, and $B(G)$ is the orbit space $E(G)/G$. The space BG is called the classifying space of G.

The space $E(G)$ is the union of the images $E(G)^n$ of the spaces $\coprod_{m \leq n} G^{m+1} \times \Delta_m$, and

$$E(G)^n - E(G)^{n-1} = (G^n - W) \times G \times (\Delta_n - \partial \Delta_n),$$

where $W \subset G^n$ is the "fat wedge" consisting of those points at least one of whose coordinates is the identity element e. Similarly, we have subspaces $B(G)^n$ such that

$$B(G)^n - B(G)^{n-1} = (G^n - W) \times (\Delta_n - \partial \Delta_n).$$

The map p restricts to the projection between these subspaces. Intuitively, it looks as if p should be a bundle with fiber G, and this is indeed the case if the identity element of G is a nondegenerate basepoint. This condition is enough to ensure local triviality as we glue together over the filtration $\{B(G)^n\}$. It is less intuitive, but true, that the space $E(G)$ is contractible. By the long exact homotopy sequence, these facts imply that

$$\pi_{q+1}(BG) \cong \pi_q(G)$$

for all $q \geq 0$.

For topological groups G and H, the obvious shuffle homeomorphisms

$$(G \times H)^n \cong G^n \times H^n$$

specify isomorphisms of simplicial spaces
$$E_*(G \times H) \cong E_*(G) \times E_*(H) \quad \text{and} \quad B_*(G \times H) \cong B_*(G) \times B_*(H)$$
that are compatible with the projections. Since geometric realization commutes with products, we conclude that $B(G \times H)$ is homeomorphic to $B(G) \times B(H)$. Thus B is a product-preserving functor on the category of topological groups.

Now suppose that G is a commutative topological group. Then its multiplication $G \times G \longrightarrow G$ and inverse map $G \longrightarrow G$ are homomorphisms. We conclude that $B(G)$ and $E(G)$ are again commutative topological groups. The multiplication on $B(G)$ is determined by the multiplication on G as the composite
$$B(G) \times B(G) \cong B(G \times G) \longrightarrow B(G).$$
Moreover, the map $p : E(G) \longrightarrow B(G)$ and the inclusion of G in $E(G)$ as the fiber over the basepoint (the unique point in $B_0(G)$) are homomorphisms. This allows us to iterate the construction, setting $B^0(G) = G$ and $B^n(G) = B(B^{n-1}(G))$ for $n \geq 1$. Specializing to a discrete Abelian group π, we define
$$K(\pi, n) = B^n(\pi).$$
As promised, we have
$$\pi_q(K(\pi, n)) = \pi_{q-1}(K(\pi, n-1)) = \cdots = \pi_{q-n}(K(\pi, 0)) = \begin{cases} \pi & \text{if } q = n \\ 0 & \text{if } q \neq n. \end{cases}$$

PROBLEMS

1. Let X be a space that satisfies the hypotheses used to construct a universal cover $\tilde{X}$. Let $\pi = \pi_1(X)$ and consider the action of the group π on the space $\tilde{X}$ given by the isomorphism of π with $\text{Aut}(\tilde{X})$. Let A be an Abelian group and let $\mathbb{Z}[\pi]$ act trivially on A, $a \cdot \sigma = a$ for $\sigma \in \pi$ and $a \in A$. Do one or both of the following, and convince yourself that the other choice also works.

    (a) [Cellular chains] Assume that X is a CW complex. Show that $\tilde{X}$ is a CW complex such that the action of π on $\tilde{X}$ induces an action of the group ring $\mathbb{Z}[\pi]$ on the cellular chain complex $C_*(\tilde{X})$ such that each $C_q(\tilde{X})$ is a free $\mathbb{Z}[\pi]$-module and
    $$C_*(X; A) \cong A \otimes_{\mathbb{Z}[\pi]} C_*(\tilde{X}).$$

    (b) [Singular chains] Show that the action of π on $\tilde{X}$ induces an action of $\mathbb{Z}[\pi]$ on the singular chain complex $C_*(\tilde{X})$ such that each $C_q(\tilde{X})$ is a free $\mathbb{Z}[\pi]$-module and
    $$C_*(X; A) \cong A \otimes_{\mathbb{Z}[\pi]} C_*(\tilde{X}).$$

2. Let π be a group and let $K(\pi, 1)$ be a connected CW complex such that $\pi_1(K(\pi, 1)) = \pi$ and $\pi_q(K(\pi, 1)) = 0$ for $q \neq 1$. Use Problem 1 to show that there is an isomorphism
    $$H_*(K(\pi, 1); A) \cong \text{Tor}_*^{\mathbb{Z}[\pi]}(A, \mathbb{Z}).$$

3. Let $p : Y \longrightarrow X$ be a covering space with finite fibers, say of cardinality n. Using singular chains, construct a homomorphism $t : H_*(X; A) \longrightarrow H_*(Y; A)$ such that the composite $p_* \circ t : H_*(X; A) \longrightarrow H_*(X; A)$ is multiplication by n; t is called a "transfer homomorphism."

CHAPTER 17

Some more homological algebra

The reader will by now appreciate that the calculation of homology groups, although far simpler than the calculation of homotopy groups, can still be a difficult task. In practice, one seldom uses chains explicitly; rather, one uses them to prove algebraic theorems that simplify topological calculations. Indeed, if one focuses on singular chains, then one eschews chain level computations in principle as well as in practice.

We here recall some classical results in homological algebra that explain how to calculate $H_*(X;\pi)$ from $H_*(X) \equiv H_*(X;\mathbb{Z})$ and how to calculate $H_*(X \times Y)$ from $H_*(X) \otimes H_*(Y)$. We then say a little about cochain complexes in preparation for the definition of cohomology groups.

We again work over a general commutative ring R, although the main example will be $R = \mathbb{Z}$. Tensor products are understood to be taken over R.

1. Universal coefficients in homology

Let X and Y be chain complexes over R. We think of $H_*(X) \otimes H_*(Y)$ as a graded R-module which, in degree n, is $\sum_{p+q=n} H_p(X) \otimes H_q(Y)$. We define

$$\alpha : H_*(X) \otimes H_*(Y) \longrightarrow H_*(X \otimes Y)$$

by $\alpha([x] \otimes [y]) = [x \otimes y]$ for cycles x and y that represent homology classes $[x]$ and $[y]$. As a special case, for an R-module M we have

$$\alpha : H_*(X) \otimes M \longrightarrow H_*(X \otimes M).$$

We omit the proof of the following standard result, but we shall shortly give the quite similar proof of a cohomological analogue. Recall that an R-module M is said to be flat if the functor $M \otimes N$ is exact (that is, preserves exact sequences in the variable N). We say that a graded R-module is flat if each of its terms is flat.

We assume that the reader has seen torsion products, which measure the failure of tensor products to be exact functors. For a principal ideal domain (PID) R, the only torsion product is the first one, denoted $\text{Tor}_1^R(M,N)$. It can be computed by constructing a short exact sequence

$$0 \longrightarrow F_1 \longrightarrow F_0 \longrightarrow M \longrightarrow 0$$

and tensoring with N to obtain an exact seqence

$$0 \longrightarrow \text{Tor}_1^R(M,N) \longrightarrow F_1 \otimes N \longrightarrow F_0 \otimes N \longrightarrow M \otimes N \longrightarrow 0,$$

where F_1 and F_0 are free R-modules. That is, we choose an epimorphism $F_0 \longrightarrow M$ and note that, since R is a PID, its kernel F_1 is also free.

THEOREM (Universal coefficient). *Let R be a PID and let X be a flat chain complex over R. Then, for each n, there is a natural short exact sequence*

$$0 \longrightarrow H_n(X) \otimes M \xrightarrow{\alpha} H_n(X \otimes M) \xrightarrow{\beta} \mathrm{Tor}_1^R(H_{n-1}(X), M) \longrightarrow 0.$$

The sequence splits, so that

$$H_n(X \otimes M) \cong (H_n(X) \otimes M) \oplus \mathrm{Tor}_1^R(H_{n-1}(X), M),$$

but the splitting is not natural.

In Chapter 20 §3, we shall see an important class of examples in which the splitting is very far from being natural.

COROLLARY. *If R is a field, then*

$$\alpha : H_*(X) \otimes M \longrightarrow H_*(X; M)$$

is a natural isomorphism.

2. The Künneth theorem

The universal coefficient theorem in homology is a special case of the Künneth theorem.

THEOREM (Künneth). *Let R be a PID and let X be a flat chain complex and Y be any chain complex. Then, for each n, there is a natural short exact sequence*

$$0 \longrightarrow \sum_{p+q=n} H_p(X) \otimes H_q(Y) \xrightarrow{\alpha} H_n(X \otimes Y) \xrightarrow{\beta} \sum_{p+q=n-1} \mathrm{Tor}_1^R(H_p(X), H_q(Y)) \longrightarrow 0.$$

The sequence splits, so that

$$H_n(X \otimes Y) \cong (\sum_{p+q=n} H_p(X) \otimes H_q(Y)) \oplus (\sum_{p+q=n-1} \mathrm{Tor}_1^R(H_p(X), H_q(Y))),$$

but the splitting is not natural.

Returning to topology for a moment, observe that this applies directly to the computation of the homology of the Cartesian product of CW complexes X and Y in view of the isomorphism

$$C_*(X \times Y) \cong C_*(X) \otimes C_*(Y).$$

COROLLARY. *If R is a field, then*

$$\alpha : H_*(X) \otimes H_*(Y) \longrightarrow H_*(X \otimes Y)$$

is a natural isomorphism.

We prove the corollary to give the idea. The general case is proved by an elaboration of the argument. In fact, in practice, algebraic topologists carry out the vast majority of their calculations using a field of coefficients, and it is then the corollary that is relevant to the study of the homology of Cartesian products. There is a simple but important technical point to make about this. Let us for the moment remember to indicate the ring over which we are taking tensor products. For chain complexes X and Y over $\mathbb{Z}$, we have

$$(X \otimes_\mathbb{Z} R) \otimes_R (Y \otimes_\mathbb{Z} R) \cong (X \otimes_\mathbb{Z} Y) \otimes_\mathbb{Z} R.$$

We can therefore use the corollary to compute $H_*(X \otimes_\mathbb{Z} Y; R)$ from $H_*(X; R)$ and $H_*(Y; R)$.

PROOF OF THE COROLLARY. Assume first that $X_i = 0$ for $i \neq p$, so that $X = X_p$ is just an R-module with no differential. The square commutes and the row and column are exact in the diagram

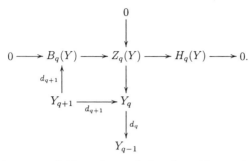

Since all modules over a field are free and thus flat, this remains true when we tensor the diagram with X_p. This proves that if $n = p + q$, then
$$Z_n(X_p \otimes Y) = X_p \otimes Z_q(Y), \quad B_n(X_p \otimes Y) = X_p \otimes B_q(Y),$$
and therefore
$$H_n(X \otimes Y) = X_p \otimes H_q(Y).$$
In the general case, regard the graded modules $Z(X)$ and $B(X)$ as chain complexes with zero differential. The exact sequences
$$0 \longrightarrow Z_p(X) \longrightarrow X_p \xrightarrow{d_p} B_{p-1}(X) \longrightarrow 0$$
of R-modules define a short exact seqence of chain complexes since $d_{p-1} \circ d_p = 0$. Define the suspension of a graded R-module N by $(\Sigma N)_{n+1} = N_n$. Tensoring with Y, we obtain a short exact sequence of chain complexes
$$0 \longrightarrow Z(X) \otimes Y \longrightarrow X \otimes Y \longrightarrow \Sigma B(X) \otimes Y \longrightarrow 0.$$
It follows from the first part and additivity that
$$H_*(Z(X) \otimes Y) = Z(X) \otimes H_*(Y) \quad \text{and} \quad H_*(\Sigma B(X) \otimes Y) = \Sigma B(X) \otimes H_*(Y).$$
Moreover, by inspection of definitions, the connecting homomorphism of the long exact sequence of homology modules associated to our short exact sequence of chain complexes is just the inclusion $B \otimes H_*(Y) \longrightarrow Z \otimes H_*(Y)$. In particular, the long exact sequence breaks up into short exact sequences
$$0 \longrightarrow B(X) \otimes H_*(Y) \longrightarrow Z(X) \otimes H_*(Y) \longrightarrow H_*(X \otimes Y) \longrightarrow 0.$$
However, since tensoring with $H_*(Y)$ is an exact functor, the cokernel of the inclusion $B \otimes H_*(Y) \longrightarrow Z \otimes H_*(Y)$ is $H_*(X) \otimes H_*(Y)$. The conclusion follows. □

3. Hom functors and universal coefficients in cohomology

For a chain complex $X = X_*$, we define the dual cochain complex X^* by setting
$$X^q = \text{Hom}(X_q, R) \quad \text{and} \quad d^q = (-1)^q \text{Hom}(d_{q+1}, \text{id}).$$
As with tensor products, we understand Hom to mean Hom_R when R is clear from the context. On elements, for an R-map $f : X_q \longrightarrow R$ and an element $x \in X_{q+1}$,
$$(d^q f)(x) = (-1)^q f(d_q(x)).$$

More generally, for an R-module M, we define a cochain complex $\operatorname{Hom}(X, M)$ in the same way. The sign is conventional and is designed to facilitate the definition of $\operatorname{Hom}(X, Y)$ for a chain complex X and cochain complex Y; however, we shall not have occasion to use the latter definition.

In analogy with the notation $H_*(X; M) = H_*(X \otimes M)$, we write
$$H^*(X; M) = H^*(\operatorname{Hom}(X, M)).$$
We have a cohomological version of the universal coefficient theorem. We assume that the reader has seen Ext modules, which measure the failure of Hom to be an exact functor. For a PID R, the only Ext module is the first one, denoted $\operatorname{Ext}^1_R(M, N)$. It can be computed by constructing a short exact sequence
$$0 \longrightarrow F_1 \longrightarrow F_0 \longrightarrow M \longrightarrow 0$$
and applying Hom to obtain an exact seqence
$$0 \longrightarrow \operatorname{Hom}(M, N) \longrightarrow \operatorname{Hom}(F_0, N) \longrightarrow \operatorname{Hom}(F_1, N) \longrightarrow \operatorname{Ext}^1_R(M, N) \longrightarrow 0,$$
where F_1 and F_0 are free R-modules.

For each n, define
$$\alpha : H^n(\operatorname{Hom}(X, M)) \longrightarrow \operatorname{Hom}(H_n(X), M)$$
by letting $\alpha[f]([x]) = f(x)$ for a cohomology class $[f]$ represented by a "cocycle" $f : X_n \longrightarrow M$ and a homology class $[x]$ represented by a cycle x. It is easy to check that $f(x)$ is independent of the choices of f and x since x is a cycle and f is a cocycle.

THEOREM (Universal coefficient). *Let R be a PID and let X be a free chain complex over R. Then, for each n, there is a natural short exact sequence*
$$0 \longrightarrow \operatorname{Ext}^1_R(H_{n-1}(X), M) \xrightarrow{\beta} H^n(X; M) \xrightarrow{\alpha} \operatorname{Hom}(H_n(X), M) \longrightarrow 0.$$
The sequence splits, so that
$$H^n(X; M) \cong \operatorname{Hom}(H_n(X), M) \oplus \operatorname{Ext}^1_R(H_{n-1}(X), M),$$
but the splitting is not natural.

COROLLARY. *If R is a field, then*
$$\alpha : H^*(X; M) \longrightarrow \operatorname{Hom}(H_*(X), M)$$
is a natural isomorphism.

Again, there is a technical point to be made here. If X is a complex of free Abelian groups and M is an R-module, such as R itself, then
$$\operatorname{Hom}_{\mathbb{Z}}(X, M) \cong \operatorname{Hom}_R(X \otimes_{\mathbb{Z}} R, M).$$
One way to see this is to observe that, if B is a basis for a free Abelian group F, then $\operatorname{Hom}_{\mathbb{Z}}(F, M)$ and $\operatorname{Hom}_R(F \otimes_{\mathbb{Z}} R, M)$ are both in canonical bijective correspondence with maps of sets $B \longrightarrow M$. More algebraically, a homomorphism $f : F \longrightarrow M$ of Abelian groups determines the corresponding map of R-modules as the composite of $f \otimes \operatorname{id}$ and the action of R on M:
$$F \otimes_{\mathbb{Z}} R \longrightarrow M \otimes_{\mathbb{Z}} R \longrightarrow M.$$

4. Proof of the universal coefficient theorem

We need two properties of Ext in the proof. First, $\text{Ext}_R^1(F, M) = 0$ for a free R-module F. Second, when R is a PID, a short exact sequence

$$0 \longrightarrow L' \longrightarrow L \longrightarrow L'' \longrightarrow 0$$

of R-modules gives rise to a six-term exact sequence

$$0 \longrightarrow \text{Hom}(L'', M) \longrightarrow \text{Hom}(L, M) \longrightarrow \text{Hom}(L', M)$$
$$\stackrel{\delta}{\longrightarrow} \text{Ext}_R^1(L'', M) \longrightarrow \text{Ext}_R^1(L, M) \longrightarrow \text{Ext}_R^1(L', M) \longrightarrow 0.$$

PROOF OF THE UNIVERSAL COEFFICIENT THEOREM. We write $B_n = B_n(X)$, $Z_n = Z_n(X)$, and $H_n = H_n(X)$ to abbreviate notation. Since each X_n is a free R-module and R is a PID, each B_n and Z_n is also free. We have short exact sequences

$$0 \longrightarrow B_n \stackrel{i_n}{\longrightarrow} Z_n \stackrel{\pi_n}{\longrightarrow} H_n \longrightarrow 0$$

and

$$0 \longrightarrow Z_n \stackrel{j_n}{\longrightarrow} X_n \mathrel{\substack{d_n \\ \longleftarrow \\ \sigma_n}} B_{n-1} \longrightarrow 0;$$

we choose a splitting σ_n of the second. Writing $f^* = \text{Hom}(f, M)$ consistently, we obtain a commutative diagram with exact rows and columns

[commutative diagram]

By inspection of the diagram, we see that the canonical map α coincides with the composite

$$H^n(X; M) = \ker d_{n+1}^* / \text{im } d_n^* = \ker i_n^* j_n^* / \text{im } d_n^* i_{n-1}^* \stackrel{j_n^*}{\longrightarrow} \text{im } \pi_n^* \stackrel{(\pi_n^*)^{-1}}{\longrightarrow} \text{Hom}(H_n, M).$$

Since j_n^* is an epimorphism, so is α. The kernel of α is $\text{im } d_n^* / \text{im } d_n^* i_{n-1}^*$, and $\delta(d_n^*)^{-1}$ maps this group isomorphically onto $\text{Ext}_R^1(H_{n-1}, M)$. The composite $\delta \sigma_n^*$ induces the required splitting. □

5. Relations between ⊗ and Hom

We shall need some observations about cochain complexes and tensor products, and we first recall some general facts about the category of R-modules. For R-modules L, M, and N, we have an adjunction

$$\text{Hom}(L \otimes M, N) \cong \text{Hom}(L, \text{Hom}(M, N)).$$

We also have a natural homomorphism
$$\mathrm{Hom}(L, M) \otimes N \longrightarrow \mathrm{Hom}(L, M \otimes N),$$
and this is an isomorphism if either L or N is a finitely generated projective R-module. Again, we have a natural map
$$\mathrm{Hom}(L, M) \otimes \mathrm{Hom}(L', M') \longrightarrow \mathrm{Hom}(L \otimes L', M \otimes M'),$$
which is an isomorphism if L and L' are finitely generated and projective or if L is finitely generated and projective and $M = R$.

We can replace L and L' by chain complexes and obtain similar maps, inserting signs where needed. For example, a chain homotopy $X \otimes \mathscr{I} \longrightarrow X'$ between chain maps $f, g : X \longrightarrow X'$ induces a chain map
$$\mathrm{Hom}(X', M) \longrightarrow \mathrm{Hom}(X \otimes \mathscr{I}, M) \cong \mathrm{Hom}(\mathscr{I}, \mathrm{Hom}(X, M)) \cong \mathrm{Hom}(X, M) \otimes \mathscr{I}^*,$$
where $\mathscr{I}^* = \mathrm{Hom}(\mathscr{I}, R)$. It should be clear that this implies that our original chain homotopy induces a homotopy of cochain maps
$$f^* \simeq g^* : \mathrm{Hom}(X', M) \longrightarrow \mathrm{Hom}(X, M).$$
If Y and Y' are cochain complexes, then we have the natural homomorphism
$$\alpha : H^*(Y) \otimes H^*(Y') \longrightarrow H^*(Y \otimes Y')$$
given by $\alpha([y] \otimes [y']) = [y \otimes y']$, exactly as for chain complexes. (In fact, by regrading, we may view this as a special case of the map for chain complexes.) The Künneth theorem applies to this map. For its flatness hypothesis, it is useful to remember that, for any Noetherian ring R, the dual $\mathrm{Hom}(F, R)$ of a free R-module is a flat R-module.

As indicated above, if $Y = \mathrm{Hom}(X, M)$ and $Y' = \mathrm{Hom}(X', M')$ for chain complexes X and X' and R-modules M and M', then we also have the map of cochain complexes
$$\omega : \mathrm{Hom}(X, M) \otimes \mathrm{Hom}(X', M') \longrightarrow \mathrm{Hom}(X \otimes X', M \otimes M')$$
specified by the formula
$$\omega(f \otimes f')(x \otimes x') = (-1)^{(\deg f')(\deg x)} f(x) \otimes f'(x').$$
We continue to write ω for the map it induces on cohomology, and we then have the composite
$$\omega \circ \alpha : H^*(X; M) \otimes H^*(X'; M') \longrightarrow H^*(X \otimes X'; M \otimes M').$$
When $M = M' = A$ is a commutative R-algebra, we may compose with the map
$$H^*(X \otimes X'; A \otimes A) \longrightarrow H^*(X \otimes X'; A)$$
induced by the multiplication of A to obtain a map
$$H^*(X; A) \otimes H^*(X'; A) \longrightarrow H^*(X \otimes X'; A).$$
We are especially interested in the case when $R = \mathbb{Z}$ and A is either $\mathbb{Z}$ or a field.

CHAPTER 18

Axiomatic and cellular cohomology theory

We give a treatment of cohomology that is precisely parallel to our treatment of homology. The essential new feature is the cup product structure that makes the cohomology of X with coefficients in a commutative ring R a commutative graded R-algebra. This additional structure ties together the cohomology groups in different degrees and is fundamentally important to most of the applications.

1. Axioms for cohomology

Fix an Abelian group π and consider pairs of spaces (X, A). We shall see that π determines a "cohomology theory on pairs (X, A)."

THEOREM. *For integers q, there exist* contravariant *functors $H^q(X, A; \pi)$ from the homotopy category of pairs of spaces to the category of Abelian groups together with natural transformations $\delta : H^q(A; \pi) \longrightarrow H^{q+1}(X, A; \pi)$, where $H^q(X; \pi)$ is defined to be $H^q(X, \emptyset; \pi)$. These functors and natural transformations satisfy and are characterized by the following axioms.*

- *DIMENSION If X is a point, then $H^0(X; \pi) = \pi$ and $H^q(X; \pi) = 0$ for all other integers.*
- *EXACTNESS The following sequence is exact, where the unlabeled arrows are induced by the inclusions $A \longrightarrow X$ and $(X, \emptyset) \longrightarrow (X, A)$:*

$$\cdots \longrightarrow H^q(X, A; \pi) \longrightarrow H^q(X; \pi) \longrightarrow H^q(A, n) \overset{\delta}{\longrightarrow} H^{q+1}(X, A; \pi) \longrightarrow \cdots.$$

- *EXCISION If $(X; A, B)$ is an excisive triad, so that X is the union of the interiors of A and B, then the inclusion $(A, A \cap B) \longrightarrow (X, B)$ induces an isomorphism*

$$H^*(X, B; \pi) \longrightarrow H^*(A, A \cap B; \pi).$$

- *ADDITIVITY If (X, A) is the disjoint union of a set of pairs (X_i, A_i), then the inclusions $(X_i, A_i) \longrightarrow (X, A)$ induce an isomorphism*

$$H^*(X, A; \pi) \longrightarrow \prod_i H^*(X_i, A_i; \pi).$$

- *WEAK EQUIVALENCE If $f : (X, A) \longrightarrow (Y, B)$ is a weak equivalence, then*

$$f^* : H^*(Y, B; \pi) \longrightarrow H^*(X, A; \pi)$$

is an isomorphism.

We write f^* instead of $H^*(f)$ or $H^q(f)$. As in homology, our approximation theorems for spaces, pairs, maps, homotopies, and excisive triads directly imply that such a theory determines and is determined by an appropriate theory defined on CW pairs, as spelled out in the following CW version of the theorem.

THEOREM. For integers q, there exist functors $H^q(X, A; \pi)$ from the homotopy category of pairs of CW complexes to the category of Abelian groups together with natural transformations $\delta : H^q(A) \longrightarrow H^{q+1}(X, A; \pi)$, where $H^q(X; \pi)$ is defined to be $H^q(X, \emptyset; \pi)$. These functors and natural transformations satisfy and are characterized by the following axioms.

- DIMENSION If X is a point, then $H^0(X; \pi) = \pi$ and $H^q(X; \pi) = 0$ for all other integers.
- EXACTNESS The following sequence is exact, where the unlabeled arrows are induced by the inclusions $A \longrightarrow X$ and $(X, \emptyset) \longrightarrow (X, A)$:

$$\cdots \longrightarrow H^q(X, A; \pi) \longrightarrow H^q(X; \pi) \longrightarrow H^q(A; \pi) \xrightarrow{\delta} H^{q+1}(X, A; \pi) \longrightarrow \cdots.$$

- EXCISION If X is the union of subcomplexes A and B, then the inclusion $(A, A \cap B) \longrightarrow (X, B)$ induces an isomorphism

$$H^*(X, B; \pi) \longrightarrow H^*(A, A \cap B; \pi).$$

- ADDITIVITY If (X, A) is the disjoint union of a set of pairs (X_i, A_i), then the inclusions $(X_i, A_i) \longrightarrow (X, A)$ induce an isomorphism

$$H^*(X, A; \pi) \longrightarrow \prod_i H^*(X_i, A_i; \pi).$$

Such a theory determines and is determined by a theory as in the previous theorem.

2. Cellular and singular cohomology

We define the cellular cochains of a CW pair (X, A) with coefficients in an Abelian group π to be

$$C^*(X, A; \pi) = \mathrm{Hom}(C_*(X, A), \pi).$$

We then define the cellular cohomology groups to be

$$H^*(X, A; \pi) = H^*(C^*(X, A; \pi)).$$

If M is a module over a commutative ring R, we have a natural identification

$$C^*(X, A; M) \cong \mathrm{Hom}_R(C_*(X, A) \otimes R, M)$$

which allows us to do homological algebra over R rather than over $\mathbb{Z}$ when convenient. In particular, if R is a field, then

$$H^*(X, A; M) \cong \mathrm{Hom}_R(H_*(X, A; R), M).$$

In general, with $R = \mathbb{Z}$, we have a natural and splittable short exact sequence

$$0 \longrightarrow \mathrm{Ext}^1_{\mathbb{Z}}(H_{n-1}(X, A), \pi) \longrightarrow H^n(X, A; \pi) \longrightarrow \mathrm{Hom}(H_n(X, A), \pi) \longrightarrow 0.$$

The verification of the axioms listed in the previous section is immediate, as in homology. The fact that cellularly homotopic maps induce the same map on cohomology uses our observations relating homotopies of chain complexes with homotopies of cochain complexes. For exactness, the fact that our chain complexes are free over $\mathbb{Z}$ implies that we have a short exact sequence of cochain complexes

$$0 \longrightarrow C^*(X, A; \pi) \longrightarrow C^*(X; \pi) \longrightarrow C^*(A; \pi) \longrightarrow 0.$$

The required natural long exact sequence follows. The rest is the same as in homology.

For general spaces X, we can use $\Gamma X = |S_*X|$ as a canonical CW approximation functor. We define the singular cochains of X to be the cellular cochains of ΓX. Then our passage from the cohomology of CW complexes to the cohomology of general spaces can be realized by taking the cohomology of singular cochain complexes.

3. Cup products in cohomology

If X and Y are CW complexes, we have an isomorphism
$$C_*(X \times Y) \cong C_*(X) \otimes C_*(Y)$$
of chain complexes and therefore, for any Abelian groups π and π', an isomorphism of cochain complexes
$$C^*(X \times Y; \pi \otimes \pi') \cong \mathrm{Hom}(C_*(X) \otimes C_*(Y), \pi \otimes \pi').$$
By our observations about cochain complexes, there results a natural homomorphism
$$H^*(X; \pi) \otimes H^*(Y; \pi') \longrightarrow H^*(X \times Y; \pi \otimes \pi').$$
If $X = Y$ and if $\pi = \pi' = R$ is a commutative ring, we can use the diagonal map $\Delta : X \longrightarrow X \times X$ and the product $R \otimes R \longrightarrow R$ to obtain a "cup product"
$$\cup : H^*(X; R) \otimes_R H^*(X; R) \longrightarrow H^*(X; R).$$
More precisely, for $p \geq 0$ and $q \geq 0$, we have a product
$$\cup : H^p(X; R) \otimes_R H^q(X; R) \longrightarrow H^{p+q}(X; R).$$
We have noted that we can use $C_*(X; R)$ instead of $C_*(X)$ and so justify tensoring over R rather than $\mathbb{Z}$. This product makes $H^*(X; R)$ into a graded unital, associative, and "commutative" R-algebra. Here commutativity is understood in the appropriate graded sense, namely
$$xy = (-1)^{pq} yx \text{ if } \deg x = p \text{ and } \deg y = q.$$
The image of $1 \in R = H^0(*; R)$ under the map $\pi^* : H^0(*; R) \longrightarrow H^0(X; R)$ induced by the unique map $\pi : X \longrightarrow \{*\}$ is the unit (= identity element) for the product. In fact, the diagrams that say that $H^*(X; R)$ is unital, associative, and commutative result by passing to cohomology from the evident commutative diagrams

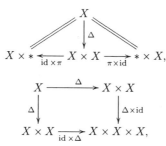

and

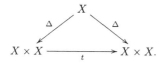

Here $t : X \times Y \longrightarrow Y \times X$ is the transposition, $t(x,y) = (y,x)$. The following diagrams commute in homology and cohomology with cofficients in R:

$$\begin{array}{ccc} H_*(X) \otimes_R H_*(Y) & \xrightarrow{\alpha} & H_*(X \times Y) \\ \tau \downarrow & & \downarrow t_* \\ H_*(Y) \otimes_R H_*(X) & \xrightarrow{\alpha} & H_*(Y \times X) \end{array}$$

and

$$\begin{array}{ccc} H^*(X) \otimes_R H^*(Y) & \xrightarrow{\alpha} & H^*(X \times Y) \\ \tau \downarrow & & \downarrow t^* \\ H^*(Y) \otimes_R H^*(X) & \xrightarrow{\alpha} & H^*(Y \times X). \end{array}$$

In both diagrams,

$$\tau(x \otimes y) = (-1)^{pq} y \otimes x \quad \text{if} \quad \deg x = p \quad \text{and} \quad \deg y = q.$$

The reason is that, on the topological level, t permutes p-cells past q-cells and, on the level of cellular chains, this involves the transposition

$$S^{p+q} = S^p \wedge S^q \longrightarrow S^q \wedge S^p = S^{p+q}.$$

We leave it as an exercise that this map has degree $(-1)^{pq}$. It is this fact that forces the cup product to be commutative in the graded sense.

In principle, the way to compute cup products is to pass to cellular chains from a cellular approximation to the diagonal map Δ. The point is that Δ fails to be cellular since it carries the n-skeleton of X to the $2n$-skeleton of $X \times X$. In practice, this does not work very well and more indirect means of computation must be used.

4. An example: $\mathbb{R}P^n$ and the Borsuk-Ulam theorem

Remember that $\mathbb{R}P^n$ is a CW complex with one q-cell for each $q \leq n$. The differential on $C_q(\mathbb{R}P^n) \cong \mathbb{Z}$ is zero if q is odd and multiplication by 2 if q is even. When we dualize to $C^*(\mathbb{R}P^n)$, we find that the differential on $C^q(\mathbb{R}P^n)$ is multiplication by 2 if q is odd and zero if q is even. We read off that

$$H^q(\mathbb{R}P^n; \mathbb{Z}) = \begin{cases} \mathbb{Z} & \text{if } q = 0 \\ \mathbb{Z}_2 & \text{if } 0 < q \leq n \quad \text{and} \quad q \text{ is even} \\ \mathbb{Z} & \text{if } q = n \text{ is odd} \\ 0 & \text{otherwise.} \end{cases}$$

If we work mod 2, taking $\mathbb{Z}_2$ as coefficient group, then the answer takes a nicer form, namely

$$H^q(\mathbb{R}P^n; \mathbb{Z}_2) = \begin{cases} \mathbb{Z}_2 & \text{if } 0 \leq q \leq n \\ 0 & \text{if } q > n. \end{cases}$$

The reader may find it instructive to compare with the calculations in homology, checking the correctness of the calculation by comparison with the universal coefficient theorem.

We shall later use Poincaré duality to give a quick proof that the cohomology algebra $H^*(\mathbb{R}P^n; \mathbb{Z}_2)$ is a truncated polynomial algebra $\mathbb{Z}_2[x]/(x^{n+1})$, where $\deg x = 1$. That is, for $1 \leq q \leq n$, the unique non-zero element of $H^q(\mathbb{R}P^n; \mathbb{Z}_2)$ is the qth power of x. This means that the elements are so tightly bound together

4. AN EXAMPLE: $\mathbb{R}P^n$ AND THE BORSUK-ULAM THEOREM

that knowledge of the cohomological behavior of a map $f : \mathbb{R}P^m \longrightarrow \mathbb{R}P^n$ on cohomology in degree one determines its behavior on cohomology in all higher degrees. We assume that $m \geq 1$ and $n \geq 1$ to avoid triviality.

PROPOSITION. *Let* $f : \mathbb{R}P^m \longrightarrow \mathbb{R}P^n$ *be a map such that* $f_* : \pi_1(\mathbb{R}P^m) \longrightarrow \pi_1(\mathbb{R}P^n)$ *is non-zero. Then* $m \leq n$.

PROOF. Since $\pi_1(\mathbb{R}P^1) = \mathbb{Z}$ and $\pi_1(\mathbb{R}P^m) = \mathbb{Z}_2$ if $m \geq 2$, the result is certainly true if $n = 1$. Thus assume that $n > 1$ and assume for a contradiction that $m > n$. By the naturality of the Hurewicz isomorphism, $f_* : H_1(\mathbb{R}P^m; \mathbb{Z}) \longrightarrow H_1(\mathbb{R}P^n; \mathbb{Z})$ is non-zero. By our universal coefficient theorems, the same is true for mod 2 homology and for mod 2 cohomology. That is, if x is the non-zero element of $H^1(\mathbb{R}P^n; \mathbb{Z}_2)$, then $f^*(x)$ is the non-zero element of $H^1(\mathbb{R}P^m; \mathbb{Z}_2)$. By the naturality of cup products

$$(f^*(x))^m = f^*(x^m).$$

However, the left side is non-zero in $H^m(\mathbb{R}P^m; \mathbb{Z}_2)$ and the right side is zero since $x^m = 0$ by our assumption that $m > n$. The contradiction establishes the conclusion. □

We use this fact together with covering space theory to prove a celebrated result known as the Borsuk-Ulam theorem. A map $g : S^m \longrightarrow S^n$ is said to be antipodal if it takes pairs of antipodal points to pairs of antipodal points. It then induces a map $f : \mathbb{R}P^m \longrightarrow \mathbb{R}P^n$ such that the following diagram commutes:

$$\begin{array}{ccc} S^m & \xrightarrow{g} & S^n \\ p_m \downarrow & & \downarrow p_n \\ \mathbb{R}P^m & \xrightarrow{f} & \mathbb{R}P^n, \end{array}$$

where p_m and p_n are the canonical coverings.

THEOREM. *If* $m > n \geq 1$, *then there exist no antipodal maps* $S^m \longrightarrow S^n$.

PROOF. Suppose given an antipodal map $g : \mathbb{R}P^m \longrightarrow \mathbb{R}P^n$. According to the proposition, $f_* : \pi_1(\mathbb{R}P^m) \longrightarrow \pi_1(\mathbb{R}P^n)$ is zero. According to the fundamental theorem of covering space theory, there is a map $\tilde{f} : \mathbb{R}P^m \longrightarrow S^n$ such that $p_n \circ \tilde{f} = f$. Let $s \in S^m$. Then $\tilde{f}(p_m(s)) = \tilde{f}(p_m(-s))$ must be either $g(s)$ or $g(-s)$, since these are the only two points in $p_n^{-1}(f(p_m(s)))$. Thus either $t = s$ or $t = -s$ satisfies $\tilde{f}(p_m(t)) = g(t)$. Therefore, by the fundamental theorem of covering space theory, the maps $\tilde{f} \circ p_m$ and g must be equal since they agree on a point. This is absurd: $\tilde{f} \circ p_m$ takes antipodal points to the same point, while g was assumed to be antipodal. □

THEOREM (Borsuk-Ulam). *For any continuous map* $f : S^n \longrightarrow \mathbb{R}^n$, *there exists* $x \in S^n$ *such that* $f(x) = f(-x)$.

PROOF. Suppose for a contradiction that $f(x) \neq f(-x)$ for all x. We could then define a continuous antipodal map $g : S^n \longrightarrow S^{n-1}$ by letting $g(x)$ be the point at which the vector from 0 through $f(x) - f(-x)$ intersects S^{n-1}. □

5. Obstruction theory

We give an outline of one of the most striking features of cohomology: the cohomology groups of a space X with coefficients in the homotopy groups of a space Y control the construction of homotopy classes of maps $X \longrightarrow Y$. As a matter of motivation, this helps explain why one is interested in general coefficient groups. It also explains why the letter π is so often used to denote coefficient groups.

DEFINITION. Fix $n \geq 1$. A connected space X is said to be n-simple if $\pi_1(X)$ is Abelian and acts trivially on the homotopy groups $\pi_q(X)$ for $q \leq n$; X is said to be simple if it is n-simple for all n.

Let (X, A) be a relative CW complex with relative skeleta X^n and let Y be an n-simple space. The assumption on Y has the effect that we need not worry about basepoints. Let $f : X^n \longrightarrow Y$ be a map. We ask when f can be extended to a map $X^{n+1} \longrightarrow Y$ that restricts to the given map on A.

If we compose the attaching maps $S^n \to X$ of cells of $X \setminus A$ with f, we obtain elements of $\pi_n(Y)$. These elements specify a well defined "obstruction cocycle"
$$c_f \in C^{n+1}(X, A; \pi_n(Y)).$$
Clearly, by considering extensions cell by cell, f extends to X^{n+1} if and only if $c_f = 0$. This is not a computable criterion. However, if we allow ourselves to modify f a little, then we can refine the criterion to a cohomological one that often is computable. If f and f' are maps $X^n \to Y$ and h is a homotopy rel A of the restrictions of f and f' to X^{n-1}, then f, f', and h together define a map
$$h(f, f') : (X \times I)^n \longrightarrow Y.$$
Applying $c_{h(f,f')}$ to cells $j \times I$, we obtain a "deformation cochain"
$$d_{f,f',h} \in C^n(X, A; \pi_n(Y))$$
such that $\delta d_{f,f',h} = c_f - c_{f'}$. Moreover, given f and d, there exists f' that coincides with f on X^{n-1} and satisfies $d_{f,f'} = d$, where the constant homotopy h is understood. This gives the following result.

THEOREM. For $f : X^n \longrightarrow Y$, the restriction of f to X^{n-1} extends to a map $X^{n+1} \to Y$ if and only if $[c_f] = 0$ in $H^{n+1}(X, A; \pi_n(Y))$.

It is natural to ask further when such extensions are unique up to homotopy, and a similar argument gives the answer.

THEOREM. Given maps $f, f' : X^n \to Y$ and a homotopy rel A of their restrictions to X^{n-1}, there is an obstruction class in $H^n(X, A; \pi_n(Y))$ that vanishes if and only if the restriction of the given homotopy to X^{n-2} extends to a homotopy $f \simeq f'$ rel A.

PROBLEMS

The first few problems here are parallel to those at the end of Chapter 16.

1. Let X be a space that satisfies the hypotheses used to construct a universal cover $\tilde{X}$ and let A be an Abelian group. Using cellular or singular chains, show that
$$C^*(X; A) \cong \mathrm{Hom}_{\mathbb{Z}[\pi]}(C_*(\tilde{X}), A).$$

2. Show that there is an isomorphism

$$H^*(K(\pi,1);A) \cong \mathrm{Ext}^*_{\mathbb{Z}[\pi]}(\mathbb{Z},A).$$

When A is a commutative ring, the Ext groups have algebraically defined products, constructed as follows. The evident isomorphism $\mathbb{Z} \cong \mathbb{Z} \otimes \mathbb{Z}$ is covered by a map of free $\mathbb{Z}[\pi]$-resolutions $P \longrightarrow P \otimes P$, where $\mathbb{Z}[\pi]$ acts diagonally on tensor products, $\alpha(x \otimes y) = \alpha x \otimes \alpha y$. This chain map is unique up to chain homotopy. It induces a map of chain complexes

$$\mathrm{Hom}_{\mathbb{Z}[\pi]}(P,A) \otimes \mathrm{Hom}_{\mathbb{Z}[\pi]}(P,A) \longrightarrow \mathrm{Hom}_{\mathbb{Z}[\pi]}(P,A)$$

and therefore an induced product on $\mathrm{Ext}^*_{\mathbb{Z}[\pi]}(\mathbb{Z},A)$. Convince yourself that the isomorphism above preserves products and explain the intuition (don't worry about technical exactitude).

3. * Now use homological algebra to determine $H^*(\mathbb{R}P^\infty;\mathbb{Z}_2)$ as a ring.
4. Use the previous problem to deduce the ring structure on $H^*(\mathbb{R}P^n;\mathbb{Z}_2)$ for each $n \geq 1$.
5. Let $p: Y \longrightarrow X$ be a covering space with finite fibers, say of cardinality n. Construct a "transfer homomorphism" $t: H^*(Y;A) \longrightarrow H^*(X;A)$ and show that $t \circ p^*: H^*(X;A) \longrightarrow H^*(X;A)$ is multiplication by n.
6. Let X and Y be CW complexes. Show that the interchange map

$$t: X \times Y \longrightarrow Y \times X$$

satisfies $t_*([i] \otimes [j]) = (-1)^{pq} [j] \otimes [i]$ for a p-cell of X and a q-cell of Y. Deduce that the cohomology ring $H^*(X)$ is commutative in the graded sense:

$$x \cup y = (-1)^{pq} y \cup x \quad \text{if} \quad \deg x = p \quad \text{and} \quad \deg y = q.$$

An "H-space" is a space X with a basepoint e and a product $\phi: X \times X \longrightarrow X$ such that the maps $\lambda: X \longrightarrow X$ and $\rho: X \longrightarrow X$ given by left and right multiplication by e are each homotopic to the identity map. Note that λ and ρ specify a map $X \vee X \longrightarrow X$ that is homotopic to the codiagonal or folding map ∇, which restricts to the identity on each wedge summand. The following two problems are optional review exercises.

7. If e is a nondegenerate basepoint for X, then ϕ is homotopic to a product ϕ' such that left and right multiplication by e under the product ϕ' are both identity maps.
8. Show that the product on $\pi_1(X,e)$ induced by the based map $\phi': X \times X \longrightarrow X$ agrees with the multiplication given by composition of paths and that both products are commutative.
9. For an H-space X, the following diagram is commutative:

$$\begin{array}{ccccc} X \times X & \xrightarrow{\Delta \times \Delta} & X \times X \times X \times X & \xrightarrow{\mathrm{id} \times t \times \mathrm{id}} & X \times X \times X \times X \\ \phi \downarrow & & & & \downarrow \phi \times \phi \\ X & & \xrightarrow{\Delta} & & X \times X \end{array}$$

(Check it: it is too trivial to write down.) Let X be $(n-1)$-connected, $n \geq 2$, and let $x \in H^n(X)$.

(a) Show that $\phi^*(x) = x \otimes 1 + 1 \otimes x$.

(b) Show that
$$(\Delta \times \Delta)^*(\mathrm{id} \times t \times \mathrm{id})^*(\phi \times \phi)^*(x \otimes x) = x^2 \otimes 1 + (1 + (-1)^n)(x \otimes x) + 1 \otimes x^2.$$
(c) Prove that, if n is even, then either $2(x \otimes x) = 0$ in $H^*(X \times X)$ or $x^2 \neq 0$. Deduce that S^n cannot be an H-space if n is even.

CHAPTER 19

Derivations of properties from the axioms

Returning to the axiomatic approach to cohomology, we assume given a theory on pairs of spaces and give some deductions from the axioms. This may be viewed as a dualized review of what we did in homology, and we generally omit the proofs. The only significant difference that we will encounter is in the computation of the cohomology of colimits. In a final section, we show the uniqueness of (ordinary) cohomology with coefficients in π.

Prior to that section, we make no use of the dimension axiom in this chapter. A "generalized cohomology theory" E^* is defined to be a system of functors $E^q(X, A)$ and natural transformations $\delta : E^q(A) \longrightarrow E^{q+1}(X, A)$ that satisfy all of our axioms except for the dimension axiom. Similarly, we have the notion of a generalized cohomology theory on CW pairs, and the following result holds.

THEOREM. *A cohomology theory E^* on pairs of spaces determines and is determined by its restriction to a cohomology theory E^* on pairs of CW complexes.*

1. Reduced cohomology groups and their properties

For a based space X, we define the reduced cohomology of X to be

$$\tilde{E}^q(X) = E^q(X, *).$$

There results a direct sum decomposition

$$E^*(X) \cong \tilde{E}^*(X) \oplus E^*(*)$$

that is natural with respect to based maps. For $* \in A \subset X$, the summand $E^*(*)$ maps isomorphically under the map $E^*(X) \longrightarrow E^*(A)$, and the exactness axiom implies that there is a reduced long exact sequence

$$\cdots \longrightarrow \tilde{E}^{q-1}(A) \xrightarrow{\delta} E^q(X, A) \longrightarrow \tilde{E}^q(X) \longrightarrow \tilde{E}^q(A) \longrightarrow \cdots.$$

The unreduced cohomology groups are recovered as the special cases

$$E^*(X) = \tilde{E}^*(X_+)$$

of reduced ones, and similarly for maps. Relative cohomology groups are also special cases of reduced ones.

THEOREM. *For any cofibration $i : A \longrightarrow X$, the quotient map $q : (X, A) \longrightarrow (X/A, *)$ induces an isomorphism*

$$\tilde{E}^*(X/A) = E^*(X/A, *) \cong E^*(X, A).$$

We may replace any inclusion $i : A \longrightarrow X$ by the canonical cofibration $A \longrightarrow Mi$ and then apply the result just given to obtain an isomorphism

$$E^*(X, A) \cong \tilde{E}^*(Ci).$$

THEOREM. *For a nondegenerately based space X, there is a natural isomorphism*
$$\Sigma : \tilde{E}^q(X) \cong \tilde{E}^{q+1}(\Sigma X).$$

COROLLARY. *Let $* \in A \subset X$, where $i : A \longrightarrow X$ is a cofibration between nondegenerately based spaces. In the long exact sequence*
$$\cdots \longrightarrow \tilde{E}^{q-1}(A) \xrightarrow{\delta} \tilde{E}^q(X/A) \longrightarrow \tilde{E}^q(X) \longrightarrow \tilde{E}^q(A) \longrightarrow \cdots$$
of the pair (X, A), the connecting homomorphism δ is the composite
$$\tilde{E}^{q-1}(A) \xrightarrow{\Sigma} \tilde{E}^q(\Sigma A) \xrightarrow{\partial^*} \tilde{E}^q(X/A).$$

COROLLARY. *For any n and q,*
$$\tilde{E}^q(S^n) \cong \tilde{E}^{q-n}(*).$$

2. Axioms for reduced cohomology

DEFINITION. A reduced cohomology theory $\tilde{E}^*$ consists of functors $\tilde{E}^q$ from the homotopy category of nondegenerately based spaces to the category of Abelian groups that satisfy the following axioms.

- EXACTNESS If $i : A \longrightarrow X$ is a cofibration, then the sequence
$$\tilde{E}^q(X/A) \longrightarrow \tilde{E}^q(X) \longrightarrow \tilde{E}^q(A)$$
is exact.
- SUSPENSION For each integer q, there is a natural isomorphism
$$\Sigma : \tilde{E}^q(X) \cong \tilde{E}^{q+1}(\Sigma X).$$
- ADDITIVITY If X is the wedge of a set of nondegenerately based spaces X_i, then the inclusions $X_i \longrightarrow X$ induce an isomorphism
$$\tilde{E}^*(X) \longrightarrow \prod_i \tilde{E}^*(X_i).$$
- WEAK EQUIVALENCE If $f : X \longrightarrow Y$ is a weak equivalence, then
$$f^* : \tilde{E}^*(Y) \longrightarrow \tilde{E}^*(X)$$
is an isomorphism.

The reduced form of the dimension axiom would read
$$\tilde{H}^0(S^0) = \pi \quad \text{and} \quad \tilde{H}^q(S^0) = 0 \text{ for } q \neq 0.$$

THEOREM. *A cohomology theory E^* on pairs of spaces determines and is determined by a reduced cohomology theory $\tilde{E}^*$ on nondegenerately based spaces.*

DEFINITION. A reduced cohomology theory $\tilde{E}^*$ on based CW complexes consists of functors $\tilde{E}^q$ from the homotopy category of based CW complexes to the category of Abelian groups that satisfy the following axioms.

- EXACTNESS If A is a subcomplex of X, then the sequence
$$\tilde{E}^q(X/A) \longrightarrow \tilde{E}^q(X) \longrightarrow \tilde{E}^q(A)$$
is exact.
- SUSPENSION For each integer q, there is a natural isomorphism
$$\Sigma : \tilde{E}^q(X) \cong \tilde{E}^{q+1}(\Sigma X).$$

- ADDITIVITY If X is the wedge of a set of based CW complexes X_i, then the inclusions $X_i \longrightarrow X$ induce an isomorphism
$$\tilde{E}^*(X) \longrightarrow \prod_i \tilde{E}^*(X_i).$$

THEOREM. *A reduced cohomology theory $\tilde{E}^*$ on nondegenerately based spaces determines and is determined by its restriction to a reduced cohomology theory on based CW complexes.*

THEOREM. *A cohomology theory E^* on CW pairs determines and is determined by a reduced cohomology theory $\tilde{E}^*$ on based CW complexes.*

3. Mayer-Vietoris sequences in cohomology

We have Mayer-Vietoris sequences in cohomology just like those in homology. The proofs are the same. Poincaré duality between the homology and cohomology of manifolds will be proved by an inductive comparison of homology and cohomology Mayer-Vietoris sequences. We record two preliminaries.

PROPOSITION. *For a triple (X, A, B), the following sequence is exact:*
$$\cdots E^{q-1}(A, B) \xrightarrow{\delta} E^q(X, A) \xrightarrow{j^*} E^q(X, B) \xrightarrow{i^*} E^q(A, B) \longrightarrow \cdots.$$
Here $i : (A, B) \longrightarrow (X, B)$ and $j : (X, B) \longrightarrow (X, A)$ are inclusions and δ is the composite
$$E^{q-1}(A, B) \longrightarrow E^{q-1}(A) \xrightarrow{\delta} E^q(X, A).$$

Now let $(X; A, B)$ be an excisive triad and set $C = A \cap B$.

LEMMA. *The map*
$$E^*(X, C) \longrightarrow E^*(A, C) \oplus E^*(B, C)$$
induced by the inclusions of (A, C) and (B, C) in (X, C) is an isomorphism.

THEOREM (Mayer-Vietoris sequence). *Let $(X; A, B)$ be an excisive triad and set $C = A \cap B$. The following sequence is exact:*
$$\cdots \longrightarrow E^{q-1}(C) \xrightarrow{\Delta^*} E^q(X) \xrightarrow{\phi^*} E^q(A) \oplus E^q(B) \xrightarrow{\psi^*} E^q(C) \longrightarrow \cdots.$$
Here, if $i : C \longrightarrow A$, $j : C \longrightarrow B$, $k : A \longrightarrow X$, and $\ell : B \longrightarrow X$ are the inclusions, then
$$\phi^*(\chi) = (k^*(\chi), \ell^*(\chi)) \quad \text{and} \quad \psi^*(\alpha, \beta) = i^*(\alpha) - j^*(\beta)$$
and Δ^* is the composite
$$E^{q-1}(C) \xrightarrow{\delta} E^q(A, C) \cong E^q(X, B) \longrightarrow E^q(X).$$

For the relative version, let X be contained in some ambient space Y.

THEOREM (Relative Mayer-Vietoris sequence). *The following sequence is exact:*
$$\cdots \longrightarrow E^{q-1}(Y, C) \xrightarrow{\Delta^*} E^q(Y, X) \xrightarrow{\phi^*} E^q(Y, A) \oplus E^q(Y, B) \xrightarrow{\psi^*} E^q(Y, C) \longrightarrow \cdots.$$
Here, if $i : (Y, C) \longrightarrow (Y, A)$, $j : (Y, C) \longrightarrow (Y, B)$, $k : (Y, A) \longrightarrow (Y, X)$, and $\ell : (Y, B) \longrightarrow (Y, X)$ are the inclusions, then
$$\phi^*(\chi) = (k^*(\chi), \ell^*(\chi)) \quad \text{and} \quad \psi^*(\alpha, \beta) = i^*(\alpha) - j^*(\beta)$$

and Δ^* is the composite

$$E^{q-1}(Y,C) \longrightarrow E^{q-1}(A,C) \cong E^{q-1}(X,B) \xrightarrow{\delta} E^q(Y,X).$$

COROLLARY. *The absolute and relative Mayer-Vietoris sequences are related by the following commutative diagram:*

$$\begin{array}{ccccccc}
E^{q-1}(C) & \xrightarrow{\Delta^*} & E^q(X) & \xrightarrow{\phi^*} & E^q(A) \oplus E^q(B) & \xrightarrow{\psi^*} & E^q(C) \\
\delta \downarrow & & \delta \downarrow & & \delta+\delta \downarrow & & \delta \downarrow \\
E^q(Y,C) & \xrightarrow{\Delta^*} & E^{q+1}(Y,X) & \xrightarrow{\phi^*} & E^{q+1}(Y,A) \oplus E^{q+1}(Y,B) & \xrightarrow{\psi^*} & E^{q+1}(Y,C).
\end{array}$$

4. Lim^1 and the cohomology of colimits

In this section, we let X be the union of an expanding sequence of subspaces X_i, $i \geq 0$. We shall use the additivity and weak equivalence axioms and the Mayer-Vietoris sequence to explain how to compute $E^*(X)$. The answer is more subtle than in homology because, algebraically, limits are less well behaved than colimits: they are not exact functors from diagrams of Abelian groups to Abelian groups. Rather than go into the general theory, we simply display how the "first right derived functor" lim^1 of an inverse sequence of Abelian groups can be computed.

LEMMA. *Let $f_i : A_{i+1} \longrightarrow A_i$, $i \geq 1$, be a sequence of homomorphisms of Abelian groups. Then there is an exact sequence*

$$0 \longrightarrow \lim A_i \xrightarrow{\beta} \prod_i A_i \xrightarrow{\alpha} \prod_i A_i \longrightarrow \lim{}^1 A_i \longrightarrow 0,$$

where α is the difference of the identity map and the map with coordinates f_i and β is the map whose projection to A_i is the canonical map given by the definition of a limit.

That is, we may as well define lim$^1 A_i$ to be the displayed cokernel. We then have the following result.

THEOREM. *For each q, there is a natural short exact sequence*

$$0 \longrightarrow \lim{}^1 E^{q-1}(X_i) \longrightarrow E^q(X) \xrightarrow{\pi} \lim E^q(X_i) \longrightarrow 0,$$

where π is induced by the inclusions $X_i \longrightarrow X$.

PROOF. We use the notations and constructions in the proof that homology commutes with colimits and consider the excisive triad (tel $X_i; A, B$) with $C = A \cap B$ constructed there. By the additivity axiom,

$$E^*(A) = \prod_i E^*(X_{2i}), \quad E^*(B) = \prod_i E^*(X_{2i+1}), \quad \text{and} \quad E^*(C) = \prod_i E^*(X_i).$$

We construct the following commutative diagram, whose top row is the cohomology Mayer-Vietoris sequence of the triad (tel $X_i; A, B$) and whose bottom row is an exact sequence of the sort displayed in the previous lemma.

$$\begin{array}{ccccccccc}
\cdots & \longrightarrow & E^q(\text{tel } X_i) & \longrightarrow & E^q(A) \oplus E^q(B) & \longrightarrow & E^q(C) & \longrightarrow & E^{q+1}(\text{tel } X_i) & \longrightarrow \cdots \\
 & & \cong \downarrow & & \cong \downarrow & & \cong \downarrow & & \cong \downarrow & \\
\cdots & \longrightarrow & E^q(X) & \xrightarrow{\beta'} & \prod_i E^q(X_i) & \xrightarrow{\alpha'} & \prod_i E^q(X_i) & \longrightarrow & E^{q+1}(X) & \longrightarrow \cdots \\
 & & \pi' \downarrow & & \prod(-1)^i \downarrow & & \prod_i(-1)^i \downarrow & & & \\
0 & \longrightarrow & \lim E^q(X_i) & \xrightarrow{\beta} & \prod_i E^q(X_i) & \xrightarrow{\alpha} & \prod_i E^q(X_i) & \longrightarrow & \lim{}^1 E^q(X_i) & \longrightarrow 0.
\end{array}$$

The commutativity of the bottom middle square is a comparison based on the sign used in the Mayer-Vietoris sequence. Here the map π' differs by alternating signs from the canonical map π, but this does not affect the conclusion. A chase of the diagram implies the result. □

The $\lim^1$ "error terms" are a nuisance, and it is important to know when they vanish. We say that an inverse sequence $f_i : A_{i+1} \longrightarrow A_i$ satisfies the Mittag-Leffler condition if, for each fixed i, there exists $j \geq i$ such that, for every $k > j$, the image of the composite $A_k \longrightarrow A_i$ is equal to the image of the composite $A_j \longrightarrow A_i$. For example, this holds if all but finitely many of the f_i are epimorphisms or if the A_i are all finite. As a matter of algebra, we have the following vanishing result.

LEMMA. *If the inverse sequence $f_i : A_{i+1} \longrightarrow A_i$ satisfies the Mittag-Leffler condition, then $\lim^1 A_i = 0$.*

For example, for $q < n$, the inclusion $X^n \longrightarrow X^{n+1}$ of skeleta in a CW complex induces an isomorphism $H^q(X^{n+1}; \pi) \longrightarrow H^q(X^n; \pi)$ and we conclude that the canonical map
$$H^q(X; \pi) \longrightarrow H^q(X^n; \pi)$$
is an isomorphism for $q < n$. This is needed in the proof of the uniqueness of ordinary cohomology.

5. The uniqueness of the cohomology of CW complexes

As with homology, one reason for defining ordinary cohomology with coefficients in an Abelian group π in terms of cellular cochains is the inevitability of the definition. If we assume given a theory that satisfies the axioms, we see that the cochains with coefficients in π of a CW complex X can be redefined by
$$C^n(X; \pi) = H^n(X^n, X^{n-1}; \pi),$$
with differential
$$d : C^n(X; \pi) \longrightarrow C^{n+1}(X; \pi)$$
the composite
$$H^n(X^n, X^{n-1}; \pi) \longrightarrow H^n(X^n) \xrightarrow{\delta} H^{n+1}(X^{n+1}, X^n).$$
That is, the following result holds.

THEOREM. *$C^*(X; \pi)$ as just defined is isomorphic to $\mathrm{Hom}(C_*(X), \pi)$.*

We define the reduced cochains $\tilde{C}^*(X; \pi)$ of a based space X to be the kernel of the map $C^*(X; \pi) \longrightarrow C^*(*; \pi)$ induced by the inclusion $\{*\} \longrightarrow X$. For a CW pair (X, A), we define $C^*(X, A; \pi)$ to be the kernel of the epimorphism
$$C^*(X; \pi) \longrightarrow C^*(A; \pi)$$
induced by the inclusion $A \longrightarrow X$. The analogue of the previous result for reduced and relative cochains follows directly. This leads to a uniqueness theorem for cohomology just like that for homology.

THEOREM. *There is a natural isomorphism*
$$H^*(X, A; \pi) \cong H^*(C^*(X, A; \pi))$$

under which the natural transformation δ agrees with the natural transformation induced by the connecting homomorphisms associated to the short exact sequences

$$0 \longrightarrow C^*(X, A; \pi) \longrightarrow C^*(X; \pi) \longrightarrow C^*(A; \pi) \longrightarrow 0.$$

PROOF. It suffices to obtain a natural isomorphism of reduced theories on based CW complexes X. We have seen that $\tilde{H}^q(X; \pi) \cong \tilde{H}^q(X^n; \pi)$ for $q < n$, and we obtain a diagram dual to that used in the proof of the analogue in homology by arrow reversal. We leave further details as an exercise for the reader. □

PROBLEMS

1. Complete the proof of the uniqueness theorem for cohomology.

In the following sequence of problems, we take cohomology with coefficients in a commutative ring R and we write $\otimes$ for $\otimes_R$.

2. Let A and B be subspaces of a space X. Construct a relative cup product

$$H^p(X, A) \otimes H^q(X, B) \longrightarrow H^{p+q}(X, A \cup B)$$

and show that the following diagram is commutative:

$$\begin{array}{ccc} H^p(X, A) \otimes H^q(X, B) & \longrightarrow & H^{p+q}(X, A \cup B) \\ \downarrow & & \downarrow \\ H^p(X) \otimes H^q(X) & \longrightarrow & H^{p+q}(X). \end{array}$$

The horizontal arrows are cup products; the vertical arrows are induced from $X \longrightarrow (X, A)$, and so forth.

3. Let X have a basepoint $* \in A \cap B$. Deduce a commutative diagram

$$\begin{array}{ccc} H^p(X, A) \otimes H^q(X, B) & \longrightarrow & H^{p+q}(X, A \cup B) \\ \downarrow & & \downarrow \\ \tilde{H}^p(X) \otimes \tilde{H}^q(X) & \longrightarrow & \tilde{H}^{p+q}(X). \end{array}$$

4. Let $X = A \cup B$, where A and B are contractible and $A \cap B \neq \emptyset$. Deduce that the cup product

$$\tilde{H}^p(X) \otimes \tilde{H}^q(X) \longrightarrow \tilde{H}^{p+q}(X)$$

is the zero homomorphism.

5. Let $X = \Sigma Y = Y \wedge S^1$. Deduce that the cup product

$$\tilde{H}^p(X) \otimes \tilde{H}^q(X) \longrightarrow \tilde{H}^{p+q}(X)$$

is the zero homomorphism.

Commentary: Additively, cohomology groups are "stable," in the sense that

$$\tilde{H}^p(Y) \cong \tilde{H}^{p+1}(\Sigma Y).$$

Cup products are "unstable," in the sense that they vanish on suspensions. This is an indication of how much more information they carry than the mere additive groups. The proof given by this sequence of exercises actually applies to any "multiplicative" cohomology theory, by which we understand a theory that has suitable cup products.

CHAPTER 20

The Poincaré duality theorem

The crucial starting point for applications of algebraic topology to geometric topology is the Poincaré duality theorem. It gives a tight algebraic constraint on the homology and cohomology groups of compact manifolds.

1. Statement of the theorem

It is apparent that there is a kind of duality relating the construction of homology and cohomology. In its simplest form, this is reflected by the fact that evaluation of cochains on chains gives a natural homomorphism

$$C^p(X;\pi) \otimes C_p(X;\rho) \longrightarrow \pi \otimes \rho.$$

This passes to homology and cohomology to give an evaluation pairing

$$H^p(X;\pi) \otimes H_p(X;\rho) \longrightarrow \pi \otimes \rho.$$

Taking $\pi = \rho$ to be a commutative ring R and using its product, there results a pairing

$$H^p(X;R) \otimes_R H_p(X;R) \longrightarrow R.$$

It is usually written $\langle \alpha, x \rangle$ for $\alpha \in H^p(X;R)$ and $x \in H_p(X;R)$. When R is a field and the $H_p(X;R)$ are finite dimensional vector spaces, the adjoint of this pairing is an isomorphism

$$H^p(X;R) \cong \mathrm{Hom}_R(H_p(X;R), R).$$

That is, the cohomology groups of X are the vector space duals of the homology groups of X.

Now let M be a compact manifold of dimension n. We shall study manifolds without boundary in this chapter, turning to manifolds with boundary in the next. We do not assume that M is differentiable. It is known that M can be given the structure of a finite CW complex, and its homology and cohomology groups are therefore finitely generated. When M is differentiable, it is not hard to prove this using Morse theory, but it is a deep theorem in the general topological case. We shall not go into the proof but shall take the result as known.

We have the cup product

$$H^p(M;R) \otimes H^{n-p}(M;R) \longrightarrow H^n(M;R).$$

If R is a field and M is "R-orientable," then there is an "R-fundamental class" $z \in H_n(M;R)$. The composite of the cup product and evaluation on z gives a cup product pairing

$$H^p(M;R) \otimes H^{n-p}(M;R) \longrightarrow R.$$

One version of the Poincaré duality theorem asserts that this pairing is nonsingular, so that its adjoint is an isomorphism

$$H^p(M;R) \cong \mathrm{Hom}_R(H^{n-p}(M;R), R) \cong H_{n-p}(M;R).$$

In fact, Poincaré duality does not require the commutative ring R to be a field, and it is useful to allow R-modules π as coefficients in our homology and cohomology groups. We shall gradually make sense of and prove the following theorem.

THEOREM (Poincaré duality). *Let M be a compact R-oriented n-manifold. Then, for an R-module π, there is an isomorphism*

$$D : H^p(M;\pi) \longrightarrow H_{n-p}(M;\pi).$$

We shall define the notion of an R-orientation and an R-fundamental class of a manifold in §3, and we shall later prove the following result.

PROPOSITION. *If M is a compact n-manifold, then an R-orientation of M determines and is determined by an R-fundamental class $z \in H_n(M;R)$.*

The isomorphism D is given by the adjoint of the cup product pairing determined by z, but it is more convenient to describe it in terms of the "cap product." For any space X and R-module π, there is a cap product

$$\cap : H^p(X;\pi) \otimes_R H_n(X;R) \longrightarrow H_{n-p}(X;\pi).$$

We shall define it in the next section. The isomorphism D is specified by

$$D(\alpha) = \alpha \cap z.$$

When $\pi = R$, we shall prove that the cap product, cup product, and evaluation pairing are related by the fundamental identity

$$\langle \alpha \cup \beta, x \rangle = \langle \beta, \alpha \cap x \rangle.$$

Taking $x = z$, this shows that in this case D is adjoint to the cup product pairing determined by z.

We explain a few consequences before beginning to fill in the details and proofs. Let M be a connected compact oriented ($=$ $\mathbb{Z}$-oriented) n-manifold. Taking integer coefficients, we have $D : H^p(M) \cong H_{n-p}(M)$. With $p = 0$, this shows that $H_n(M) \cong \mathbb{Z}$ with generator the fundamental class z. With $p = n$, it shows that $H^n(M) \cong \mathbb{Z}$ with generator ζ dual to z, $\langle \zeta, z \rangle = 1$. The relation between $\cup$ and $\cap$ has the following consequence.

COROLLARY. *Let $T_p \subset H^p(M)$ be the torsion subgroup. The cup product pairing $\alpha \otimes \beta \longrightarrow \langle \alpha\beta, z \rangle$ induces a nonsingular pairing*

$$H^p(M)/T_p \otimes H^{n-p}(M)/T_{n-p} \longrightarrow \mathbb{Z}.$$

PROOF. If $\alpha \in T_p$, say $r\alpha = 0$, and $\beta \in H^{n-p}(M)$, then $r(\alpha \cup \beta) = 0$ and therefore $\alpha \cup \beta = 0$ since $H^n(M) = \mathbb{Z}$. Thus the pairing vanishes on torsion elements. Since $\text{Ext}^1_{\mathbb{Z}}(\mathbb{Z}_r, \mathbb{Z}) = \mathbb{Z}_r$ and each $H_p(M)$ is finitely generated, $\text{Ext}^1_{\mathbb{Z}}(H_*(M), \mathbb{Z})$ is a torsion group. By the universal coefficient theorem, this implies that

$$H^p(M)/T_p = \text{Hom}(H_p(M), \mathbb{Z}).$$

Thus, if $\alpha \in H^p(M)$ projects to a generator of the free Abelian group $H^p(M)/T_p$, then there exists $a \in H_p(M)$ such that $\langle \alpha, a \rangle = 1$. By Poincaré duality, there exists $\beta \in H^{n-p}(M)$ such that $\beta \cap z = a$. Then

$$\langle \beta \cup \alpha, z \rangle = \langle \alpha, \beta \cap z \rangle = 1. \quad \square$$

We shall see that any simply connected manifold, such as $\mathbb{C}P^n$, is orientable. The previous result allows us to compute the cup products of $\mathbb{C}P^n$.

COROLLARY. As a graded ring, $H^*(\mathbb{C}P^n)$ is the truncated polynomial algebra $\mathbb{Z}[\alpha]/(\alpha^{n+1})$, where $\deg \alpha = 2$. That is, $H^{2q}(\mathbb{C}P^n)$ is the free Abelian group with generator α^q for $1 \leq q \leq n$.

PROOF. We know that $\mathbb{C}P^n$ is a CW complex with one $2q$-cell for each q, $0 \leq q \leq n$. Therefore the conclusion is correct additively: $H^{2q}(\mathbb{C}P^n)$ is a free Abelian group on one generator for $0 \leq q \leq n$. Moreover $\mathbb{C}P^{n-1}$ is the $(2n-1)$-skeleton of $\mathbb{C}P^n$, and the inclusion $\mathbb{C}P^{n-1} \longrightarrow \mathbb{C}P^n$ therefore induces an isomorphism $H^{2q}(\mathbb{C}P^n) \longrightarrow H^{2q}(\mathbb{C}P^{n-1})$ for $q < n$. We proceed by induction on n, the conclusion being obvious for $\mathbb{C}P^1 \cong S^2$. The induction hypothesis implies that if α generates $H^2(\mathbb{C}P^n)$, then α^q generates $H^{2q}(\mathbb{C}P^n)$ for $q < n$. By the previous result, there exists $\beta \in H^{2n-2}(\mathbb{C}P^n)$ such that $\langle \alpha \cup \beta, z \rangle = 1$. Clearly β must be a generator, so that $\beta = \pm \alpha^{n-1}$, and therefore α^n must generate $H^{2n}(\mathbb{C}P^n)$. □

In the presence of torsion in the cohomology of M, it is convenient to work with coefficients in a field. We shall see that an oriented manifold is R-oriented for any commutative ring R. The same argument as for integer coefficients gives the following more convenient nonsingular pairing result.

COROLLARY. Let M be a connected compact R-oriented n-manifold, where R is a field. Then $\alpha \otimes \beta \longrightarrow \langle \alpha \cup \beta, z \rangle$ defines a nonsingular pairing

$$H^p(M;R) \otimes_R H^{n-p}(M;R) \longrightarrow R.$$

We shall see that every manifold is $\mathbb{Z}_2$-oriented, and an argument exactly like that for $\mathbb{C}P^n$ allows us to compute the cup products in $H^*(\mathbb{R}P^n; \mathbb{Z}_2)$. We used this information in our proof of the Borsuk-Ulam theorem.

COROLLARY. As a graded ring, $H^*(\mathbb{R}P^n; \mathbb{Z}_2)$ is the truncated polynomial algebra $\mathbb{Z}_2[\alpha]/(\alpha^{n+1})$, where $\deg \alpha = 1$. That is, α^q is the non-zero element of $H^q(\mathbb{R}P^n; \mathbb{Z}_2)$ for $1 \leq q \leq n$.

2. The definition of the cap product

To define the cap product, we may as well assume that X is a CW complex, by CW approximation. The diagonal map $\Delta : X \longrightarrow X \times X$ is not cellular, but it is homotopic to a cellular map Δ'. Thus we have a chain map

$$\Delta'_* : C_*(X) \longrightarrow C_*(X \times X) \cong C_*(X) \otimes C_*(X).$$

It carries $C_n(X)$ to $\sum C_p(X) \otimes C_{n-p}(X)$. Tensoring over a commutative ring R and using that $R \cong R \otimes_R R$, we obtain

$$\Delta'_* : C_*(X;R) \longrightarrow C_*(X \times X; R) \cong C_*(X;R) \otimes_R C_*(X;R).$$

For an R-module π, we define

$$\cap : C^*(X;\pi) \otimes_R C_*(X;R) \longrightarrow C_*(X;\pi)$$

to be the composite

$$C^*(X;\pi) \otimes_R C_*(X;R)$$
$$\downarrow \text{id} \otimes \Delta'_*$$
$$C^*(X;\pi) \otimes_R C_*(X;R) \otimes_R C_*(X;R)$$
$$\downarrow \varepsilon \otimes \text{id}$$
$$\pi \otimes_R C_*(X;R) \cong C_*(X;\pi).$$

Here ε evaluates cochains on chains. Precisely, it must be interpreted as zero on $C^p(X;\pi) \otimes_R C_q(X;R)$ if $p \neq q$ and the evident evaluation map

$$\text{Hom}_R(C_p(X;R),\pi) \otimes_R C_p(X;R) \longrightarrow \pi$$

if $p = q$. Therefore the cap product is given degreewise by maps

$$\cap : C^p(X;\pi) \otimes_R C_n(X;R) \longrightarrow C_{n-p}(X;\pi).$$

To understand this, it makes sense to think in terms of $C^*(X;\pi)$ regraded by negative degrees and so thought of as a chain complex rather than a cochain complex. Our convention on cochains that $(d\alpha)(x) = (-1)^{p+1}\alpha(dx)$ for $\alpha \in C^p(X;\pi)$ and $x \in C_{p+1}(X;R)$ means that $\varepsilon \circ d = 0$, where d is the tensor product differential on the chain complex $C^*(X;\pi) \otimes_R C_*(X;R)$. That is, ε is a map of chain complexes, where π is thought of as a chain complex concentrated in degree zero, with zero differential. It follows that $\cap$ is a chain map. That is,

$$d(\alpha \cap x) = (d\alpha) \cap x + (-1)^{\deg \alpha} \alpha \cap dx.$$

Using the evident natural map from the tensor product of homologies to the homology of a tensor product, we see that $\cap$ passes to homology to induce a pairing

$$\cap : H^*(X;\pi) \otimes_R H_*(X;R) \longrightarrow H_*(X;\pi).$$

To relate the cap and cup products, recall that the latter is induced by

$$\Delta'^* : C^*(X;R) \otimes_R C^*(X;R) \cong C^*(X \times X;R) \longrightarrow C^*(X;R).$$

It is trivial that the following diagram commutes:

$$\begin{array}{ccc} C^*(X \times X;R) \otimes_R C_*(X;R) & \xrightarrow{\text{id} \otimes \Delta'_*} & C^*(X \times X;R) \otimes_R C_*(X \times X;R) \\ {\scriptstyle \Delta'^* \otimes \text{id}} \downarrow & & \downarrow {\scriptstyle \varepsilon} \\ C^*(X;R) \otimes_R C_*(X;R) & \xrightarrow{\varepsilon} & R. \end{array}$$

We may identify the chains and cochains of $X \times X$ on the top row with tensor products of chains and cochains of X. After this identification, the right-hand map ε becomes the composite

$$C^*(X;R) \otimes_R C^*(X;R) \otimes_R C_*(X;R) \otimes_R C_*(X;R)$$
$$\downarrow \text{id} \otimes t \otimes \text{id}$$
$$C^*(X;R) \otimes_R C_*(X;R) \otimes_R C^*(X;R) \otimes_R C_*(X;R)$$
$$\downarrow \varepsilon \otimes \varepsilon$$
$$R \otimes_R R \cong R.$$

Noting the agreement of signs introduced by the two maps t, we see that this composite is the same as the composite

$$C^*(X;R) \otimes_R C^*(X;R) \otimes_R C_*(X;R) \otimes_R C_*(X;R)$$
$$\downarrow t \otimes \mathrm{id} \otimes \mathrm{id}$$
$$C^*(X;R) \otimes_R C^*(X;R) \otimes_R C_*(X;R) \otimes_R C_*(X;R)$$
$$\downarrow \mathrm{id} \otimes \varepsilon \otimes \mathrm{id}$$
$$C^*(X;R) \otimes_R C_*(X;R)$$
$$\downarrow \varepsilon$$
$$R.$$

Inspecting definitions, we see that, on elements, these observations prove the fundamental identity

$$\langle \alpha \cup \beta, x \rangle = \langle \beta, \alpha \cap x \rangle.$$

For use in the proof of the Poincaré duality theorem, we observe that the cap product generalizes to relative cap products

$$\cap : H^p(X,A;\pi) \otimes_R H_n(X,A;R) \longrightarrow H_{n-p}(X;\pi)$$

and

$$\cap : H^p(X;\pi) \otimes_R H_n(X,A;R) \longrightarrow H_{n-p}(X,A;\pi)$$

for pairs (X,A). Indeed, we may assume that (X,A) is a CW pair and that Δ' restricts to a map $A \longrightarrow A \times A$ that is homotopic to the diagonal of A. Via the quotient map $X \longrightarrow X/A$, Δ' induces relative diagonal approximations

$$\Delta'_* : C_*(X,A;R) \longrightarrow C_*(X,A;R) \otimes C_*(X;R)$$

and

$$\Delta'_* : C_*(X,A;R) \longrightarrow C_*(X;R) \otimes C_*(X,A;R).$$

These combine with the evident evaluation maps to give the required relative cap products.

3. Orientations and fundamental classes

Let M be an n-manifold, not necessarily compact; the extra generality will be crucial to our proof of the Poincaré duality theorem. For $x \in M$, we can choose a coordinate chart $U \cong \mathbb{R}^n$ with $x \in U$. By excision, exactness, and homotopy invariance, we have isomorphisms

$$H_i(M, M - x) \cong H_i(U, U - x) \cong \tilde{H}_{i-1}(U - x) \cong \tilde{H}_{i-1}(S^{n-1}).$$

This holds with any coefficient group, but we agree to take coefficients in a given commutative ring R. Thus $H_i(M, M - x) = 0$ if $i \neq n$ and $H_n(M, M - x) \cong R$. We think of $H_n(M, M - x)$ as a free R-module on one generator, but the generator (which corresponds to a unit of the ring R) is unspecified. Intuitively, an R-orientation of M is a consistent choice of generators.

DEFINITION. An R-fundamental class of M at a subspace X is an element $z \in H_n(M, M - X)$ such that, for each $x \in X$, the image of z under the map

$$H_n(M, M - X) \longrightarrow H_n(M, M - x)$$

induced by the inclusion $(M, M - X) \longrightarrow (M, M - x)$ is a generator. If $X = M$, we refer to $z \in H_n(M)$ as a fundamental class of M. An R-orientation of M is an open cover $\{U_i\}$ and R-fundamental classes z_i of M at U_i such that if $U_i \cap U_j$ is non-empty, then u_i and u_j map to the same element of $H_n(M, M - U_i \cap U_j)$.

We say that M is R-orientable if it admits an R-orientation. When $R = \mathbb{Z}$, we refer to orientations and orientability. There are various equivalent ways of formulating these notions. We leave it as an exercise for the reader to reconcile the present definition of orientability with any other definition he or she may have seen.

Clearly an R-fundamental class z determines an R-orientation: given any open cover $\{U_i\}$, we take z_i to be the image of z in $H_n(M, M - U_i)$. The converse holds when M is compact. To show this, we need the following vanishing theorem, which we shall prove in the next section.

THEOREM (Vanishing). Let M be an n-manifold. For any coefficient group π, $H_i(M; \pi) = 0$ if $i > n$, and $\tilde{H}_n(M; \pi) = 0$ if M is connected and is not compact.

We can use this together with Mayer-Vietoris sequences to construct R-fundamental classes at compact subspaces from R-orientations. To avoid trivialities, we tacitly assume that $n > 0$. (The trivial case $n = 0$ forced the use of reduced homology in the statement; where arguments use reduced homology below, it is only to ensure that what we write is correct in dimension zero.)

THEOREM. Let K be a compact subset of M. Then, for any coefficient group π, $H_i(M, M - K; \pi) = 0$ if $i > n$, and an R-orientation of M determines an R-fundamental class of M at K. In particular, if M is compact, then an R-orientation of M determines an R-fundamental class of M.

PROOF. First assume that K is contained in a coordinate chart $U \cong \mathbb{R}^n$. By excision and exactness, we then have

$$H_i(M, M - K; \pi) \cong H_i(U, U - K; \pi) \cong \tilde{H}_{i-1}(U - K; \pi).$$

Since $U - K$ is open in U, the vanishing theorem implies that $\tilde{H}_{i-1}(U - K; \pi) = 0$ for $i > n$. In fact, a lemma used in the proof of the vanishing theorem will prove this directly. In this case, an R-fundamental class in $H_n(M, M - U)$ maps to an R-fundamental class in $H_n(M, M - K)$. A general compact subset K of M can be written as the union of finitely many compact subsets, each of which is contained in a coordinate chart. By induction, it suffices to prove the result for $K \cup L$ under the assumption that it holds for K, L, and $K \cap L$. With any coefficients, we have the Mayer-Vietoris sequence

$$\cdots \longrightarrow H_{i+1}(M, M - K \cap L) \xrightarrow{\Delta} H_i(M, M - K \cup L)$$
$$\xrightarrow{\psi} H_i(M, M - K) \oplus H_i(M, M - L) \xrightarrow{\phi} H_i(M, M - K \cap L) \longrightarrow \cdots.$$

The vanishing of $H_i(M, M - K \cup L; \pi)$ for $i > n$ follows directly. Now take $i = n$ and take coefficients in R. Then ψ is a monomorphism. The R-fundamental classes $z_K \in H_n(M, M - K)$ and $z_L \in H_n(M, M - L)$ determined by a given R-orientation

both map to the R-fundamental class $z_{K \cap L} \in H_n(M, M - K \cap L)$ determined by the given R-orientation. Therefore
$$\phi(z_K, z_L) = z_{K \cap L} - z_{K \cap L} = 0$$
and there exists a unique $z_{K \cup L} \in H_n(M, M - K \cup L)$ such that
$$\psi(z_{K \cup L}) = (z_K, z_L).$$
Clearly $z_{K \cup L}$ is an R-fundamental class of M at $K \cup L$. □

The vanishing theorem also implies the following dichotomy, which we have already noticed in our examples of explicit calculations.

COROLLARY. *Let M be a connected compact n-manifold, $n > 0$. Then either M is not orientable and $H_n(M; \mathbb{Z}) = 0$ or M is orientable and the map*
$$H_n(M; \mathbb{Z}) \longrightarrow H_n(M, M - x; \mathbb{Z}) \cong \mathbb{Z}$$
is an isomorphism for every $x \in M$.

PROOF. Since $M - x$ is connected and not compact, $H_n(M - x; \pi) = 0$ and thus
$$H_n(M; \pi) \longrightarrow H_n(M, M - x; \pi) \cong \pi$$
is a monomorphism for all coefficient groups π. In particular, by the universal coefficient theorem,
$$H_n(M; \mathbb{Z}) \otimes \mathbb{Z}_q \longrightarrow H_n(M, M - x; \mathbb{Z}) \otimes \mathbb{Z}_q \cong \mathbb{Z}_q$$
is a monomorphism for all positive integers q. If $H_n(M; \mathbb{Z}) \neq 0$, then $H_n(M; \mathbb{Z}) \cong \mathbb{Z}$ with generator mapped to some multiple of a generator of $H_n(M, M - x; \mathbb{Z})$. By the mod q monomorphism, the coefficient must be ± 1. □

As an aside, the corollary leads to a striking example of the failure of the naturality of the splitting in the universal coefficient theorem. Consider a connected, compact, non-orientable n-manifold M. Let $x \in M$ and write M_x for the pair $(M, M - x)$. Since M is $\mathbb{Z}_2$-orientable, the middle vertical arrow in the following diagram is an isomorphism between copies of $\mathbb{Z}_2$:

$$\begin{array}{ccccccccc}
0 & \longrightarrow & H_n(M) \otimes \mathbb{Z}_2 & \longrightarrow & H_n(M; \mathbb{Z}_2) & \longrightarrow & \mathrm{Tor}_1^{\mathbb{Z}}(H_{n-1}(M), \mathbb{Z}_2) & \longrightarrow & 0 \\
& & \downarrow 0 & & \downarrow \cong & & \downarrow 0 & & \\
0 & \longrightarrow & H_n(M_x) \otimes \mathbb{Z}_2 & \longrightarrow & H_n(M_x; \mathbb{Z}_2) & \longrightarrow & \mathrm{Tor}_1^{\mathbb{Z}}(H_{n-1}(M_x), \mathbb{Z}_2) & \longrightarrow & 0.
\end{array}$$

Clearly $H_{n-1}(M, M - x) = 0$, and the corollary gives that $H_n(M) = 0$. Thus the left and right vertical arrows are zero. If the splittings of the rows were natural, this would imply that the middle vertical arrow is also zero.

4. The proof of the vanishing theorem

Let M be an n-manifold, $n > 0$. Take all homology groups with coefficients in a given Abelian group π in this section. We must prove the intuitively obvious statement that $H_i(M) = 0$ for $i > n$ and the much more subtle statement that $H_n(M) = 0$ if M is connected and is not compact. The last statement is perhaps the technical heart of our proof of the Poincaré duality theorem.

We begin with the general observation that homology is "compactly supported" in the sense of the following result.

LEMMA. *For any space X and element $x \in H_q(X)$, there is a compact subspace K of X and an element $k \in H_q(K)$ that maps to x.*

PROOF. Let $\gamma : Y \longrightarrow X$ be a CW approximation of X and let $x = \gamma_*(y)$. If y is represented by a cycle $z \in C_q(Y)$, then z, as a finite linear combination of q-cells, is an element of $C_q(L)$ for some finite subcomplex L of Y. Let $K = \gamma(L)$ and let k be the image of the homology class represented by z. Then K is compact and k maps to x. □

We need two lemmas about open subsets of $\mathbb{R}^n$ to prove the vanishing theorem, the first of which is just a special case.

LEMMA. *If U is open in $\mathbb{R}^n$, then $H_i(U) = 0$ for $i \geq n$.*

PROOF. Let $s \in H_i(U)$, $i \geq n$. There is a compact subspace K of U and an element $k \in H_i(K)$ that maps to s. We may decompose $\mathbb{R}^n$ as a CW complex whose n-cells are small n-cubes in such a way that there is a finite subcomplex L of $\mathbb{R}^n$ with $K \subset L \subset U$. (To be precise, use a cubical grid with small enough mesh.) For $i > 0$, the connecting homomorphisms ∂ are isomorphisms in the commutative diagram

$$\begin{array}{ccc} H_{i+1}(\mathbb{R}^n, L) & \longrightarrow & H_{i+1}(\mathbb{R}^n, U) \\ \partial \downarrow & & \downarrow \partial \\ H_i(L) & \longrightarrow & H_i(U). \end{array}$$

Since $(\mathbb{R}^n, L)$ has no relative q-cells for $q > n$, the groups on the left are zero for $i \geq n$. Since s is in the image of $H_i(L)$, $s = 0$. □

LEMMA. *Let U be open in $\mathbb{R}^n$. Suppose that $t \in H_n(\mathbb{R}^n, U)$ maps to zero in $H_n(\mathbb{R}^n, \mathbb{R}^n - x)$ for all $x \in \mathbb{R}^n - U$. Then $t = 0$.*

PROOF. We prove the equivalent statement that if $s \in \tilde{H}_{n-1}(U)$ maps to zero in $\tilde{H}_{n-1}(\mathbb{R}^n - x)$ for all $x \in \mathbb{R}^n - U$, then $s = 0$. Choose a compact subspace K of U such that s is in the image of $\tilde{H}_{n-1}(K)$. Then K is contained in an open subset V whose closure $\bar{V}$ is compact and contained in U, hence s is the image of an element $r \in \tilde{H}_{n-1}(V)$. We claim that r maps to zero in $\tilde{H}_{n-1}(U)$, so that $s = 0$. Of course, r maps to zero in $\tilde{H}_{n-1}(\mathbb{R}^n - x)$ if $x \notin U$. Let T be an open contractible subset of $\mathbb{R}^n$ such that $\bar{V} \subset T$ and $\bar{T}$ is compact. For example, T could be a large enough open cube. Let $L = T - (T \cap U)$. For each $x \in \bar{L}$, choose a closed cube D that contains x and is disjoint from V. A finite set $\{D_1, \ldots, D_q\}$ of these cubes covers $\bar{L}$. Let $C_i = D_i \cap T$ and observe that $(\mathbb{R}^n - D_i) \cap T = T - C_i$. We see by induction on p that r maps to zero in $\tilde{H}_{n-1}(T - (C_1 \cup \cdots \cup C_p))$ for $0 \leq p \leq q$. This is clear if $p = 0$. For the inductive step, observe that

$$T - (C_1 \cup \cdots \cup C_p) = (T - (C_1 \cup \cdots \cup C_{p-1})) \cap (\mathbb{R}^n - D_p)$$

and that $H_n((T - (C_1 \cup \cdots \cup C_{p-1})) \cup (\mathbb{R}^n - D_p)) = 0$ by the previous lemma. Therefore the map

$$\tilde{H}_{n-1}(T - (C_1 \cup \cdots \cup C_p)) \longrightarrow \tilde{H}_{n-1}(T - (C_1 \cup \cdots \cup C_{p-1})) \oplus \tilde{H}_{n-1}(\mathbb{R}^n - D_p)$$

in the Mayer-Vietoris sequence is a monomorphism. Since $r \in \tilde{H}_{n-1}(V)$ maps to zero in the two right-hand terms, by the induction hypothesis and the contractibility

of D_p to a point $x \notin U$, it maps to zero in the left-hand term. Since
$$V \subset T - (C_1 \cup \cdots \cup C_q) \subset T \cap U \subset U,$$
this implies our claim that r maps to zero in $\tilde{H}_{n-1}(U)$. □

PROOF OF THE VANISHING THEOREM. Let $s \in H_i(M)$. We must prove that $s = 0$ if $i > n$ and if $i = n$ when M is connected and not compact. Choose a compact subspace K of M such that s is in the image of $H_i(K)$. Then K is contained in some finite union $U_1 \cup \cdots \cup U_q$ of coordinate charts, and it suffices to prove that $H_i(U_1 \cup \cdots \cup U_q)$ for the specified values of i. Inductively, using that $H_i(U) = 0$ for $i \geq n$ when U is an open subset of a coordinate chart, it suffices to prove that $H_i(U \cup V) = 0$ for the specified values of i when U is a coordinate chart and V is an open subspace of M such that $H_i(V) = 0$ for the specified values of i. We have the Mayer-Vietoris sequence
$$H_i(U) \oplus H_i(V) \longrightarrow H_i(U \cup V) \longrightarrow \tilde{H}_{i-1}(U \cap V) \longrightarrow \tilde{H}_{i-1}(U) \oplus \tilde{H}_{i-1}(V).$$
If $i > n$, the vanishing of $H_i(U \cup V)$ follows immediately. Thus assume that M is connected and not compact and consider the case $i = n$. We have $H_n(U) = 0$, $H_n(V) = 0$, and $\tilde{H}_{n-1}(U) = 0$. It follows that $H_n(U \cup V) = 0$ if and only if $i_* : \tilde{H}_{n-1}(U \cap V) \longrightarrow \tilde{H}_{n-1}(V)$ is a monomorphism, where $i : U \cap V \longrightarrow V$ is the inclusion.

We claim first that $H_n(M) \longrightarrow H_n(M, M - y)$ is the zero homomorphism for any $y \in M$. If $x \in M$ and L is a path in M connecting x to y, then the diagram

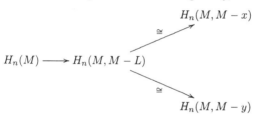

shows that if $s \in H_n(M)$ maps to zero in $H_n(M, M - x)$, then it maps to zero in $H_n(M, M - y)$. If s is in the image of $H_n(K)$ where K is compact, we may choose a point $x \in M - K$. Then the map $K \longrightarrow M \longrightarrow (M, M - x)$ factors through $(M - x, M - x)$ and therefore s maps to zero in $H_n(M, M - x)$. This proves our claim.

Now consider the following diagram, where $y \in U - U \cap V$:

Let $r \in \ker i_*$. Since $\tilde{H}_{n-1}(U) = 0$, the bottom map ∂ is an epimorphism and there exists $s \in H_n(U, U \cap V)$ such that $\partial(s) = r$. We claim that s maps to zero in $H_n(U, U - y)$ for every $y \in U - (U \cap V)$. By the previous lemma, this will imply that $s = 0$ and thus $r = 0$, so that i_* is indeed a monomorphism. Since $i_*(r) = 0$, there exists $t \in H_n(V, U \cap V)$ such that $\partial(t) = r$. Let s' and t' be the images of s and t in $H_n(U \cup V, U \cap V)$. Then $\partial(s' - t') = 0$, hence there exists $w \in H_n(U \cup V)$ that maps to $s' - t'$. Since w maps to zero in $H_n(M, M - y)$, so does $s' - t'$. Since the map $(V, U \cap V) \longrightarrow (M, M - y)$ factors through $(M - y, M - y)$, t and thus also t' maps to zero in $H_n(M, M - y)$. Therefore s' maps to zero in $H_n(M, M - y)$ and thus s maps to zero in $H_n(U, U - y)$, as claimed. □

5. The proof of the Poincaré duality theorem

Let M be an R-oriented n-manifold, not necessarily compact. Unless otherwise specified, we take homology and cohomology with coefficients in a given R-module π in this section. Remember that homology is a covariant functor with compact supports. Cohomology is a contravariant functor, and it does not have compact supports. We would like to prove the Poincaré duality theorem by inductive comparisons of Mayer-Vietoris sequences, and the opposite variance of homology and cohomology makes it unclear how to proceed. To get around this, we introduce a variant of cohomology that does have compact supports and has enough covariant functoriality to allow us to proceed by comparisons of Mayer-Vietoris sequences.

Consider the set $\mathscr{K}$ of compact subspaces K of M. This set is directed under inclusion; to conform with our earlier discussion of colimits, we may view $\mathscr{K}$ as a category whose objects are the compact subspaces K and whose maps are the inclusions between them. We define

$$H_c^q(M) = \operatorname{colim} H^q(M, M - K),$$

where the colimit is taken with respect to the homomorphisms

$$H^q(M, M - K) \longrightarrow H^q(M, M - L)$$

induced by the inclusions $(M, M - L) \subset (M, M - K)$ for $K \subset L$. This is the cohomology of M with compact supports. Intuitively, thinking in terms of singular cohomology, its elements are represented by cocycles that vanish off some compact subspace.

A map $f : M \longrightarrow N$ is said to be proper if $f^{-1}(L)$ is compact in M when L is compact in N. This holds, for example, if f is the inclusion of a closed subspace. For such f, we obtain an induced homomorphism $f^* : H_c^*(N) \longrightarrow H_c^*(M)$ in an evident way. However, we shall make no use of this contravariant functoriality.

What we shall use is a kind of covariant functoriality that will allow us to compare long exact sequences in homology and cohomology. Explicitly, for an open subspace U of M, we obtain a homomorphism $H_c^q(U) \longrightarrow H_c^q(M)$ by passage to colimits from the excision isomorphisms

$$H^q(U, U - K) \longrightarrow H^q(M, M - K)$$

for compact subspaces K of U.

For each compact subspace K of M, the R-orientation of M determines a fundamental class $z_K \in H_n(M, M - K; R)$. Taking the relative cap product with z_K, we obtain a duality homomorphism

$$D_K : H^p(M, M - K) \longrightarrow H_{n-p}(M).$$

5. THE PROOF OF THE POINCARÉ DUALITY THEOREM

If $K \subset L$, the following diagram commutes:

$$\begin{array}{ccc} H^p(M, M-K) & \longrightarrow & H^p(M, M-L) \\ & \searrow {\scriptstyle D_K} \quad \swarrow {\scriptstyle D_L} & \\ & H_{n-p}(M). & \end{array}$$

We may therefore pass to colimits to obtain a duality homomorphism

$$D : H^p_c(M) \longrightarrow H_{n-p}(M).$$

If U is open in M and is given the induced R-orientation, then the following naturality diagram commutes:

$$\begin{array}{ccc} H^p_c(U) & \xrightarrow{D} & H_{n-p}(U) \\ \downarrow & & \downarrow \\ H^p_c(M) & \xrightarrow[D]{} & H_{n-p}(M). \end{array}$$

If M itself is compact, then M is cofinal among the compact subspaces of M. Therefore $H^p_c(M) = H^p(M)$, and the present duality map D coincides with that of the Poincaré duality theorem as originally stated. We shall prove a generalization to not necessarily compact manifolds.

THEOREM (Poincaré duality). *Let M be an R-oriented n-manifold. Then $D : H^p_c(M) \longrightarrow H_{n-p}(M)$ is an isomorphism.*

PROOF. We shall prove that $D : H^p_c(U) \longrightarrow H_{n-p}(U)$ is an isomorphism for every open subspace U of M. The proof proceeds in five steps.

STEP 1. *The result holds for any coordinate chart U.*
We may take $U = M = \mathbb{R}^n$. The compact cubes K are cofinal among the compact subspaces of $\mathbb{R}^n$. For such K and for $x \in K$,

$$H^p(\mathbb{R}^n, \mathbb{R}^n - K) \cong H^p(\mathbb{R}^n, \mathbb{R}^n - x) \cong \tilde{H}^{p-1}(S^{n-1}) \cong \tilde{H}^p(S^n).$$

The maps of the colimit system defining $H^p_c(\mathbb{R}^n)$ are clearly isomorphisms. By the definition of the cap product, we see that $D : H^n(\mathbb{R}^n, \mathbb{R}^n - x) \longrightarrow H_0(\mathbb{R}_n)$ is an isomorphism. Therefore D_K is an isomorphism for every compact cube K and so $D : H^n_c(\mathbb{R}^n) \longrightarrow H_0(\mathbb{R}^n)$ is an isomorphism. □

STEP 2. *If the result holds for open subspaces U and V and their intersection, then it holds for their union.*
Let $W = U \cap V$ and $Z = U \cup V$. The compact subspaces of Z that are unions of a compact subspace K of U and a compact subspace L of V are cofinal among all of the compact subspaces of Z. For such K and L, we have the following commutative diagram with exact rows. We let $J = K \cap L$ and $N = K \cup L$, and we

write $U_K = (U, U - K)$, and so on, to abbreviate notation.

$$
\begin{array}{ccccccccc}
\longrightarrow & H^p(Z_J) & \longrightarrow & H^p(Z_K) \oplus H^p(Z_L) & \longrightarrow & H^p(Z_N) & \longrightarrow & H^{p+1}(Z_J) & \longrightarrow \\
 & \downarrow \cong & & \downarrow \cong & & \| & & \downarrow \cong & \\
\longrightarrow & H^p(W_J) & \longrightarrow & H^p(U_K) \oplus H^p(V_L) & \longrightarrow & H^p(Z_N) & \longrightarrow & H^{p+1}(W_J) & \longrightarrow \\
 & \downarrow D & & \downarrow D \oplus D & & \downarrow D & & \downarrow D & \\
\longrightarrow & H_{n-p}(W) & \longrightarrow & H_{n-p}(U) \oplus H_{n-p}(V) & \longrightarrow & H_{n-p}(Z) & \longrightarrow & H_{n-p-1}(W) & \longrightarrow
\end{array}
$$

The top row is the relative Mayer-Vietoris sequence of the triad $(Z; Z - K, Z - L)$. The middle row results from the top row by excision isomorphisms. The bottom row is the absolute Mayer-Vietoris sequence of the triad $(Z; U, V)$. The left two squares commute by naturality. The right square commutes by a diagram chase from the definition of the cap product. The entire diagram is natural with respect to pairs (K, L). We obtain a commutative diagram with exact rows on passage to colimits, and the conclusion follows by the five lemma. □

STEP 3. *If the result holds for each U_i in a totally ordered set of open subspaces $\{U_i\}$, then it holds for the union U of the U_i.*
Any compact subspace K of U is contained in a finite union of the U_i and therefore in one of the U_i. Since homology is compactly supported, it follows that $\text{colim}\, H_{n-p}(U_i) \cong H_{n-p}(U)$. On the cohomology side, we have

$$
\begin{aligned}
\text{colim}_i\, H_c^p(U_i) &= \text{colim}_i \text{colim}_{\{K | K \subset U_i\}} H^p(U_i, U_i - K) \\
&\cong \text{colim}_{\{K \subset U\}} \text{colim}_{\{i | K \subset U_i\}} H^p(U_i, U_i - K) \\
&\cong \text{colim}_{\{K \subset U\}} H^p(U, U - K) = H_c^p(U).
\end{aligned}
$$

Here the first isomorphism is an (algebraic) interchange of colimits isomorphism: both composite colimits are isomorphic to $\text{colim}\, H_c^p(U_i, U_i - K)$, where the colimit runs over the pairs (K, i) such that $K \subset U_i$. The second isomorphism holds since $\text{colim}_{\{i | K \subset U_i\}} H^p(U_i, U_i - K) \cong H^p(U, U - K)$ because the colimit is taken over a system of inverses of excision isomorphisms. The conclusion follows since a colimit of isomorphisms is an isomorphism. □

STEP 4. *The result holds if U is an open subset of a coordinate neighborhood.*
We may take $M = \mathbb{R}^n$. If U is a convex subset of $\mathbb{R}^n$, then U is homeomorphic to $\mathbb{R}^n$ and Step 1 applies. Since the intersection of two convex sets is convex, it follows by induction from Step 2 that the conclusion holds for any finite union of convex open subsets of $\mathbb{R}^n$. Any open subset U of $\mathbb{R}^n$ is the union of countably many convex open subsets. By ordering them and letting U_i be the union of the first i, we see that the conclusion for U follows from Step 3. □

STEP 5. *The result holds for any open subset U of M.*
We may as well take $M = U$. By Step 3, we may apply Zorn's lemma to conclude that there is a maximal open subset V of M for which the conclusion holds. If V is not all of M, say $x \notin V$, we may choose a coordinate chart U such that $x \in U$. By Steps 2 and 4, the result holds for $U \cup V$, contradicting the maximality of V. □

This completes the proof of the Poincaré duality theorem. □

6. The orientation cover

There is an orientation cover of a manifold that helps illuminate the notion of orientability. For the moment, we relax the requirement that the total space of a cover be connected. Here we take homology with integer coefficients.

PROPOSITION. *Let M be a connected n-manifold. Then there is a 2-fold cover $p : \tilde{M} \longrightarrow M$ such that $\tilde{M}$ is connected if and only if M is not orientable.*

PROOF. Define $\tilde{M}$ to be the set of pairs (x, α), where $x \in M$ and where $\alpha \in H_n(M, M - x) \cong \mathbb{Z}$ is a generator. Define $p(x, \alpha) = x$. If $U \subset M$ is open and $\beta \in H_n(M, M - U)$ is a fundamental class of M at U, define

$$\langle U, \beta \rangle = \{(x, \alpha) | x \in U \text{ and } \beta \text{ maps to } \alpha\}.$$

The sets $\langle U, \beta \rangle$ form a base for a topology on $\tilde{M}$. In fact, if $(x, \alpha) \in \langle U, \beta \rangle \cap \langle V, \gamma \rangle$, we can choose a coordinate neighborhood $W \subset U \cap V$ such that $x \in W$. There is a unique class $\alpha' \in H_n(M, M - W)$ that maps to α, and both β and γ map to α'. Therefore

$$\langle W, \alpha' \rangle \subset \langle U, \beta \rangle \cap \langle V, \gamma \rangle.$$

Clearly p maps $\langle U, \beta \rangle$ homeomorphically onto U and

$$p^{-1}(U) = \langle U, \beta \rangle \cup \langle U, -\beta \rangle.$$

Therefore $\tilde{M}$ is an n-manifold and p is a 2-fold cover. Moreover, $\tilde{M}$ is oriented. Indeed, if U is a coordinate chart and $(x, \alpha) \in \langle U, \beta \rangle$, then the following maps all induce isomorphisms on passage to homology:

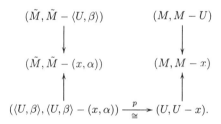

Via the diagram, $\beta \in H_n(M, M - U)$ specifies an element $\tilde{\beta} \in H_n(\tilde{M}, \tilde{M} - \langle U, \beta \rangle)$, and $\tilde{\beta}$ is independent of the choice of (x, α). These classes are easily seen to specify an orientation of $\tilde{M}$. Essentially by definition, an orientation of M is a cross section $s : M \longrightarrow \tilde{M}$: if $s(U) = \langle U, \beta \rangle$, then these β specify an orientation. Given one section s, changing the signs of the β gives a second section $-s$ such that $\tilde{M} = \text{im}(s) \amalg \text{im}(-s)$, showing that $\tilde{M}$ is not connected if M is oriented. □

The theory of covering spaces gives the following consequence.

COROLLARY. *If M is simply connected, or if $\pi_1(M)$ contains no subgroup of index 2, then M is orientable. If M is orientable, then M admits exactly two orientations.*

PROOF. If M is not orientable, then $p_*(\pi_1(\tilde{M}))$ is a subgroup of $\pi_1(M)$ of index 2. This implies the first statement, and the second statement is clear. □

We can use homology with coefficients in a commutative ring R to construct an analogous R-orientation cover. It depends on the units of R. For example, if $R = \mathbb{Z}_2$, then the R-orientation cover is the identity map of M since there is a unique unit in R. This reproves the obvious fact that any manifold is $\mathbb{Z}_2$-oriented. The evident ring homomorphism $\mathbb{Z} \longrightarrow R$ induces a natural homomorphism $H_*(X;\mathbb{Z}) \longrightarrow H_*(X;R)$, and we see immediately that an orientation of M induces an R-orientation of M for any R.

PROBLEMS

1. Prove: there is no homotopy equivalence $f : \mathbb{C}P^{2n} \longrightarrow \mathbb{C}P^{2n}$ that reverses orientation (induces multiplication by -1 on $H_{4n}(\mathbb{C}P^{2n})$).

In the problems below, M is assumed to be a compact connected n-manifold (without boundary), where $n \geq 2$.

2. Prove that if M is a Lie group, then M is orientable.
3. Prove that if M is orientable, then $H_{n-1}(M;\mathbb{Z})$ is a free Abelian group.
4. Prove that if M is not orientable, then the torsion subgroup of $H_{n-1}(M;\mathbb{Z})$ is cyclic of order 2 and $H_n(M;\mathbb{Z}_q)$ is zero if q is odd and is cyclic of order 2 if q is even. (Hint: use universal coefficients and the transfer homomorphism of the orientation cover.)
5. Let M be oriented with fundamental class z. Let $f : S^n \longrightarrow M$ be a map such that $f_*(i_n) = qz$, where $i_n \in H_n(S^n;\mathbb{Z})$ is the fundamental class and $q \neq 0$.
   (a) Show that $f_* : H_*(S^n;\mathbb{Z}_p) \longrightarrow H_*(M;\mathbb{Z}_p)$ is an isomorphism if p is a prime that does not divide q.
   (b) Show that multiplication by q annihilates $H_i(M;\mathbb{Z})$ if $1 \leq i \leq n-1$.
6. (a) Let M be a compact n-manifold. Suppose that M is homotopy equivalent to ΣY for some connected based space Y. Deduce that M has the same integral homology groups as S^n. (Hint: use the vanishing of cup products on $\tilde{H}^*(\Sigma Y)$ and Poincaré duality, treating the cases M orientable and M non-orientable separately.)
   (b) Deduce that M is homotopy equivalent to S^n. Does it follow that Y is homotopy equivalent to S^{n-1}?
7. * Essay: The singular cohomology $H^*(M;\mathbb{R})$ is isomorphic to the de Rham cohomology of M. Why is this plausible? Sketch proof?

CHAPTER 21

The index of manifolds; manifolds with boundary

The Poincaré duality theorem imposes strong constraints on the Euler characteristic of a manifold. It also leads to new invariants, most notably the index. Moreover, there is a relative version of Poincaré duality in the context of manifolds with boundary, and this leads to necessary algebraic conditions on the cohomology of a manifold that must be satisfied if it is to be a boundary. In particular, the index of a compact oriented $4n$-manifold M is zero if M is a boundary. We shall later outline the theory of cobordism, which leads to necessary *and sufficient* algebraic conditions for a manifold to be a boundary.

1. The Euler characteristic of compact manifolds

The Euler characteristic $\chi(X)$ of a space with finitely generated homology is defined by
$$\chi(X) = \sum_i (-1)^i \text{ rank } H_i(X; \mathbb{Z}).$$
The universal coefficient theorem implies that
$$\chi(X) = \sum_i (-1)^i \dim H_i(X; F)$$
for *any* field of coefficients F. Examination of the relevant short exact sequences shows that
$$\chi(X) = \sum_i (-1)^i \text{ rank } C_i(X; \mathbb{Z})$$
for *any* decomposition of X as a finite CW complex. The verifications of these statements are immediate from earlier exercises.

Now consider a compact oriented n-manifold. Recall that we take it for granted that M can be decomposed as a finite CW complex, so that each $H_i(M; \mathbb{Z})$ is finitely generated. By the universal coefficient theorem and Poincaré duality, we have
$$H_i(M; F) \cong H^i(M; F) \cong H_{n-i}(M; F)$$
for any field F. We may take $F = \mathbb{Z}_2$, and so dispense with the requirement that M be oriented. If n is odd, the summands of $\chi(M)$ cancel in pairs, and we obtain the following conclusion.

PROPOSITION. *If M is a compact manifold of odd dimension, then $\chi(M) = 0$.*

If $n = 2m$ and M is oriented, then
$$\chi(M) = \sum_{i=0}^{m-1} (-1)^i 2 \dim H_i(M) + (-1)^m \dim H_m(M)$$
for any field F of coefficients. Let us take $F = \mathbb{Q}$. Of course, we can replace homology by cohomology in the definition and formulas for $\chi(M)$. The middle dimensional cohomology group $H^m(M)$ plays a particularly important role. Recall that we have the cup product pairing
$$\phi : H^m(M) \otimes H^m(M) \longrightarrow \mathbb{Q}$$

specified by $\phi(\alpha,\beta) = \langle \alpha \cup \beta, z \rangle$. This pairing is nonsingular. Since $\alpha \cup \beta = (-1)^m \beta \cup \alpha$, it is skew symmetric if m is odd and is symmetric if m is even. When m is odd, we obtain the following conclusion.

PROPOSITION. *If M is a compact oriented n-manifold, where $n \equiv 2 \mod 4$, then $\chi(M)$ is even.*

PROOF. It suffices to prove that $\dim H^m(M)$ is even, where $n = 2m$, and this is immediate from the following algebraic observation. □

LEMMA. *Let F be a field of characteristic $\neq 2$, V be a finite dimensional vector space over F, and $\phi: V \times V \longrightarrow F$ be a nonsingular skew symmetric bilinear form. Then V has a basis $\{x_1, \ldots, x_r, y_1, \ldots, y_r\}$ such that $\phi(x_i, y_i) = 1$ for $1 \leq i \leq r$ and $\phi(z,w) = 0$ for all other pairs of basis elements (z,w). Therefore the dimension of V is even.*

PROOF. We proceed by induction on $\dim V$, and we may assume that $V \neq 0$. Since $\phi(x,y) = -\phi(y,x)$, $\phi(x,x) = 0$ for all $x \in V$. Choose $x_1 \neq 0$. Certainly there exists y_1 such that $\phi(x_1, y_1) = 1$, and x_1 and y_1 are then linearly independent. Define
$$W = \{x | \phi(x,x_1) = 0 \text{ and } \phi(x,y_1) = 0\} \subset V.$$
That is, W is the kernel of the homomorphism $\psi: V \longrightarrow F \times F$ specified by $\psi(x) = (\phi(x,x_1), \phi(x,y_1))$. Since $\psi(x_1) = (0,1)$ and $\psi(y_1) = (-1,0)$, ψ is an epimorphism. Thus $\dim W = \dim V - 2$. Since ϕ restricts to a nonsingular skew symmetric bilinear form on W, the conclusion follows from the induction hypothesis. □

2. The index of compact oriented manifolds

To study manifolds of dimension $4k$, we consider an analogue for symmetric bilinear forms of the previous algebraic lemma. Since we will need to take square roots, we will work over $\mathbb{R}$.

LEMMA. *Let V be a finite dimensional real vector space and $\phi: V \times V \longrightarrow \mathbb{R}$ be a nonsingular symmetric bilinear form. Define $q(x) = \phi(x,x)$. Then V has a basis $\{x_1, \ldots, x_r, y_1, \ldots, y_s\}$ such that $\phi(z,w) = 0$ for all pairs (z,w) of distinct basis elements, $q(x_i) = 1$ for $1 \leq i \leq r$ and $q(y_j) = -1$ for $1 \leq j \leq s$. The number $r - s$ is an invariant of ϕ, called the signature of ϕ.*

PROOF. We proceed by induction on $\dim V$, and we may assume that $V \neq 0$. Clearly $q(rx) = r^2 q(x)$. Since we can take square roots in $\mathbb{R}$, we can choose $x_1 \in V$ such that $q(x_1) = \pm 1$. Define $\psi: V \longrightarrow \mathbb{R}$ by $\psi(x) = \phi(x,x_1)$ and let $W = \ker \psi$. Since $\psi(x_1) = \pm 1$, ψ is an epimorphism and $\dim W = \dim V - 1$. Since ϕ restricts to a nonsingular symmetric bilinear form on W, the existence of a basis as specified follows directly from the induction hypothesis. Invariance means that the integer $r - s$ is independent of the choice of basis on which q takes values ± 1, and we leave the verification to the reader. □

DEFINITION. Let M be a compact oriented n-manifold. If $n = 4k$, define the index of M, denoted $I(M)$, to be the signature of the cup product form $H^{2k}(M;\mathbb{R}) \otimes H^{2k}(M;\mathbb{R}) \longrightarrow \mathbb{R}$. If $n \not\equiv 0 \mod 4$, define $I(M) = 0$.

The Euler characteristic and index are related by the following congruence.

PROPOSITION. *For any compact oriented n-manifold, $\chi(M) \equiv I(M) \mod 2$.*

2. THE INDEX OF COMPACT ORIENTED MANIFOLDS

PROOF. If n is odd, then $\chi(M) = 0$ and $I(M) = 0$. If $n \equiv 2 \bmod 4$, then $\chi(M)$ is even and $I(M) = 0$. If $n = 4k$, then $I(M) = r - s$, where $r + s = \dim H^{2k}(M;\mathbb{R}) \equiv \chi(M) \bmod 2$. □

Observe that the index of M changes sign if the orientation of M is reversed. We write $-M$ for M with the reversed orientation, and then $I(-M) = -I(M)$. We also have the following algebraic identities. Write $H^*(M) = H^*(M;\mathbb{R})$.

LEMMA. *If M and M' are compact oriented n-manifolds, then*
$$I(M \amalg M') = I(M) + I(M'),$$
where $M \amalg M'$ is given the evident orientation induced from those of M and M'.

PROOF. There is nothing to prove unless $n = 4k$, in which case
$$H^{2k}(M \amalg M') = H^{2k}(M) \times H^{2k}(M').$$
Clearly the cup product of an element of $H^*(M)$ with an element of $H^*(M')$ is zero, and the cup product form on $H^{2k}(M \amalg M')$ is given by
$$\phi((x,x'),(y,y')) = \phi(x,y) + \phi(x',y')$$
for $x,y \in H^{2k}(M)$ and $x',y' \in H^{2k}(M')$. The conclusion follows since the signature of a sum of forms is the sum of the signatures. □

LEMMA. *Let M be a compact oriented m-manifold and N be a compact oriented n-manifold. Then*
$$I(M \times N) = I(M) \cdot I(N),$$
where $M \times N$ is given the orientation induced from those of M and N.

PROOF. We must first make sense of the induced orientation on $M \times N$. For CW pairs (X,A) and (Y,B), we have an identification of CW complexes
$$(X \times Y)/(X \times B \cup A \times Y) \cong (X/A) \wedge (Y/B)$$
and therefore an isomorphism
$$C_*(X \times Y, X \times B \cup A \times Y) \cong C_*(X,A) \otimes C_*(Y,B).$$
This implies a relative Künneth theorem for arbitrary pairs (X,A) and (Y,B). For subspaces $K \subset M$ and $L \subset N$,
$$(M \times N, M \times N - K \times L) = (M \times N, M \times (N - L) \cup (M - K) \times N).$$
In particular, for points $x \in M$ and $y \in Y$,
$$(M \times N, M \times N - (x,y)) = (M \times N, M \times (N - y) \cup (M - x) \times N).$$
Therefore fundamental classes z_K of M at K and z_L of N at L determine a fundamental class $z_{K \times L}$ of $M \times N$ at $K \times L$. In particular, the image under $H_m(M) \otimes H_n(N) \longrightarrow H_{m+n}(M \times N)$ of the tensor product of fundamental classes of M and N is a fundamental class of $M \times N$.

Turning to the claimed product formula, we see that there is nothing to prove unless $m + n = 4k$, in which case
$$H^{2k}(M \times N) = \sum_{i+j=2k} H^i(M) \otimes H^j(N).$$
The cup product form is given by
$$\phi(x \otimes y, x' \otimes y') = (-1)^{(\deg y)(\deg x') + mn} \langle x \cup x', z_M \rangle \langle y \cup y', z_N \rangle$$

The cup product form is given by

$$\phi(x \otimes y, x' \otimes y') = (-1)^{(\deg y)(\deg x') + mn} \langle x \cup x', z_M \rangle \langle y \cup y', z_N \rangle$$

for $x, x' \in H^*(M)$ and $y, y' \in H^*(N)$. If m and n are odd, then the signature of this form is zero. If m and n are even, then this form is the sum of the tensor product of the cup product forms on the middle dimensional cohomology groups of M and N and a form whose signature is zero. Here, if m and n are congruent to 2 mod 4, the signature is zero since the lemma of the previous section implies that the signature of the tensor product of two skew symmetric forms is zero. When m and n are congruent to 0 mod 4, the conclusion holds since the signature of the tensor product of two symmetric forms is the product of their signatures. We leave the detailed verifications of these algebraic statements as exercises for the reader. □

3. Manifolds with boundary

Let $\mathbb{H}^n = \{(x_1, \ldots, x_n) | x_n \geq 0\}$ be the upper half-plane in $\mathbb{R}^n$. Recall that an n-manifold with boundary is a Hausdorff space M having a countable basis of open sets such that every point of M has a neighborhood homeomorphic to an open subset of $\mathbb{H}^n$. A point x is an interior point if it has a neighborhood homeomorphic to an open subset of $\mathbb{H}^n - \partial \mathbb{H}^n \cong \mathbb{R}^n$; otherwise it is a boundary point. It is a fact called "invariance of domain" that if U and V are homeomorphic subspaces of $\mathbb{R}^n$ and U is open, then V is open. Therefore, a homeomorphism of an open subspace of $\mathbb{H}^n$ onto an open subspace of $\mathbb{H}^n$ carries boundary points to boundary points.

We denote the boundary of an n-manifold M by ∂M. Thus M is a manifold without boundary if ∂M is empty; M is said to be closed if, in addition, it is compact. The space ∂M is an $(n-1)$-manifold without boundary.

It is a fundamental question in topology to determine which closed manifolds are boundaries. The question makes sense with varying kinds of extra structure. For example, we can ask whether or not a smooth (= differentiable) closed manifold is the boundary of a smooth manifold (with the induced smooth structure). Numerical invariants in algebraic topology give criteria. One such criterion is given by the following consequence of the Poincaré duality theorem. Remember that $\chi(M) = 0$ if M is a closed manifold of odd dimension.

PROPOSITION. *If $M = \partial W$, where W is a compact $(2m+1)$-manifold, then $\chi(M) = 2\chi(W)$.*

PROOF. The product $W \times I$ is a $(2m+2)$-manifold with

$$\partial(W \times I) = (W \times \{0\}) \cup (M \times I) \cup (W \times \{1\}).$$

Let $U = \partial(W \times I) - (W \times \{1\})$ and $V = \partial(W \times I) - (W \times \{0\})$. Then U and V are open subsets of $\partial(W \times I)$. Clearly U and V are both homotopy equivalent to W and $U \cap V$ is homotopy equivalent to M. We have the Mayer-Vietoris sequence

$$\begin{array}{ccccccc}
H_{i+1}(U \cup V) & \longrightarrow & H_i(U \cap V) & \longrightarrow & H_i(U) \oplus H_i(V) & \longrightarrow & H_i(U \cup V) \\
\| & & \downarrow \cong & & \downarrow \cong & & \| \\
H_{i+1}(\partial(W \times I)) & \longrightarrow & H_i(M) & \longrightarrow & H_i(W) \oplus H_i(W) & \longrightarrow & H_i(\partial(W \times I)).
\end{array}$$

Therefore $2\chi(W) = \chi(M) + \chi(\partial(W \times I))$. However, $\chi(\partial(W \times I)) = 0$ since $\partial(W \times I)$ is a closed manifold of odd dimension. □

COROLLARY. *If $M = \partial W$ for a compact manifold W, then $\chi(M)$ is even.*

For example, since $\chi(\mathbb{R}P^{2m}) = 1$ and $\chi(\mathbb{C}P^n) = n+1$, this criterion shows that $\mathbb{R}P^{2m}$ and $\mathbb{C}P^{2m}$ cannot be boundaries. Notice that we have proved that these are not boundaries of topological manifolds, let alone of smooth ones.

4. Poincaré duality for manifolds with boundary

The index gives a more striking criterion: if a closed oriented $4k$-manifold M is the boundary of a (topological) manifold, then $I(M) = 0$. To prove this, we must first obtain a relative form of the Poincaré duality theorem applicable to manifolds with boundary.

We let M be an n-manifold with boundary, $n > 0$, throughout this section, and we let R be a given commutative ring. We say that M is R-orientable (or orientable if $R = \mathbb{Z}$) if its interior $\mathring{M} = M - \partial M$ is R-orientable; similarly, an R-orientation of M is an R-orientation of its interior. To study these notions, we shall need the following result, which is intuitively clear but is somewhat technical to prove. In the case of smooth manifolds, it can be seen in terms of inward-pointing unit vectors of the normal line bundle of the embedding $\partial M \longrightarrow M$.

THEOREM (Topological collaring). *There is an open neighborhood V of ∂M in M such that the identification $\partial M = \partial M \times \{0\}$ extends to a homeomorphism $V \cong \partial M \times [0,1)$.*

It follows that the inclusion $\mathring{M} \longrightarrow M$ is a homotopy equivalence and the inclusion $\partial M \longrightarrow M$ is a cofibration. We take homology with coefficients in R in the next two results.

PROPOSITION. *An R-orientation of M determines an R-orientation of ∂M.*

PROOF. Consider a coordinate chart U of a point $x \in \partial M$. If $\dim M = n$, then U is homeomorphic to an open half-disk in $\mathbb{H}^n$. Let $V = \partial U = U \cap \partial M$ and let $y \in \mathring{U} = U - V$. We have the following chain of isomorphisms:

$$\begin{aligned}
H_n(\mathring{M}, \mathring{M} - \mathring{U}) &\cong H_n(\mathring{M}, \mathring{M} - y) \\
&\cong H_n(M, M - y) \\
&\cong H_n(M, M - \mathring{U}) \\
&\xrightarrow{\partial} H_{n-1}(M - \mathring{U}, M - U) \\
&\cong H_{n-1}(M - \mathring{U}, (M - \mathring{U}) - x) \\
&\cong H_{n-1}(\partial M, \partial M - x) \\
&\cong H_{n-1}(\partial M, \partial M - V).
\end{aligned}$$

The first and last isomorphisms are restrictions of the sort that enter into the definition of an R-orientation, and the third isomorphism is similar. We see by use of a small boundary collar that the inclusion $(\mathring{M}, \mathring{M} - y) \longrightarrow (M, M - y)$ is a homotopy equivalence, and that gives the second isomorphism. The connecting homomorphism is that of the triple $(M, M - \mathring{U}, M - U)$ and is an isomorphism since $H_*(M, M - U) \cong H_*(M, M) = 0$. The isomorphism that follows comes from the observation that the inclusion $(M - \mathring{U}) - x \longrightarrow M - U$ is a homotopy equivalence, and the next to last isomorphism is given by excision of $\mathring{M} - \mathring{U}$. The conclusion is an easy consequence of these isomorphisms. □

PROPOSITION. *If M is compact and R-oriented and $z_{\partial M} \in H_{n-1}(\partial M)$ is the fundamental class determined by the induced R-orientation on ∂M, then there is a unique element $z \in H_n(M, \partial M)$ such that $\partial z = z_{\partial M}$; z is called the R-fundamental class determined by the R-orientation of M.*

PROOF. Since $\overset{\circ}{M}$ is a non-compact manifold without boundary and $\overset{\circ}{M} \longrightarrow M$ is a homotopy equivalence, $H_n(M) \cong H_n(\overset{\circ}{M}) = 0$ by the vanishing theorem. Therefore $\partial : H_n(M, \partial M) \longrightarrow H_{n-1}(\partial M)$ is a monomorphism. Let V be a boundary collar and let $N = M - V$. Then N is a closed subspace and a deformation retract of the R-oriented open manifold $\overset{\circ}{M}$, and we have

$$H_n(\overset{\circ}{M}, \overset{\circ}{M} - N) \cong H_n(M, M - \overset{\circ}{M}) = H_n(M, \partial M).$$

Since M is compact, N is a compact subspace of $\overset{\circ}{M}$. Therefore the R-orientation of $\overset{\circ}{M}$ determines a fundamental class in $H_n(\overset{\circ}{M}, \overset{\circ}{M} - N)$. Let z be its image in $H_n(M, \partial M)$. Then z restricts to a generator of $H_n(M, M - y) \cong H_n(\overset{\circ}{M}, \overset{\circ}{M} - y)$ for every $y \in \overset{\circ}{M}$. Via naturality diagrams and the chain of isomorphisms in the previous proof, we see that ∂z restricts to a generator of $H_{n-1}(\partial M, \partial M - x)$ for all $x \in \partial M$ and is the fundamental class determined by the R-orientation of ∂M. □

THEOREM (Relative Poincaré duality). *Let M be a compact R-oriented n-manifold with R-fundamental class $z \in H_n(M, \partial M; R)$. Then, with coefficients taken in any R-module π, capping with z specifies duality isomorphisms*

$$D : H^p(M, \partial M) \longrightarrow H_{n-p}(M) \quad \text{and} \quad D : H^p(M) \longrightarrow H_{n-p}(M, \partial M).$$

PROOF. The following diagram commutes by inspection of definitions:

$$\begin{array}{ccccccc}
H^{p-1}(\partial M) & \longrightarrow & H^p(M, \partial M) & \longrightarrow & H^p(M) & \longrightarrow & H^p(\partial M) \\
\downarrow D & & \downarrow D & & \downarrow D & & \downarrow D \\
H_{n-p}(\partial M) & \longrightarrow & H_{n-p}(M) & \longrightarrow & H_{n-p}(M, \partial M) & \longrightarrow & H_{n-p-1}(\partial M).
\end{array}$$

Here D for ∂M is obtained by capping with ∂z and is an isomorphism. By the five lemma, it suffices to prove that $D : H^p(M) \longrightarrow H_{n-p}(M, \partial M)$ is an isomorphism. To this end, let $N = M \cup_{\partial M} M$ be the "double" of M and let M_1 and M_2 be the two copies of M in N. Clearly N is a compact manifold without boundary, and it is easy to see that N inherits an R-orientation from the orientation on M_1 and the negative of the orientation on M_2. Of course, $\partial M = M_1 \cap M_2$. If U is the union of M_1 and a boundary collar in M_2 and V is the union of M_2 and a boundary collar in M_1, then we have a Mayer-Vietoris sequence for the triad $(N; U, V)$. Using the evident equivalences of U with M_1, V with M_2, and $U \cap V$ with ∂M, this gives the exact sequence in the top row of the following commutative diagram. The bottom row is the exact sequence of the pair $(N, \partial M)$, and the isomorphism results from the homeomorphism $N/\partial M \cong (M_1/\partial M) \vee (M_2/\partial M)$; we abbreviate $N_1 = (M_1, \partial M)$

and $N_2 = (M_2, \partial M)$:

$$\begin{array}{ccccccc}
H^p(N) & \longrightarrow & H^p(M_1) \oplus H^p(M_2) & \xrightarrow{\psi} & H^p(\partial M) & \xrightarrow{\Delta} & H^{p+1}(N) \\
\downarrow D & & \downarrow D\oplus D & & \downarrow D & & \downarrow D \\
H_{n-p}(N) & \longrightarrow & H_{n-p}(N_1) \oplus H_{n-p}(N_2) & \longrightarrow & H_{n-p-1}(\partial M) & \longrightarrow & H_{n-p-1}(N) \\
\| & & \downarrow \cong & & \| & & \| \\
H_{n-p}(N) & \longrightarrow & H_{n-p}(N, \partial M) & \longrightarrow & H_{n-p-1}(\partial M) & \longrightarrow & H_{n-p-1}(N).
\end{array}$$

The top left square commutes by naturality. In the top middle square, we have $\psi(x,y) = i_1^*(x) - i_2^*(y)$, where $i_1 : \partial M \longrightarrow M_1$ and $i_2 : \partial M \longrightarrow M_2$ are the inclusions. Since D for M_2 is the negative of D for M_1 under the identifications with M, the commutativity of this square follows from the relation $D \circ i^* = \partial \circ D :$ $H^p(M) \longrightarrow H_{n-p-1}(\partial M)$, $i : \partial M \longrightarrow M$, which holds by inspection of definitions. For the top right square, Δ is the top composite in the diagram

$$\begin{array}{ccccc}
H^p(\partial M) & \xrightarrow{\delta} & H^{p+1}(M_1, \partial M) \cong H^{p+1}(N, M_2) & \longrightarrow & H^{p+1}(N) \\
\downarrow D & & \downarrow D & & \downarrow D \\
H_{n-p-1}(\partial M) & \xrightarrow{i_{1*}} & H_{n-p-1}(M_1) & \longrightarrow & H_{n-p-1}(N).
\end{array}$$

The right square commutes by naturality, and $D \circ \delta = i_{1*} \circ D$ by inspection of definitions. By the five lemma, since the duality maps D for N and ∂M are isomorphisms, both maps D between direct summands must be isomorphisms. The conclusion follows. $\square$

5. The index of manifolds that are boundaries

We shall prove the following theorem.

THEOREM. *If M is the boundary of a compact oriented $(4k+1)$-manifold, then $I(M) = 0$.*

We first give an algebraic criterion for the vanishing of the signature of a form and then show that the cup product form on the middle dimensional cohomology of M satisfies the criterion.

LEMMA. *Let W be a n-dimensional subspace of a 2n-dimensional real vector space V. Let $\phi : V \times V \longrightarrow \mathbb{R}$ be a nonsingular symmetric bilinear form such that $\phi : W \times W \longrightarrow \mathbb{R}$ is identically zero. Then the signature of ϕ is zero.*

PROOF. Let r and s be as in the definition of the signature. Then $r + s = 2n$ and we must show that $r = s$. We prove that $r \geq n$. Applied to the form $-\phi$, this will also give that $s \geq n$, implying the conclusion. We proceed by induction on n. Let $\{x_1, \ldots, x_n, z_1, \ldots, z_n\}$ be a basis for V, where $\{x_1, \ldots, x_n\}$ is a basis for W. Define $\theta : V \longrightarrow \mathbb{R}^n$ and $\psi : V \longrightarrow \mathbb{R}^n$ by

$$\theta(x) = (\phi(x, x_1), \ldots, \phi(x, x_n)) \quad \text{and} \quad \psi(x) = (\phi(x, z_1), \ldots, \phi(x, z_n)).$$

Since ϕ is nonsingular, $\ker \theta \cap \ker \psi = 0$. Since $\ker \theta$ and $\ker \psi$ each have dimension at least n, neither can have dimension more than n and θ and ψ must both be epimorphisms. Choose y_1 such that $\theta(y_1) = (1, 0, \ldots, 0)$. Let $q(x) = \phi(x, x)$ and note that $q(x) = 0$ if $x \in W$. Since $q(x_1) = 0$ and $\phi(x_1, y_1) = 1$, $q(ax_1 + y_1) = 2a + q(y_1)$ for $a \in \mathbb{R}$. Taking $a = (1 - q(y_1))/2$, we find $q(ax_1 + y_1) = 1$. If $n = 1$, this gives $r \geq 1$ and completes the proof. If $n > 1$, define $\omega : V \longrightarrow \mathbb{R}^2$ by $\omega(x) = (\phi(x, x_1), \phi(x, y_1))$. Since $\omega(x_1) = (0, 1)$ and $\omega(y_1) = (1, q(y_1))$, ω is an epimorphism. Let $V' = \ker \omega$ and let $W' \subset V'$ be the span of $\{x_2, \ldots, x_n\}$. The restriction of ϕ to V' satisfies the hypothesis of the lemma, and the induction hypothesis together with the construction just given imply that $r \geq n$. □

Take homology and cohomology with coefficients in $\mathbb{R}$.

LEMMA. *Let $M = \partial W$, where W is a compact oriented $(4k+1)$-manifold, and let $i : M \longrightarrow W$ be the inclusion. Let $\phi : H^{2k}(M) \otimes H^{2k}(M) \longrightarrow \mathbb{R}$ be the cup product form. Then the image of $i^* : H^{2k}(W) \longrightarrow H^{2k}(M)$ is a subspace of half the dimension of $H^{2k}(M)$ on which ϕ is identically zero.*

PROOF. Let $z \in H_{4k+1}(W, M)$ be the fundamental class. For $\alpha, \beta \in H^{2k}(W)$,
$$\phi(i^*(\alpha), i^*(\beta)) = \langle i^*(\alpha \cup \beta), \partial z \rangle = \langle \alpha \cup \beta, i_* \partial z \rangle = 0$$
since $i_* \partial = 0$ by the long exact sequence of the pair (W, M). Thus ϕ is identically zero on $\operatorname{im} i^*$. The commutative diagram with exact rows

$$\begin{array}{ccccc} H^{2k}(W) & \xrightarrow{i^*} & H^{2k}(M) & \xrightarrow{\delta} & H^{2k+1}(W, M) \\ {\scriptstyle D}\downarrow & & {\scriptstyle D}\downarrow & & {\scriptstyle D}\downarrow \\ H_{2k+1}(W, M) & \xrightarrow{\partial} & H_{2k}(M) & \xrightarrow{i_*} & H_{2k}(W) \end{array}$$

implies that $H^{2k}(M) \cong \operatorname{im} i^* \oplus \operatorname{im} \delta \cong \operatorname{im} i^* \oplus \operatorname{im} i_*$. Since i^* and i_* are dual homomorphisms, $\operatorname{im} i^*$ and $\operatorname{im} i_*$ are dual vector spaces and thus have the same dimension. □

PROBLEMS

Let M be a compact connected n-manifold with boundary ∂M, where $n \geq 2$.

1. Prove: ∂M is not a retract of M.
2. Prove: if M is contractible, then ∂M has the homology of a sphere.
3. Assume that M is orientable. Let $n = 2m + 1$ and let K be the kernel of the homomorphism $H_m(\partial M) \longrightarrow H_m(M)$ induced by the inclusion, where homology is taken with coefficients in a field. Prove: $\dim H_m(\partial M) = 2 \dim K$.

Let $n = 3$ in the rest of the problems.

4. Prove: if M is orientable, ∂M is empty, and $H_1(M; \mathbb{Z}) = 0$, then M has the same homology groups as a 3-sphere.
5. Prove: if M is nonorientable and ∂M is empty, then $H_1(M; \mathbb{Z})$ is infinite.

(Hint for the last three problems: use the standard classification of closed 2-manifolds and think about first homology groups.)

6. Prove: if M is orientable and $H_1(M; \mathbb{Z}) = 0$, then ∂M is a disjoint union of 2-spheres.
7. Prove: if M is orientable, $\partial M \neq \phi$, and ∂M contains no 2-spheres, then $H_1(M; \mathbb{Z})$ is infinite.
8. Prove: if M is nonorientable and ∂M contains no 2-spheres and no projective planes, then $H_1(M; \mathbb{Z})$ is infinite.

CHAPTER 22

Homology, cohomology, and $K(\pi,n)$s

We have given an axiomatic definition of ordinary homology and cohomology, and we have shown how to realize the axioms by means of either cellular or singular chain and cochain complexes. We here give a homotopical way of constructing ordinary theories that makes no use of chains, whether cellular or singular. We also show how to construct cup and cap products homotopically. This representation of homology and cohomology in terms of Eilenberg-MacLane spaces is the starting point of the modern approach to homology and cohomology theory, and we shall indicate how theories that do not satisfy the dimension axiom can be represented. We shall also describe Postnikov systems, which give a way to approximate general (simple) spaces by weakly equivalent spaces built up out of Eilenberg-MacLane spaces. This is conceptually dual to the way that CW complexes allow the approximation of spaces by weakly equivalent spaces built up out of spheres. Finally, we present the important notion of cohomology operations and relate them to the cohomology of Eilenberg-MacLane spaces.

1. $K(\pi,n)$s and homology

Recall that a reduced homology theory on based CW complexes is a sequence of functors $\tilde{E}_q$ from the homotopy category of based CW complexes to the category of Abelian groups. Each $\tilde{E}_q$ must satisfy the exactness and additivity axioms, and there must be a natural suspension isomorphism. Up to isomorphism, ordinary reduced homology with coefficients in π is characterized as the unique such theory that satisfies the dimension axiom: $\tilde{E}_0(S^0) = \pi$ and $\tilde{E}_q(S^0) = 0$ if $q \neq 0$. We proceed to construct such a theory homotopically.

For based spaces X and Y, we let $[X,Y]$ denote the set of based homotopy classes of based maps $X \longrightarrow Y$. Recall that we require Eilenberg-MacLane spaces $K(\pi,n)$ to have the homotopy types of CW complexes and that, up to homotopy equivalence, there is a unique such space for each n and π. By a result of Milnor, if X has the homotopy type of a CW complex, then so does ΩX. By the Whitehead theorem, we therefore have a homotopy equivalence

$$\tilde{\sigma} : K(\pi,n) \longrightarrow \Omega K(\pi,n+1).$$

This map is the adjoint of a map

$$\sigma : \Sigma K(\pi,n) \longrightarrow K(\pi,n+1).$$

We may take the smash product of the map σ with a based CW complex X and use the suspension homomorphism on homotopy groups to obtain maps

$$\pi_{q+n}(X \wedge K(\pi,n)) \xrightarrow{\Sigma} \pi_{q+n+1}(\Sigma(X \wedge K(\pi,n)))$$
$$= \pi_{q+n+1}(X \wedge \Sigma K(\pi,n)) \xrightarrow{(\mathrm{id}\wedge\sigma)_*} \pi_{q+n+1}(X \wedge K(\pi,n+1)).$$

THEOREM. *For CW complexes X, Abelian groups π and integers $n \geq 0$, there are natural isomorphisms*

$$\tilde{H}_q(X;\pi) \cong \operatorname{colim}_n \pi_{q+n}(X \wedge K(\pi,n)).$$

It suffices to verify the axioms, and the dimension axiom is clear. If $X = S^0$, then $X \wedge K(\pi,n) = K(\pi,n)$. Here the homotopy groups in the colimit system are zero if $q \neq 0$, and, if $q = 0$, the colimit runs over a sequence of isomorphisms between copies of π.

The verifications of the rest of the axioms are exercises in the use of the homotopy excision and Freudenthal suspension theorems, and it is worthwhile to carry out these exercises in greater generality.

DEFINITION. A prespectrum is a sequence of based spaces T_n, $n \geq 0$, and based maps $\sigma : \Sigma T_n \longrightarrow T_{n+1}$.

The example at hand is the Eilenberg-Mac Lane prespectrum $\{K(\pi,n)\}$. Another example is the "suspension prespectrum" $\{\Sigma^n X\}$ of a based space X; the required maps $\Sigma(\Sigma^n X) \longrightarrow \Sigma^{n+1} X$ are the evident identifications. When $X = S^0$, this is called the sphere prespectrum.

THEOREM. *Let $\{T_n\}$ be a prespectrum such that T_n is $(n-1)$-connected and of the homotopy type of a CW complex for each n. Define*

$$\tilde{E}_q(X) = \operatorname{colim}_n \pi_{q+n}(X \wedge T_n),$$

where the colimit is taken over the maps

$$\pi_{q+n}(X \wedge T_n) \xrightarrow{\Sigma} \pi_{q+n+1}(\Sigma(X \wedge T_n)) \cong \pi_{q+n+1}(X \wedge \Sigma T_n) \xrightarrow{\operatorname{id} \wedge \sigma} \pi_{q+n+1}(X \wedge T_{n+1}).$$

Then the functors $\tilde{E}_q$ define a reduced homology theory on based CW complexes.

PROOF. Certainly the $\tilde{E}$ are well defined functors from the homotopy category of based CW complexes to the category of Abelian groups. We must verify the exactness, additivity, and suspension axioms. Without loss of generality, we may take the T_n to be CW complexes with one vertex and no other cells of dimension less than n. Then $X \wedge T_n$ is a quotient complex of $X \times T_n$, and it too has one vertex and no other cells of dimension less than n. In particular, it is $(n-1)$-connected.

If A is a subcomplex of X, then the homotopy excision theorem implies that the quotient map

$$(X \wedge T_n, A \wedge T_n) \longrightarrow ((X \wedge T_n)/(A \wedge T_n), *) \cong ((X/A) \wedge T_n, *)$$

is a $(2n-1)$-equivalence. We may restrict to terms with $n > q - 1$ in calculating $\tilde{E}_q(X)$, and, for such q, the long exact sequence of homotopy groups of the pair $(X \wedge T_n, A \wedge T_n)$ gives that the sequence

$$\pi_{q+n}(A \wedge T_n) \longrightarrow \pi_{q+n}(X \wedge T_n) \longrightarrow \pi_{q+n}((X/A) \wedge T_n)$$

is exact. Since passage to colimits preserves exact sequences, this proves the exactness axiom.

We need some preliminaries to prove the additivity axiom.

DEFINITION. Define the weak product $\prod_i^w Y_i$ of a set of based spaces Y_i to be the subspace of $\prod_i Y_i$ consisting of those points all but finitely many of whose coordinates are basepoints.

LEMMA. *For a set of based spaces $\{Y_i\}$, the canonical map*
$$\sum_i \pi_q(Y_i) \longrightarrow \pi_q(\prod_i^w Y_i)$$
is an isomorphism.

PROOF. The homotopy groups of $\prod_i^w Y_i$ are the colimits of the homotopy groups of the finite subproducts of the Y_i, and the conclusion follows. □

LEMMA. *If $\{Y_i\}$ is a set of based CW complexes, then $\prod_i^w Y_i$ is a CW complex whose cells are the cells of the finite subproducts of the Y_i. If each Y_i has a single vertex and no q-cells for $q < n$, then the $(2n-1)$-skeleton of $\prod_i^w Y_i$ coincides with the $(2n-1)$-skeleton of $\bigvee_i Y_i$. Therefore the inclusion $\bigvee_i Y_i \longrightarrow \prod_i^w Y_i$ is a $(2n-1)$-equivalence.*

Returning to the proof of the additivity axiom, suppose given based CW complexes X_i and consider the natural map
$$(\bigvee_i X_i) \wedge T_n \cong \bigvee_i (X_i \wedge T_n) \longrightarrow \prod_i^w (X_i \wedge T_n).$$
It induces isomorphisms on π_{q+n} for $q < n-1$, and the additivity axiom follows.

Finally, we must prove the suspension axiom. We have the suspension map
$$\pi_{q+n}(X \wedge T_n) \xrightarrow{\Sigma} \pi_{q+n+1}(\Sigma(X \wedge T_n)) \cong \pi_{q+n+1}((\Sigma X) \wedge T_n).$$
By the Freudenthal suspension theorem, it is an isomorphism for $q < n-1$. Keeping track of suspension coordinates and their permutation, we easily check that these maps commute with the maps defining the colimit systems for X and for ΣX. Therefore they induce a natural suspension isomorphism
$$\tilde{E}_q(X) \cong \tilde{E}_{q+1}(\Sigma X).$$
This completes the proof of the theorem. □

EXAMPLE. Applying the theorem to the sphere prespectrum, we find that the stable homotopy groups $\pi_q^s(X)$ give the values of a reduced homology theory; it is called "stable homotopy theory."

2. $K(\pi, n)$s and cohomology

The homotopical description of ordinary cohomology theories is both simpler and more important to the applications than the homotopical description of ordinary homology theories.

THEOREM. *For CW complexes X, Abelian groups π, and integers $n \geq 0$, there are natural isomorphisms*
$$\tilde{H}^n(X; \pi) \cong [X, K(\pi, n)].$$

The dimension axiom is built into the definition of $K(\pi, n)$, as we see by taking $X = S^0$. As in homology, it is worthwhile to carry out the verification of the remaining axioms in greater generality. We first state some properties of the functor $[-, Z]$ on based CW complexes that is "represented" by a based space Z.

LEMMA. *For any based space Z, the functor $[X, Z]$ from based CW complexes X to pointed sets satisfies the following properties.*
- *HOMOTOPY If $f \simeq g : X \longrightarrow Y$, then $f^* = g^* : [Y, Z] \longrightarrow [X, Z]$.*

- EXACTNESS   If A is a subcomplex of X, then the sequence
$$[X/A, Z] \longrightarrow [X, Z] \longrightarrow [A, Z]$$
is exact.
- ADDITIVITY   If X is the wedge of a set of based CW complexes X_i, then the inclusions $X_i \longrightarrow X$ induce an isomorphism
$$[X, Z] \longrightarrow \prod [X_i, Z].$$

If Z has a multiplication $\phi : Z \times Z \longrightarrow Z$ such that the basepoint $*$ of Z is a two-sided unit up to homotopy, so that Z is an "H-space," then ϕ induces an "addition"
$$[X, Z] \times [X, Z] \longrightarrow [X, Z].$$
The trivial map $X \longrightarrow Z$ acts as zero. If Z is homotopy associative, in the sense that there is a homotopy between the maps given on elements by $(xy)z$ and by $x(yz)$, then the addition is associative. If, further, Z is homotopy commutative, in the sense that there is a homotopy between the maps given on elements by xy and by yx, then this addition is commutative. We say that Z is "grouplike" if there is a map $\chi : Z \longrightarrow Z$ such that $\phi(\text{id} \times \chi)\Delta : Z \longrightarrow Z$ is homotopic to the trivial map, and then $\chi_* : [X, Z] \longrightarrow [X, Z]$ sends an element $x \in [X, Z]$ to x^{-1}.

LEMMA.   If Z is a grouplike homotopy associative and commutative H-space, then the functor $[X, Z]$ takes values in Abelian groups.

Actually, the existence of inverses can be deduced if Z is only "grouplike" in the weaker sense that $\pi_0(X)$ is a group, but we shall not need the extra generality. Now consider the multiplication on a loop space ΩY given by composition of loops. Our proof that $\pi_1(Y)$ is a group and $\pi_2(Y)$ is an Abelian group amounts to a proof of the following result.

LEMMA.   For any based space Y, ΩY is a grouplike homotopy associative H-space and $\Omega^2 Y$ is a grouplike homotopy associative and commutative H-space.

Recall too that we have
$$[\Sigma X, Y] \cong [X, \Omega Y]$$
for any based spaces X and Y.

DEFINITION.   An Ω-prespectrum is a sequence of based spaces T_n and weak homotopy equivalences $\tilde{\sigma} : T_n \longrightarrow \Omega T_{n+1}$.

It is usual, but unnecessary, to require the T_n to have the homotopy types of CW complexes, in which case the $\tilde{\sigma}$ are homotopy equivalences. Specialization of the observations above leads to the following fundamental fact.

THEOREM.   Let $\{T_n\}$ be an Ω-prespectrum. Define
$$\tilde{E}^q(X) = \begin{cases} [X, T_q] & \text{if } q \geq 0 \\ [X, \Omega^{-q} T_0] & \text{if } q < 0. \end{cases}$$
Then the functors $\tilde{E}^q$ define a reduced cohomology theory on based CW complexes.

PROOF.   We have already verified the exactness and additivity axioms, and the weak equivalences $\tilde{\sigma}$ induce the suspension isomorphisms:
$$\tilde{E}^q(X) = [X, T_q] \longrightarrow [X, \Omega T_{q+1}] \cong [\Sigma X, T_{q+1}] = \tilde{E}^{q+1}(\Sigma X). \quad \square$$

It is a consequence of a general result called the Brown representability theorem that every reduced cohomology theory is represented in this fashion by an Ω-prespectrum.

3. Cup and cap products

Changing notations, let A and B be Abelian groups and X and Y be based spaces. We have an external product

$$\tilde{H}^p(X;A) \otimes \tilde{H}^q(Y;B) \longrightarrow \tilde{H}^{p+q}(X \wedge Y; A \otimes B).$$

Indeed, if X and Y are based CW complexes, then we have an isomorphism of cellular chain complexes

$$\tilde{C}_*(X) \otimes \tilde{C}_*(Y) \cong \tilde{C}_*(X \wedge Y).$$

On passage to cochains with coefficients in A, B, and $A \otimes B$, this induces a homomorphism of cochain complexes

$$\tilde{C}^*(X;A) \otimes \tilde{C}^*(Y;B) \longrightarrow \tilde{C}^*(X \wedge Y; A \otimes B).$$

In turn, this induces the cited product on passage to cohomology. With $X = Y$, we can apply the diagonal $\Delta : X \longrightarrow X \wedge X$ and any homomorphism $A \otimes B \longrightarrow C$ to obtain a cup product

$$\tilde{H}^p(X;A) \otimes \tilde{H}^q(X;B) \longrightarrow \tilde{H}^{p+q}(X;C).$$

When $X = X'_+$ for an unbased space X', this gives the cup product on the unreduced cohomology of X'.

We can obtain these external products and therefore their induced cup products homotopically. The smash product of maps gives a pairing

$$[X, K(A,p)] \otimes [Y, K(B,q)] \longrightarrow [X \wedge Y, K(A,p) \wedge K(B,q)].$$

Therefore, to obtain an external product, we need only obtain a suitable map

$$\phi_{p,q} : K(A,p) \wedge K(B,q) \longrightarrow K(A \otimes B, p+q).$$

Such a map may be interpreted as an element of $\tilde{H}^{p+q}(K(A,p) \wedge K(B,q); A \otimes B)$. Since the space $K(A,p) \wedge K(B,q)$ is $(p+q-1)$-connected, the universal coefficient, Künneth, and Hurewicz theorems give isomorphisms

$$\begin{aligned}\tilde{H}^{p+q}(K(A,p) \wedge K(B,q); A \otimes B) &\cong \mathrm{Hom}(\tilde{H}_{p+q}(K(A,p) \wedge K(B,q)), A \otimes B) \\ &\cong \mathrm{Hom}(\tilde{H}_p(K(A,p)) \otimes \tilde{H}_q(K(B,q)), A \otimes B) \\ &\cong \mathrm{Hom}(\pi_p(K(A,p)) \otimes \pi_q(K(B,q)), A \otimes B) \\ &= \mathrm{Hom}(A \otimes B, A \otimes B).\end{aligned}$$

Therefore the identity homomorphism on the group $A \otimes B$ gives rise to the required map $\phi_{p,q}$. Arguing similarly, it is easy to check that the system of maps $\{\phi_{p,q}\}$ is associative, commutative, and unital in the sense that the following diagrams are homotopy commutative. Indeed, translating back along isomorphisms of the form just displayed, each of the diagrams translates to an elementary algebraic identity.

$$\begin{array}{ccc} K(A,p) \wedge K(B,q) \wedge K(C,r) & \xrightarrow{\phi \wedge \mathrm{id}} & K(A \otimes B, p+q) \wedge K(C,r) \\ {\scriptstyle \mathrm{id} \wedge \phi} \downarrow & & \downarrow {\scriptstyle \phi} \\ K(A,p) \wedge K(B \otimes C, q+r) & \xrightarrow{\phi} & K(A \otimes B \otimes C, p+q+r), \end{array}$$

$$K(A,p) \wedge K(B,q) \xrightarrow{\phi} K(A \otimes B, p+q)$$
$$t \downarrow \qquad\qquad \downarrow K(t,p+q)$$
$$K(B,q) \wedge K(A,p) \xrightarrow{\phi} K(B \otimes A, p+q),$$

and

$$S^0 \wedge K(A,p) = K(A,p)$$
$$i \wedge \mathrm{id} \downarrow \qquad\qquad \|$$
$$K(\mathbb{Z},0) \wedge K(A,p) \xrightarrow{\phi} K(\mathbb{Z} \otimes A, p),$$

where $i: S^0 \longrightarrow \mathbb{Z} = K(\mathbb{Z},0)$ takes 0 to 0 and 1 to 1. The associativity, graded commutativity, and unital properties of the cup product follow.

The cup products on cohomology defined in terms of cellular cochains and in terms of the homotopical representation of cohomology agree. To see this, observe that the identity homomorphism of A specifies a fundamental class

$$\iota_p \in \tilde{H}^p(K(A,p), A)$$

via the isomorphisms

$$\mathrm{Hom}(A,A) \cong \mathrm{Hom}(\pi_p(K(A,p)), A) \cong \mathrm{Hom}(\tilde{H}_p(K(A,p)), A) \cong \tilde{H}^p(K(A,p); A).$$

A moment's thought shows that the two cup products will agree on arbitrary pairs of cohomology classes if they agree when applied to $\iota_p \otimes \iota_q$ for all p and q. We may take our Eilenberg-Mac Lane spaces to be CW complexes and give their smash product the induced CW structure. Considering representative cycles for generators of our groups as images under the Hurewicz homomorphism of representative maps $S^p \longrightarrow K(A,p)$, we find that the required agreement follows from the canonical identifications $S^p \wedge S^q \cong S^{p+q}$.

We can also construct cap products homotopically. To do so, it is convenient to bring function spaces into play, using the obvious isomorphisms

$$[X,Y] \cong \pi_0 F(X,Y)$$

and evaluation maps

$$\varepsilon : F(X,Y) \wedge X \longrightarrow Y.$$

We wish to construct the cap product

$$\tilde{H}^p(X;A) \otimes \tilde{H}_n(X;B) \longrightarrow \tilde{H}_{n-p}(X; A \otimes B),$$

and it is equivalent to construct

$$\pi_0(F(X, K(A,p))) \otimes \mathrm{colim}_q \pi_{n+q}(X \wedge K(B,q)) \longrightarrow \mathrm{colim}_r \pi_{n-p+r}(X \wedge K(A \otimes B, r)).$$

Changing the variable of the second colimit by setting $r = p + q$ and recalling the algebraic fact that tensor products commute with colimits, we can rewrite this as

$$\mathrm{colim}_q(\pi_0(F(X, K(A,p))) \otimes \pi_{n+q}(X \wedge K(B,q))) \longrightarrow \mathrm{colim}_q \pi_{n+q}(X \wedge K(A \otimes B, p+q)).$$

Thus it suffices to define maps

$$\pi_0(F(X, K(A,p))) \otimes \pi_{n+q}(X \wedge K(B,q)) \longrightarrow \pi_{n+q}(X \wedge K(A \otimes B, p+q)).$$

3. CUP AND CAP PRODUCTS

These are given by the following composites:

$$\pi_0(F(X, K(A,p))) \otimes \pi_{n+q}(X \wedge K(B,q))$$
$$\downarrow \wedge$$
$$\pi_{n+q}(F(X, K(A,p)) \wedge X \wedge K(B,q))$$
$$\downarrow (\text{id} \wedge \Delta \wedge \text{id})_*$$
$$\pi_{n+q}(F(X, K(A,p)) \wedge X \wedge X \wedge K(B,q))$$
$$\downarrow (\varepsilon \wedge \text{id})_*$$
$$\pi_{n+q}(K(A,p) \wedge X \wedge K(B,q))$$
$$\downarrow (\text{id} \wedge \phi_{p,q})_*(\text{id} \wedge t)_*$$
$$\pi_{n+q}(X \wedge K(A \otimes B, p+q)).$$

Similarly, we can construct the evaluation pairing

$$\tilde{H}^n(X; A) \otimes \tilde{H}_n(X; B) \longrightarrow A \otimes B$$

homotopically. It is obtained by passage to colimits over q from the composites

$$\pi_0(F(X, K(A,n))) \otimes \pi_{n+q}(X \wedge K(B,q))$$
$$\downarrow \wedge$$
$$\pi_{n+q}(F(X, K(A,n)) \wedge X \wedge K(B,q))$$
$$\downarrow (\varepsilon \wedge \text{id})_*$$
$$\pi_{n+q}(K(A,n) \wedge K(B,q))$$
$$\downarrow (\phi_{n,q})_*$$
$$\pi_{n+q}(K(A \otimes B, n+q)) = A \otimes B.$$

The following formula relating the cup and cap products to the evaluation pairing was central to our discussion of Poincaré duality:

$$\langle \alpha \cup \beta, x \rangle = \langle \beta, \alpha \cap x \rangle \in R,$$

where $\alpha \in \tilde{H}^p(X; R)$, $\beta \in \tilde{H}^q(X; R)$, and $x \in \tilde{H}_{p+q}(X; R)$ for a commutative ring R. It is illuminating to rederive this from our homotopical descriptions of these products. In fact, a straightforward diagram chase shows that this formula is a direct consequence of the following elementary facts, where X, Y, and Z are based spaces. First, the following diagram commutes:

$$F(X,Y) \wedge F(X,Z) \wedge X \xrightarrow{\text{id} \wedge \text{id} \wedge \Delta} F(X,Y) \wedge F(X,Z) \wedge X \wedge X$$

$$\downarrow (\wedge) \wedge \text{id} \qquad\qquad\qquad\qquad\qquad\qquad \downarrow (\wedge) \wedge \text{id} \wedge \text{id}$$

$$F(X \wedge X, Y \wedge Z) \wedge X \xrightarrow{\text{id} \wedge \Delta} F(X \wedge X, Y \wedge Z) \wedge X \wedge X$$

$$\downarrow F(\Delta,\text{id}) \wedge \text{id} \qquad\qquad\qquad\qquad\qquad \downarrow \varepsilon$$

$$F(X, Y \wedge Z) \wedge X \xrightarrow{\qquad \varepsilon \qquad} Y \wedge Z.$$

Second, the right vertical composite in the diagram coincides with the common composite in the commutative diagram

$$F(X,Y) \wedge F(X,Z) \wedge X \wedge X \xrightarrow{t \wedge \text{id}} F(X,Z) \wedge F(X,Y) \wedge X \wedge X$$

$$\downarrow \text{id} \wedge t \wedge \text{id} \qquad\qquad\qquad\qquad\qquad \downarrow \text{id} \wedge \varepsilon \wedge \text{id}$$

$$F(X,Y) \wedge X \wedge F(X,Z) \wedge X \qquad\qquad\qquad F(X,Z) \wedge Y \wedge X$$

$$\downarrow \varepsilon \wedge \varepsilon \qquad\qquad\qquad\qquad\qquad\qquad\qquad \downarrow \text{id} \wedge t$$

$$Y \wedge Z \xleftarrow{\quad t \quad} Z \wedge Y \xleftarrow{\varepsilon \wedge \text{id}} F(X,Z) \wedge X \wedge Y.$$

The observant reader will see a punch line here: everything in this section applies equally well to the homology and cohomology theories represented by Ω-prespectra. A little more precisely, thinking of the case when $A = B = C$ is a commutative ring in the discussion above, we see by use of the product on A that we have a well behaved system of product maps

$$\phi_{p,q} : K(A,p) \wedge K(A,q) \longrightarrow K(A, p+q).$$

We have analogous cup and cap products and an evaluation pairing for the theories represented by any Ω-prespectrum $\{T_n\}$ with such a system of product maps

$$\phi_{p,q} : T_p \wedge T_q \longrightarrow T_{p+q}.$$

4. Postnikov systems

We have implicitly studied the represented functors $k(X) = [X, Y]$ by decomposing X into cells. This led in particular to the calculation of ordinary represented cohomology $[X, K(\pi, n)]$ by means of cellular chains. There is an Eckmann-Hilton dual way of studying $[X, Y]$ by decomposing Y into "cocells." We briefly describe this decomposition of spaces into their "Postnikov systems" here.

This decomposition answers a natural question: how close are the homotopy groups of a CW complex X to being a complete set of invariants for its homotopy type? Since $\prod_n K(\pi_n(X), n)$ has the same homotopy groups as X but is generally not weakly homotopy equivalent to it, some added information is needed. If X is simple, it turns out that the homotopy groups together with an inductively defined sequence of cohomology classes give a complete set of invariants.

Recall that a connected space X is said to be simple if $\pi_1(X)$ is Abelian and acts trivially on $\pi_n(X)$ for $n \geq 2$. A Postnikov system for a simple based space X consists of based spaces X_n together with based maps

$$\alpha_n : X \longrightarrow X_n \quad \text{and} \quad p_{n+1} : X_{n+1} \longrightarrow X_n,$$

$n \geq 1$, such that $p_{n+1} \circ \alpha_{n+1} = \alpha_n$, X_1 is an Eilenberg-Mac Lane space $K(\pi_1(X), 1)$, p_{n+1} is the fibration induced from the path space fibration over an Eilenberg-Mac Lane space $K(\pi_{n+1}(X), n+2)$ by a map
$$k^{n+2} : X_n \longrightarrow K(\pi_{n+1}(X), n+2),$$
and α_n induces an isomorphism $\pi_q(X) \to \pi_q(X_n)$ for $q \leq n$. It follows that $\pi_q(X_n) = 0$ for $q > n$. The system can be displayed diagrammatically as follows:

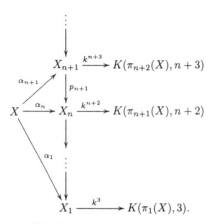

Our requirement that Eilenberg-Mac Lane spaces have the homotopy types of CW complexes implies (by a result of Milnor) that each X_n has the homotopy type of a CW complex. The maps α_n induce a weak equivalence $X \to \lim X_n$, but the inverse limit generally will not have the homotopy type of a CW complex. The "k-invariants" $\{k^{n+2}\}$ that specify the system are to be regarded as cohomology classes
$$k^{n+2} \in H^{n+2}(X_n; \pi_{n+1}(X)).$$
These classes together with the homotopy groups $\pi_n(X)$ specify the weak homotopy type of X. We outline the proof of the following theorem.

THEOREM. *A simple space X of the homotopy type of a CW complex has a Postnikov system.*

PROOF. Assume inductively that $\alpha_n : X \to X_n$ has been constructed. A consequence of the homotopy excision theorem shows that the cofiber $C(\alpha_n)$ is $(n+1)$-connected and satisfies
$$\pi_{n+2}(C(\alpha_n)) = \pi_{n+1}(X).$$
More precisely, the canonical map $\eta : F(\alpha_n) \to \Omega C(\alpha_n)$ induces an isomorphism on π_q for $q \leq n+1$. We construct
$$j : C(\alpha_n) \to K(\pi_{n+1}(X), n+2)$$
by inductively attaching cells to $C(\alpha_n)$ to kill its higher homotopy groups. We take the composite of j and the inclusion $X_n \subset C(\alpha_n)$ to be the k-invariant
$$k^{n+2} : X_n \longrightarrow K(\pi_{n+1}(X), n+2).$$

By our definition of a Postnikov system, we must define X_{n+1} to be the homotopy fiber of k^{n+2}. Thus its points are pairs (ω, x) consisting of a path $\omega : I \to K(\pi_{n+1}(X), n+2)$ and a point $x \in X_n$ such that $\omega(0) = *$ and $\omega(1) = k^{n+2}(x)$. The map $p_{n+1} : X_{n+1} \to X_n$ is given by $p_{n+1}(\omega, x) = x$, and the map $\alpha_{n+1} : X \to X_{n+1}$ is given by $\alpha_{n+1}(x) = (\omega(x), \alpha_n(x))$, where $\omega(x)(t) = j(x, 1-t)$, $(x, 1-t)$ being a point on the cone $CX \subset C(\alpha_n)$. Clearly $p_{n+1} \circ \alpha_{n+1} = \alpha_n$. It is evident that α_{n+1} induces an isomorphism on π_q for $q \leq n$, and a diagram chase shows that this also holds for $q = n+1$. □

5. Cohomology operations

Consider a "represented functor" $k(X) = [X, Z]$ and another contravariant functor k' from the homotopy category of based CW complexes to the category of sets. The following simple observation actually applies to represented functors on arbitrary categories. We shall use it to describe cohomology operations, but it also applies to describe many other invariants in algebraic topology, such as the characteristic classes of vector bundles.

LEMMA (Yoneda). *There is a canonical bijection between natural transformations* $\Phi : k \longrightarrow k'$ *and elements* $\phi \in k'(Z)$.

PROOF. Given Φ, we define ϕ to be $\Phi(\mathrm{id})$, where $\mathrm{id} \in k(Z) = [Z, Z]$ is the identity map. Given ϕ, we define $\Phi : k(X) \longrightarrow k'(X)$ by the formula $\Phi(f) = f^*(\phi)$. Here f is a map $X \longrightarrow Z$, and it induces $f^* = k'(f) : k'(Z) \longrightarrow k'(X)$. It is simple to check that these are inverse bijections. □

We are interested in the case when k' is also represented, say $k'(X) = [X, Z']$.

COROLLARY. *There is a canonical bijection between natural transformations* $\Phi : [-, Z] \longrightarrow [-, Z']$ *and elements* $\phi \in [Z, Z']$.

DEFINITION. Suppose given cohomology theories $\tilde{E}^*$ and $\tilde{F}^*$. A cohomology operation of type q and degree n is a natural transformation $\tilde{E}^q \longrightarrow \tilde{F}^{q+n}$. A stable cohomology operation of degree n is a sequence $\{\Phi^q\}$ of cohomology operations of type q and degree n such that the following diagram commutes for each q and each based space X:

$$\begin{array}{ccc} \tilde{E}^q(X) & \xrightarrow{\Phi^q} & \tilde{E}^{q+n}(X) \\ \Sigma \downarrow & & \downarrow \Sigma \\ \tilde{E}^{q+1}(\Sigma X) & \xrightarrow[\Phi^{q+1}]{} & \tilde{E}^{q+1+n}(\Sigma X). \end{array}$$

We generally abbreviate notation by setting $\Phi^q = \Phi$.

In general, cohomology operations are only natural transformations of set-valued functors. However, stable operations are necessarily homomorphisms of cohomology groups, as the reader is encouraged to check.

THEOREM. *Cohomology operations* $\tilde{H}^q(-; \pi) \longrightarrow \tilde{H}^{q+n}(-; \rho)$ *are in canonical bijective correspondence with elements of* $\tilde{H}^{q+n}(K(\pi, q); \rho)$.

PROOF. Translate to the represented level, apply the previous corollary, and translate back. □

This seems very abstract, but it has very concrete consequences. To determine all cohomology operations, we need only compute the cohomology of all Eilenberg-MacLane spaces. We have described an explicit construction of these spaces as topological Abelian groups in Chapter 16 §5, and this construction leads to an inductive method of computation. We briefly indicate a key example of how this works, without proofs.

THEOREM. *For $n \geq 0$, there are stable cohomology operations*

$$Sq^n : H^q(X; \mathbb{Z}_2) \longrightarrow H^{q+n}(X; \mathbb{Z}_2),$$

called the Steenrod operations. They satisfy the following properties.
(i) Sq^0 *is the identity operation.*
(ii) $Sq^n(x) = x^2$ *if $n = \deg x$ and $Sq^n(x) = 0$ if $n > \deg x$.*
(iii) *The Cartan formula holds:*

$$Sq^n(xy) = \sum_{i+j=n} Sq^i(x) Sq^j(y).$$

In fact, the Steenrod operations are uniquely characterized by the stated properties. There are also formulas, called the Adem relations, describing $Sq^i Sq^j$, as a linear combination of operations $Sq^{i+j-k} Sq^k$, $2k \leq i$, when $0 < i < 2j$; explicitly,

$$Sq^i Sq^j = \sum_{0 \leq k \leq [i/2]} \binom{j-k-1}{i-2k} Sq^{i+j-k} Sq^k.$$

It turns out that the Steenrod operations generate all mod 2 cohomology operations. In fact, the identity map of $K(\mathbb{Z}_2, q)$ specifies a fundamental class $\iota_q \in H^q(K(\mathbb{Z}_2, q); \mathbb{Z}_2)$, and the following theorem holds.

THEOREM. $H^*(K(\mathbb{Z}_2, q); \mathbb{Z}_2)$ *is a polynomial algebra whose generators are certain iterates of Steenrod operations applied to the fundamental class ι_q. Explicitly, writing $Sq^I = Sq^{i_1} \cdots Sq^{i_j}$ for a sequence of positive integers $I = \{i_1, \ldots, i_j\}$, the generators are the $Sq^I \iota_q$ for those sequences I such that $i_r \geq 2 i_{r+1}$ for $1 \leq r < j$ and $i_1 < i_2 + \cdots + i_j + q$.*

PROBLEMS

1. For Abelian groups π and ρ, show that $[K(\pi, n), K(\rho, n)] \cong \mathrm{Hom}(\pi, \rho)$. (Hint: use the natural isomorphism $[X, K(\rho, n)] \cong \tilde{H}^n(X; \rho)$ and universal coefficients.)
2. (a) Let $f : \pi \longrightarrow \rho$ be a homomorphism of Abelian groups. Construct cohomology operations $f^* : H^q(X; \pi) \longrightarrow H^q(X; \rho)$ for all q.
   (b) Let $0 \longrightarrow \pi \xrightarrow{f} \rho \xrightarrow{g} \sigma \longrightarrow 0$ be an exact sequence of Abelian groups. Construct cohomology operations $\beta : H^q(X; \sigma) \longrightarrow H^{q+1}(X; \pi)$ for all q such that the following is a long exact sequence:

$$\cdots \longrightarrow H^q(X; \pi) \xrightarrow{f^*} H^q(X; \rho) \xrightarrow{g^*} H^q(X; \sigma) \xrightarrow{\beta} H^{q+1}(X; \pi) \longrightarrow \cdots.$$

   The β are called Bockstein operations.
3. Using the calculation of $H^*(K(\mathbb{Z}, 2); \mathbb{Z}_2)$ stated in the text, prove that $Sq^1 : H^q(X; \mathbb{Z}_2) \longrightarrow H^{q+1}(X; \mathbb{Z}_2)$ coincides with the Bockstein operation associated to the short exact sequence $0 \longrightarrow \mathbb{Z}_2 \longrightarrow \mathbb{Z}_4 \longrightarrow \mathbb{Z}_2 \longrightarrow 0$.

4. Prove that each Φ^q of a stable cohomology operation $\{\Phi^q\}$ is a natural homomorphism.
5. Write $H^*(\mathbb{R}P^\infty; \mathbb{Z}_2) = \mathbb{Z}_2[\alpha]$, $\deg \alpha = 1$. Compute $Sq^i(\alpha^j)$ for all i and j.

CHAPTER 23

Characteristic classes of vector bundles

Some of the most remarkable applications of algebraic topology result from the translation of problems in geometric topology into problems in homotopy theory. The essential intermediary in many of these translations is the theory of vector bundles. We here explain the classification theorem for vector bundles and its relationship to the theory of characteristic classes. The reader is assumed to be familiar with the tangent and normal bundles of smooth manifolds and to be reasonably well acquainted with the definitions and elementary properties of vector bundles in general.

1. The classification of vector bundles

Let E be a (real) vector bundle over a base space B. Thus we are given a projection $p : E \longrightarrow B$ such that, for each $b \in B$, the fiber $p^{-1}(b)$ is a copy of $\mathbb{R}^n$ for some n. In the case of non-connected base spaces, the fibers over points in different components may have different dimension. We say that p is an n-plane bundle if all fibers have dimension n. For each U in some open cover of B, there is a homeomorphism (a "coordinate chart") $\phi_U : U \times \mathbb{R}^n \longrightarrow p^{-1}(U)$ over U that restricts to a linear isomorphism on each fiber. We shall require our open covers to be numerable, as can always be arranged when B is paracompact. For a second vector bundle $q : D \longrightarrow A$, a map $(g, f) : D \longrightarrow E$ of vector bundles is a pair of maps $f : A \longrightarrow B$ and $g : D \longrightarrow E$ such that $p \circ g = f \circ q$ and $g : q^{-1}(a) \longrightarrow p^{-1}(f(a))$ is linear for all $a \in A$. This gives the category of vector bundles. A map (g, f) of vector bundles is an isomorphism if and only if f is a homeomorphism and g restricts to an isomorphism on each fiber. We are mainly interested in the subcategories of n-plane bundles and maps that are linear isomorphisms on fibers. We say that two vector bundles over B are equivalent if they are isomorphic over B, so that there is an isomorphism (g, id) between them. We let $\mathscr{E}_n(B)$ denote the set of equivalence classes of n-plane bundles over B. (That this really is a well defined set will emerge shortly.) If $f : A \longrightarrow B$ is a continuous map, then the pullback of f and a vector bundle $p : E \longrightarrow B$ is a vector bundle f^*E over A. Moreover, a bundle D over A is equivalent to f^*E if and only if there is a map $(g, f) : D \longrightarrow E$ that is an isomorphism on fibers.

Thus we have a contravariant set-valued functor $\mathscr{E}_n(-)$ on spaces. Vector bundles should be thought of as rather rigid geometric objects, and the equivalence relation between them preserves that rigidity. Nevertheless, equivalence classes of n-plane bundles can be classified homotopically. This is a crucial starting point for the translation of geometric problems to homotopical ones. In turn, the starting point of the classification theorem is the observation that the functor $\mathscr{E}_n(-)$, like homology and cohomology, is homotopy invariant in the sense that it factors

183

through the homotopy category $h\mathscr{U}$. In less fancy language, this amounts to the following result.

PROPOSITION. *The pullbacks of an n-plane bundle $p : E \longrightarrow B$ along homotopic maps $f_0, f_1 : A \longrightarrow B$ are equivalent.*

SKETCH PROOF. Let $h : A \times I \longrightarrow B$ be a homotopy $f_0 \simeq f_1$. Then the restrictions of h^*E over $A \times \{0\}$ and $A \times \{1\}$ can be identified with f_0^*E and f_1^*E. Thus we change our point of view and consider a general n-plane bundle $p : E \longrightarrow B \times I$. It suffices to show that the restrictions E_0 and E_1 of E over $B \times \{0\}$ and $B \times \{1\}$ are equivalent. Define $r : B \times I \longrightarrow B \times I$ by $r(b, t) = (b, 1)$. We claim that there is a map $g : E \longrightarrow E$ such that (g, r) is a map of vector bundles. It follows that E is equivalent to r^*E, and it is easy to see that r^*E is isomorphic to the bundle $E_1 \times I$. The restriction of g to E_0 will be an equivalence to E_1. To construct g, one first proves, using the compactness of I, that there is a numerable open cover $\mathscr{O}$ of B such that the restriction of E to $U \times I$ is trivial for all $U \in \mathscr{O}$. One then uses trivializations $\phi_U : U \times I \times \mathbb{R}^n \longrightarrow p^{-1}(U \times I)$ together with functions $\lambda_U : B \longrightarrow I$ such that $\lambda_U^{-1}(0, 1] = U$ to construct g by gradually pushing the bundle to the right along neighborhoods where it is trivial. □

It can be verified on general abstract nonsense grounds, using Brown's representability theorem, that the functor $\mathscr{E}_n(-)$ is representable in the form $[-, BO(n)]$ for some space $BO(n)$. It is far more useful to have an explicit concrete construction of the relevant "classifying space" $BO(n)$. More precisely, we think of "$BO(n)$" as specifying a homotopy type of spaces, and we want an explicit representative of the homotopy type. Here $[X, Y]$ denotes unbased homotopy classes of maps. We construct a particular n-plane bundle $\gamma_n : E_n \longrightarrow BO(n)$, called the "universal n-plane bundle." By pulling back γ_n along (homotopy classes of) maps $f : B \longrightarrow BO(n)$, we obtain a natural transformation of functors $[-, BO(n)] \longrightarrow \mathscr{E}_n(-)$. We show that this natural transformation is a natural isomorphism of functors by showing how to construct a map (g, f), unique up to homotopy, from any given n-plane bundle E over any space B to the universal n-plane bundle E_n; it is in this sense that E_n is "universal."

Let $V_n(\mathbb{R}^q)$ be the Stiefel variety of orthonormal n-frames in $\mathbb{R}^q$. Its points are n-tuples of orthonormal vectors in $\mathbb{R}^q$, and it is topologized as a subspace of $(\mathbb{R}^q)^n$ or, equivalently, as a subspace of $(S^{q-1})^n$. It is a compact manifold. Let $G_n(\mathbb{R}^q)$ be the Grassmann variety of n-planes in $\mathbb{R}^q$. Its points are the n-dimensional subspaces of $\mathbb{R}^q$. Sending an n-tuple of orthonormal vectors to the n-plane they span gives a surjective function $V_n(\mathbb{R}^q) \longrightarrow G_n(\mathbb{R}^q)$, and we topologize $G_n(\mathbb{R}^q)$ as a quotient space of $V_n(\mathbb{R}^q)$. It too is a compact manifold. For example, $V_1(\mathbb{R}^q) = S^{q-1}$ and $G_1(\mathbb{R}^q) = \mathbb{R}P^{q-1}$. The standard inclusion of $\mathbb{R}^q$ in $\mathbb{R}^{q+1}$ induces inclusions $V_n(\mathbb{R}^q) \subset V_n(\mathbb{R}^{q+1})$ and $G_n(\mathbb{R}^q) \subset G_n(\mathbb{R}^{q+1})$. We define $V_n(\mathbb{R}^\infty)$ and $G_n(\mathbb{R}^\infty)$ to be the unions of the $V_n(\mathbb{R}^q)$ and $G_n(\mathbb{R}^q)$, with the topology of the union. We define the classifying space $BO(n)$ to be $G_n(\mathbb{R}^\infty)$.

Let E_n^q be the subbundle of the trivial bundle $G_n(\mathbb{R}^q) \times \mathbb{R}^q$ whose points are the pairs (x, v) such that v is a vector in the plane x; denote the projection of E_n^q by γ_n^q, so that $\gamma_n^q(x, v) = x$. When $n = 1$, γ_1^q is called the "canonical line bundle" over $\mathbb{R}P^{q-1}$. We may let q go to infinity. We let $E_n = E_n^\infty$ and let $\gamma_n = \gamma_n^\infty$: $E_n \longrightarrow BO(n)$. This is our universal bundle, and it is not hard to verify that it is indeed an n-plane bundle. We must explain why it is universal. (Technically, it is usual to assume that base spaces are paracompact, but the restriction to numerable systems of coordinate charts in our definition of vector bundles allows the use of general base spaces.)

THEOREM. *The natural transformation* $\Phi : [-, BO(n)] \longrightarrow \mathscr{E}_n(-)$ *obtained by sending the homotopy class of a map* $f : B \longrightarrow BO(n)$ *to the equivalence class of the n-plane bundle* f^*E_n *is a natural isomorphism of functors.*

SKETCH PROOF. To illustrate ideas, let M be a smooth compact n-manifold smoothly embedded in $\mathbb{R}^q$ and let $\tau(M)$ be its tangent bundle. The tangent plane τ_x at a point $x \in M \subset \mathbb{R}^q$ is then embedded as an affine plane through x in $\mathbb{R}^q$. Translating to a plane through the origin by subtracting x from each vector, we obtain a point $f(x) \in G_n(\mathbb{R}^q)$ and an isomorphism $g_x : \tau_x \longrightarrow (\gamma_n^q)^{-1}(f(x))$. The g_x glue together to give a map (g, f) of bundles from $E(\tau(M))$ to E_n^q; it is called the Gauss map of the tangent bundle of M. Similarly, using the orthogonal complements of tangent planes, we obtain the Gauss map $E(\nu) \longrightarrow E_{q-n}^q$ of the normal bundle ν of the embedding of M in $\mathbb{R}^q$.

For a general n-plane bundle $p : E \longrightarrow B$, we must construct a map $(g, f) : E \longrightarrow E_n$ of vector bundles that is an isomorphism on fibers; it will follow that E is equivalent to f^*E_n, thus showing that Φ is surjective. It suffices to construct a map $\hat{g} : E \longrightarrow \mathbb{R}^\infty$ that is a linear monomorphism on fibers, since we can then define $f(e)$ to be the image under $\hat{g}$ of the fiber through e and can define $g(e) = (f(e), \hat{g}(e))$. One first shows that one can construct a countable numerable cover of coordinate charts from a general numerable cover of coordinate charts. Using trivializations $\phi_U : U \times \mathbb{R}^n \longrightarrow p^{-1}(U)$ and functions $\lambda_U : B \longrightarrow I$ such that $U = \lambda_U^{-1}(0, 1]$, we define $\hat{g}_U : E \longrightarrow \mathbb{R}^n$ by

$$\hat{g}_U(e) = \lambda_U(p(e)) \cdot p_2(\phi_U^{-1}(e))$$

for $e \in p^{-1}(U)$, where $p_2 : U \times \mathbb{R}^n \longrightarrow \mathbb{R}^n$ is the projection, and $\hat{g}_U(e) = 0$ for $e \notin p^{-1}(U)$. Taking $\mathbb{R}^\infty$ to be the sum of countably many copies of $\mathbb{R}^n$, we then define $\hat{g} = \sum \hat{g}_U$.

To show that Φ is injective, we must show further that the resulting classifying map f is unique up to homotopy, and for this it suffices to show that any two maps (g_0, f_0) and (g_1, f_1) of vector bundles from E to E_n that are isomorphisms on fibers are bundle homotopic. These bundle maps are determined by their second coordinates $\hat{g}_0$ and $\hat{g}_1$, which are maps $E \longrightarrow \mathbb{R}^\infty$. Provided that $\hat{g}_0(e)$ is not a negative multiple of $\hat{g}_1(e)$ for any e, we obtain a homotopy $\hat{h} : \hat{g}_0 \simeq \hat{g}_1$ by setting

$$\hat{h}(e, t) = (1 - t)\hat{g}_0(e) + t\hat{g}_1(e).$$

The proviso ensures that $\hat{h}$ is a monomorphism on fibers, and $\hat{h}$ determines the required bundle homotopy $E \times I \longrightarrow E_n$. For the general case, let i_0 and i_1 be the linear isomorphisms from $\mathbb{R}^\infty$ to itself that send the qth standard basis element e_q to e_{2q} and e_{2q-1}, respectively. The composites $i_0 \circ \hat{g}_0$ and $i_1 \circ \hat{g}_1$ determine bundle maps k_0 and k_1 from E to E_n, and the construction just given applies to give bundle homotopies from g_0 to k_0, from k_0 to k_1, and from k_1 to g_1. □

2. Characteristic classes for vector bundles

DEFINITION. Let k^* be a cohomology theory, such as $H^*(-; \pi)$ for an Abelian group π. A characteristic class c of degree q for n-plane bundles is a natural assignment of a cohomology class $c(\xi) \in k^q(B)$ to bundles ξ with base space B.

Thus, if (g, f) is a map from a bundle ζ over A to a bundle ξ over B, so that ζ is equivalent to $f^*\xi$, then $f^*c(\xi) = c(\zeta)$. Clearly $c(\xi) = c(\xi')$ if ξ is equivalent to ξ'.

Since the functor $\mathscr{E}_n$ is represented by $BO(n)$, the Yoneda lemma specializes to give the following result.

LEMMA. *Evaluation on γ_n specifies a canonical bijection between characteristic classes of n-plane bundles and elements of $k^*(BO(n))$.*

The formal similarity to the definition of cohomology operations is obvious, and we shall illustrate how to exploit this similarity in the following sections. Clearly calculation of $k^*(BO(n))$ determines all characteristic classes. Moreover, the behavior of characteristic classes with respect to operations on bundles can be determined by calculating the maps on cohomology induced by maps between classifying spaces. We are particularly interested in Whitney sums of bundles. We have the evident Cartesian product, or external sum, of an m-plane bundle over A and an n-plane bundle over B; it is an $(m + n)$-plane bundle over $A \times B$. The internal sum, or Whitney sum, of two bundles over the same base space B is obtained by pulling back their external sum along the diagonal map of B.

For example, let ε denote the trivial line bundle over any space. We have the operation that sends an n-plane bundle ξ over B to the $(n + 1)$-plane bundle $\xi \oplus \varepsilon$ over B. There is a classifying map

$$i_n : BO(n) \longrightarrow BO(n+1)$$

that is characterized up to homotopy by $i_n^*(\gamma_{n+1}) = \gamma_n \oplus \varepsilon$. If we have a characteristic class c on $(n + 1)$-plane bundles, then

$$i_n^* c(\gamma_{n+1}) = c(\gamma_n \oplus \varepsilon),$$

and this leads by naturality to a description of $c(\xi \oplus \varepsilon)$ for general n-plane bundles ξ. To give an explicit description of i_n, we may think of $BO(n+1)$ as $G_{n+1}(\mathbb{R}^\infty \oplus \mathbb{R})$; precisely, we use an isomorphism between $\mathbb{R}^\infty \oplus \mathbb{R}$ and $\mathbb{R}^\infty$ to define a homeomorphism $G_{n+1}(\mathbb{R}^\infty \oplus \mathbb{R}) \cong G_{n+1}(\mathbb{R}^\infty)$, and we check that the homotopy class of this homeomorphism is independent of the choice of isomorphism. We then define i_n on $G_n(\mathbb{R}^\infty)$ by sending an n-plane x in $\mathbb{R}^\infty$ to the $(n + 1)$-plane $x \oplus \mathbb{R}$.

Similarly, we have a classifying map

$$p_{m,n} : BO(m) \times BO(n) \longrightarrow BO(m+n)$$

that is characterized up to homotopy by $p_{m,n}^*(\gamma_{m+n}) = \gamma_m \times \gamma_n$. If we have a characteristic class c on $(m + n)$-plane bundles, then

$$p_{m,n}^* c(\gamma_{m+n}) = c(\gamma_m \times \gamma_n),$$

and this leads by naturality to a description of $c(\zeta \times \xi)$ for general m-plane bundles ζ and n-plane bundles ξ. To give an explicit description of $p_{m,n}$, we may think of $BO(m + n)$ as $G_{m+n}(\mathbb{R}^\infty \oplus \mathbb{R}^\infty)$; precisely, we use an isomorphism between $\mathbb{R}^\infty \oplus \mathbb{R}^\infty$ and $\mathbb{R}^\infty$ to define a homeomorphism $G_{m+n}(\mathbb{R}^\infty \oplus \mathbb{R}^\infty) \cong G_{m+n}(\mathbb{R}^\infty)$, and we check that the homotopy class of this homeomorphism is independent of the choice of isomorphism. We then define $p_{m,n}$ on $G_m(\mathbb{R}^\infty) \times G_n(\mathbb{R}^\infty)$ by sending (x,y) to $x \oplus y$, where x is an m-plane in $\mathbb{R}^\infty$ and y is an n-plane in $\mathbb{R}^\infty$.

We bring this down to earth by describing all characteristic classes in mod 2 cohomology. In fact, we have the following equivalent pair of theorems. The first uses the language of characteristic classes, while the second describes $H^*(BO(n); \mathbb{Z}_2)$ together with the induced maps i_n^* and $p_{m,n}^*$.

THEOREM. *For n-plane bundles ξ over base spaces B, $n \geq 0$, there are characteristic classes $w_i(\xi) \in H^i(B; \mathbb{Z}_2)$, $i \geq 0$, called the Stiefel-Whitney classes. They satisfy and are uniquely characterized by the following axioms.*

1. $w_0(\xi) = 1$ and $w_i(\xi) = 0$ if $i > \dim \xi$.
2. $w_1(\gamma_1) \neq 0$, where γ_1 is the universal line bundle over $\mathbb{R}P^\infty$.
3. $w_i(\xi \oplus \varepsilon) = w_i(\xi)$.
4. $w_i(\zeta \oplus \xi) = \sum_{j=0}^{i} w_j(\zeta) \cup w_{i-j}(\xi)$.

Every mod 2 characteristic class for n-plane bundles can be written uniquely as a polynomial in the Stiefel-Whitney classes $\{w_1, \ldots, w_n\}$.

THEOREM. *For $n \geq 1$, there are elements $w_i \in H^i(BO(n); \mathbb{Z}_2)$, $i \geq 0$, called the Stiefel-Whitney classes. They satisfy and are uniquely characterized by the following axioms.*

1. $w_0 = 1$ and $w_i = 0$ if $i > n$.
2. $w_1 \neq 0$ when $n = 1$.
3. $i_n^*(w_i) = w_i$.
4. $p_{m,n}^*(w_i) = \sum_{j=0}^{i} w_j \otimes w_{i-j}$.

The mod 2 cohomology $H^*(BO(n); \mathbb{Z}_2)$ is the polynomial algebra $\mathbb{Z}_2[w_1, \ldots, w_n]$.

For the uniqueness, suppose given another collection of classes w_i' for all $n \geq 1$ that satisfy the stated properties. Since $BO(1) = \mathbb{R}P^\infty$, $w_1 = w_1'$ is the unique non-zero element of $H^1(\mathbb{R}P^\infty; \mathbb{Z}_2)$. Therefore $w_i = w_i'$ for all i when $n = 1$, and we assume that this is true for all $m < n$. Visibly i_{n-1}^* is an isomorphism in degrees less than n, and this implies that $w_i = w_i'$ in $H^i(BO(n); \mathbb{Z}_2)$ for $i < n$. It is less visible but easily checked that the $p_{m,n}^*$ are all monomorphisms in all degrees. Since $p_{1,n-1}^*(w_n) = p_{1,n-1}^*(w_n')$, this implies that $w_n = w_n'$.

3. Stiefel-Whitney classes of manifolds

It is convenient to consider $H^{**}(X) = \prod_i H^i(X)$ and to write its elements as formal sums $\sum x_i$, $\deg x_i = i$. In practice, we usually impose conditions that guarantee that the sum is finite. We define the total Stiefel-Whitney class $w(\xi)$ of a vector bundle ξ to be $\sum w_i(\xi)$; here the sum is clearly finite. Note in particular that $w(\varepsilon^q) = 1$, where ε^q is the trivial q-plane bundle. With this notation, we have the formula

$$w(\zeta \oplus \xi) = w(\zeta) \cup w(\xi).$$

It is usual to write $w_i(M) = w_i(\tau(M))$ and $w(M) = w(\tau(M))$ for a smooth compact manifold M. Suppose that M immerses in $\mathbb{R}^q$ with normal bundle ν. Then $\tau(M) \oplus \nu \cong \varepsilon^q$ and we have the "Whitney duality formula"

$$w(M) \cup w(\nu) = 1,$$

which shows how to calculate tangential Stiefel-Whitney classes in terms of normal Stiefel-Whitney classes, and conversely. This formula can be used to prove non-immersion results when we know $w(M)$. If M has dimension n, then ν has dimension $q - n$ and must satisfy $w_i(\nu) = 0$ if $i > q - n$. Calculation of $w_i(\nu)$ from the Whitney duality formula can lead to a contradiction if q is too small.

One calculation is immediate. Since the normal bundle of the standard embedding $S^q \longrightarrow \mathbb{R}^{q+1}$ is trivial, $w(S^q) = 1$. A manifold is said to be parallelizable if its tangent bundle is trivial. For some manifolds M, we can show that M is not parallelizable by showing that one of its Stiefel-Whitney classes is non-zero, but this strategy fails for $M = S^q$.

We describe some standard computations in the cohomology of projective spaces that give less trivial examples. Write ζ_q for the canonical line bundle over $\mathbb{R}P^q$ in this section. (We called it γ_1^{q+1} before.) The total space of ζ_q consists of pairs (x, v), where x is a line in $\mathbb{R}^{q+1}$ and v is a point on that line. This is a subbundle of the trivial $(q + 1)$-plane bundle ε^{q+1}, and we write $\zeta_q^\perp$ for the complementary bundle whose points are pairs (x, w) such that w is orthogonal to the line x. Thus

$$\zeta_q \oplus \zeta_q^\perp \cong \varepsilon^{q+1}.$$

Write $H^*(\mathbb{R}P^q; \mathbb{Z}_2) = \mathbb{Z}_2[\alpha]/(\alpha^{q+1})$, $\deg \alpha = 1$. Thus $\alpha = w_1(\zeta_q)$. Since ζ_q is a line bundle, $w_i(\zeta_q) = 0$ for $i > 1$. The formula $w(\zeta_q) \cup w(\zeta_q^\perp) = 1$ implies that

$$w(\zeta_q^\perp) = 1 + \alpha + \cdots + \alpha^q.$$

We can describe $\tau(\mathbb{R}P^q)$ in terms of ζ_q. Consider a point $x \in S^q$ and write (x, v) for a typical vector in the tangent plane of S^q at x. Then x is orthogonal to v in $\mathbb{R}^{q+1}$ and (x, v) and $(-x, -v)$ have the same image in $\tau(\mathbb{R}P^q)$. If L_x is the line through x, then this image point determines and is determined by the linear map $f : L_x \longrightarrow L_x^\perp$ that sends x to v. Starting from this, it is easy to check that $\tau(\mathbb{R}P^q)$ is isomorphic to the bundle $\mathrm{Hom}(\zeta_q, \zeta_q^\perp)$. As for any line bundle, we have $\mathrm{Hom}(\zeta_q, \zeta_q) \cong \varepsilon$ since the identity homomorphisms of the fibers specify a cross-section. Again, as for any bundle over a smooth manifold, a choice of Euclidean metric determines an isomorphism $\mathrm{Hom}(\zeta_q, \varepsilon) \cong \zeta_q$. These facts give the following calculation of $\tau(\mathbb{R}P^q) \oplus \varepsilon$:

$$\begin{aligned}\tau(\mathbb{R}P^q) \oplus \varepsilon &\cong \mathrm{Hom}(\zeta_q, \zeta_q^\perp) \oplus \mathrm{Hom}(\zeta_q, \zeta_q) \\ &\cong \mathrm{Hom}(\zeta_q, \zeta_q^\perp \oplus \zeta_q) \cong \mathrm{Hom}(\zeta_q, \varepsilon^{q+1}) \\ &\cong (q+1)\mathrm{Hom}(\zeta_q, \varepsilon) \cong (q+1)\zeta_q.\end{aligned}$$

Therefore

$$w(\mathbb{R}P^q) = w((q+1)\zeta_q) = w(\zeta_q)^{q+1} = (1+\alpha)^{q+1} = \sum_{0 \le i \le q} \binom{q+1}{i} \alpha^i.$$

Explicit computations are obtained by computing mod 2 binomial coefficients.

For example, $w(\mathbb{R}P^q) = 1$ if and only if $q = 2^k - 1$ for some k (as the reader should check) and therefore $\mathbb{R}P^q$ can be parallelizable only if q is of this form. If $\mathbb{R}^{q+1}$ admits a bilinear product without zero divisors, then it is not hard to prove that $\tau(\mathbb{R}P^q) \cong \mathrm{Hom}(\zeta_q, \zeta_q^\perp)$ admits q linearly independent cross-sections and is therefore trivial. We conclude that $\mathbb{R}^{q+1}$ can admit such a product only if $q+1 = 2^k$ for some k. The real numbers, complex numbers, quaternions, and Cayley numbers show that there is such a product for $q + 1 = 1, 2, 4$, and 8. As we shall explain in the next chapter, these are in fact the only q for which $\mathbb{R}^{q+1}$ admits such a product.

While the calculation of $w(\mathbb{R}P^q)$ just given is quite special, there is a remarkable general recipe, called the "Wu formula," for the computation of $w(M)$ in terms of Poincaré duality and the Steenrod operations in $H^*(M; \mathbb{Z}_2)$. In analogy with $w(M)$, we define the total Steenrod square of an element x by $Sq(x) = \sum_i Sq^i(x)$.

THEOREM (Wu formula). *Let M be a smooth closed n-manifold with fundamental class $z \in H^n(M; \mathbb{Z}_2)$. Then the total Stiefel-Whitney class $w(M)$ is equal* to $Sq(v)$, where $v = \sum v_i \in H^{**}(M; \mathbb{Z}_2)$ is the unique cohomology class such that

$$\langle v \cup x, z \rangle = \langle Sq(x), z \rangle$$

for all $x \in H^*(M; \mathbb{Z}_2)$. Thus, for $k \geq 0$, $v_k \cup x = Sq^k(x)$ for all $x \in H^{n-k}(M; \mathbb{Z}_2)$, and

$$w_k(M) = \sum_{i+j=k} Sq^i(v_j).$$

Here the existence and uniqueness of v is an easy exercise from the Poincaré duality theorem. The basic reason that such a formula holds is that the Stiefel-Whitney classes can be defined in terms of the Steenrod operations, as we shall see shortly. The Wu formula implies that the Stiefel-Whitney classes are homotopy invariant: if $f : M \longrightarrow M'$ is a homotopy equivalence between smooth closed n-manifolds, then $f^* : H^*(M'; \mathbb{Z}_2) \longrightarrow H^*(M; \mathbb{Z}_2)$ satisfies $f^*(w(M')) = w(M)$. In fact, the conclusion holds for any map f, not necessarily a homotopy equivalence, that induces an isomorphism in mod 2 cohomology. Since the tangent bundle of M depends on its smooth structure, this is rather surprising.

4. Characteristic numbers of manifolds

Characteristic classes determine important numerical invariants of manifolds, called their characteristic numbers.

DEFINITION. Let M be a smooth closed R-oriented n-manifold with fundamental class $z \in H_n(M; R)$. For a characteristic class c of degree n, define the tangential characteristic number $c[M] \in R$ by $c[M] = \langle c(\tau(M)), z \rangle$. Similarly, define the normal characteristic number $c[\nu(M)]$ by $c[\nu(M)] = \langle c(\nu(M)), z \rangle$, where $\nu(M)$ is the normal bundle associated to an embedding of M in $\mathbb{R}^q$ for q sufficiently large. (These numbers are well defined because any two embeddings of M in $\mathbb{R}^q$ for large q are isotopic and have equivalent normal bundles.)

In particular, if r_i are integers such that $\sum i r_i = n$, then the monomial $w_1^{r_1} \cdots w_n^{r_n}$ is a characteristic class of degree n, and all mod 2 characteristic classes of degree n are linear combinations of these. Different manifolds can have the same Stiefel-Whitney numbers. In fact, we have the following observation.

LEMMA. *If M is the boundary of a smooth compact $(n+1)$-manifold W, then all tangential Stiefel-Whitney numbers of M are zero.*

PROOF. Using a smooth tubular neighborhood, we see that there is an inward-pointing normal vector field along M that spans a trivial bundle ε such that

$$\tau(W)|_M \cong \tau(M) \oplus \varepsilon.$$

Therefore, if $i : M \longrightarrow W$ is the inclusion, then $i^*(w_j(W)) = w_j(M)$. Let f be a polynomial in the w_j of degree n. Recall that the fundamental class of M is ∂z, where $z \in H_{n+1}(W, M)$ is the fundamental class of the pair (W, M). We have

$$\langle f(M), \partial z \rangle = \langle i^* f(W), \partial z \rangle = \langle f(W), i_* \partial z \rangle = 0$$

since $i_* \partial = 0$ by the long exact homology sequence of the pair. □

LEMMA. *All tangential Stiefel-Whitney numbers of a smooth closed manifold M are zero if and only if all normal Stiefel-Whitney numbers of M are zero.*

PROOF. The Whitney duality formula implies that every $w_i(M)$ is a polynomial in the $w_i(\nu(M))$ and every $w_i(\nu(M))$ is a polynomial in the $w_i(M)$. □

We shall explain the following amazing result of Thom in the last chapter.

THEOREM (Thom). *If M is a smooth closed n-manifold all of whose normal Stiefel-Whitney numbers are zero, then M is the boundary of a smooth $(n+1)$-manifold.*

Thus we need only compute the Stiefel-Whitney numbers of M to determine whether or not it is a boundary. By Wu's formula, the computation only requires knowledge of the mod 2 cohomology of M, with its Steenrod operations. In practice, it might be fiendishly difficult to actually construct a manifold with boundary M geometrically.

5. Thom spaces and the Thom isomorphism theorem

There are several ways to construct the Stiefel-Whitney classes. The most illuminating one depends on a simple, but fundamentally important, construction on vector bundles, namely their "Thom spaces." This construction will also be at the heart of the proof of Thom's theorem in the last chapter.

DEFINITION. Let $\xi : E \longrightarrow B$ be an n-plane bundle. Apply one-point compactification to each fiber of ξ to obtain a new bundle $Sph(E)$ over B whose fibers are spheres S^n with given basepoints, namely the points at ∞. These basepoints specify a cross-section $B \longrightarrow Sph(E)$. Define the Thom space $T\xi$ to be the quotient space $T(\xi) = Sph(E)/B$. That is, $T(\xi)$ is obtained from E by applying fiberwise one-point compactification and then identifying all of the points at ∞ to a single basepoint (denoted ∞). Observe that this construction is functorial with respect to maps of vector bundles.

REMARK. If we give the bundle ξ a Euclidean metric and let $D(E)$ and $S(E)$ denote its unit disk bundle and unit sphere bundle, then there is an evident homeomorphism between $T\xi$ and the quotient space $D(E)/S(E)$. In turn, $D(E)/S(E)$ is homotopy equivalent to the cofiber of the inclusion $S(E) \longrightarrow D(E)$ and therefore to the cofiber of the projection $S(E) \longrightarrow B$.

If the bundle ξ is trivial, so that $E = B \times \mathbb{R}^n$, then $Sph(E) = B \times S^n$. Quotienting out B amounts to the same thing as giving B a disjoint basepoint and then forming the smash product $B_+ \wedge S^n$. That is, in this case the Thom complex is $\Sigma^n B_+$. Therefore, for any cohomology theory k^*,

$$k^q(B) = \tilde{k}^q(B_+) \cong \tilde{k}^{n+q}(T\xi).$$

There is a conceptual way of realizing this isomorphism. For any n-plane bundle $\xi : E \longrightarrow B$, we have a projection $\xi : Sph(E) \longrightarrow B$ and a quotient map $\pi : Sph(E) \longrightarrow T\xi$. We can compose their product with the diagonal map of $Sph(E)$ to obtain a composite map

$$Sph(E) \longrightarrow Sph(E) \times Sph(E) \longrightarrow B \times T\xi.$$

This sends all points at ∞ to points of $B \times \{\infty\}$. Therefore it factors through a map

$$\Delta : T\xi \longrightarrow B_+ \wedge T\xi,$$

5. THOM SPACES AND THE THOM ISOMORPHISM THEOREM

which is called the "Thom diagonal." For a commutative ring R, we can use Δ to define a cup product

$$H^p(B;R) \otimes \tilde{H}^q(T\xi;R) \longrightarrow \tilde{H}^{p+q}(T\xi;R).$$

When the bundle ξ is trivial, we let $\mu \in \tilde{H}^n(B_+ \wedge S^n; R)$ be the suspension of the identity element $1 \in H^0(B; R)$, and we find that $x \longrightarrow x \cup \mu$ specifies the suspension isomorphism $H^q(B;R) \cong \tilde{H}^{n+q}(B_+ \wedge S^n; R) = \tilde{H}^{n+q}(T\xi; R)$.

Now consider a general bundle ξ. On neighborhoods U of B over which ξ is trivial, we have $H^q(U;R) \cong \tilde{H}^{n+q}(T(\xi|_U); R)$. The isomorphism depends on the trivialization $\phi_U : U \times \mathbb{R}^n \longrightarrow \xi^{-1}(U)$. It is natural to ask if these isomorphisms patch together to give a global isomorphism $H^q(B_+) \longrightarrow \tilde{H}^{n+q}(T\xi)$. This should look very similar to the problem of patching local fundamental classes to obtain a global one; that is, it looks like a question of orientation. This leads to the following definition and theorem. For a point $b \in B$, let S_b^n be the one-point compactification of the fiber $\xi^{-1}(b)$; since S_b^n is the Thom space of $\xi|_b$, we have a canonical map $i_b : S_b^n \longrightarrow T\xi$.

DEFINITION. Let $\xi : E \longrightarrow B$ be an n-plane bundle. An R-orientation, or Thom class, of ξ is an element $\mu \in \tilde{H}^n(T\xi; R)$ such that, for every point $b \in B$, $i_b^*(\mu)$ is a generator of the free R-module $\tilde{H}^n(S_b^n)$.

We leave it as an instructive exercise to verify that an R-orientation of a closed n-manifold M determines and is determined by an R-orientation of its tangent bundle $\tau(M)$.

THEOREM (Thom isomorphism theorem). *Let* $\mu \in \tilde{H}^n(T\xi; R)$ *be a Thom class for an n-plane bundle* $\xi : E \longrightarrow B$. *Define*

$$\Phi : H^q(B;R) \longrightarrow \tilde{H}^{n+q}(T\xi; R)$$

by $\Phi(x) = x \cup \mu$. *Then* Φ *is an isomorphism.*

SKETCH PROOF. When R is a field, this can be proved by an inductive Mayer-Vietoris sequence argument. To exploit inverse images of open subsets of B, it is convenient to observe that, by easy homotopy and excision arguments,

$$\tilde{H}^*(T\xi) \cong H^*(Sph(E), B) \cong H^*(Sph(E), Sph(E)_0) \cong H^*(E, E_0),$$

where E_0 and $Sph(E)_0$ are the subspaces of E and $Sph(E)$ obtained by deleting $\{0\}$ from each fiber. Use of a field ensures that the cohomology of the relevant direct limits is the inverse limit of the cohomologies. An alternative argument that works for general R can be obtained by first showing that one can assume that B is a CW complex, by replacing ξ by its pullback along a CW approximation of B, and then proceeding by induction over the restrictions of ξ to the skeleta of B; one point is that the restriction of ξ to any cell is trivial and another is that the cohomology of B is the inverse limit of the cohomologies of its skeleta. However, much the best proof from the point of view of anyone seriously interested in algebraic topology is to apply the Serre spectral sequence of the bundle $Sph(E)$. The Serre spectral sequence is a device for computing the cohomology of the total space E of a fibration from the cohomologies of its base B and fiber F. It measures the cohomological deviation of $H^*(E)$ from $H^*(B) \otimes H^*(F)$. In the present situation, the existence of a Thom class ensures that there is no deviation for the sphere bundle $Sph(E) \longrightarrow B$, so that

$$H^*(Sph(E); R) \cong H^*(B; R) \otimes H^*(S^n; R).$$

The section given by the points at ∞ induces an isomorphism of $H^*(B;R) \otimes H^0(S^n;R)$ with $H^*(B;R)$, and the quotient map $Sph(E) \longrightarrow T\xi$ induces an isomorphism of $\tilde{H}^*(T\xi;R)$ with $H^*(B;R) \otimes H^n(S^n;R)$. □

Just as in orientation theory for manifolds, the question of orientability depends on the structure of the units of the ring R, and this leads to the following conclusion.

PROPOSITION. *Every vector bundle admits a unique $\mathbb{Z}_2$-orientation.*

This can be proved along with the Thom isomorphism theorem by a Mayer-Vietoris argument.

6. The construction of the Stiefel-Whitney classes

We indicate two constructions of the Stiefel-Whitney classes. Each has distinct advantages over the other. First, taking the characteristic class point of view, we define the Stiefel-Whitney classes in terms of the Steenrod operations by setting

$$w_i(\xi) = \Phi^{-1} Sq^i \Phi(1) = \Phi^{-1} Sq^i \mu.$$

Naturality is obvious. Axiom 1 is immediate from the relations $Sq^0 = $ id and $Sq^i(x) = 0$ if $i > \deg x$. For axiom 2, we use the following observation.

LEMMA. *There is a homotopy equivalence $j : \mathbb{R}P^\infty \longrightarrow T\gamma_1$.*

PROOF. $T\gamma_1$ is homeomorphic to $D(\gamma_1)/S(\gamma_1)$. Here $S(\gamma_1)$ is the infinite sphere S^∞, which is the universal cover of $\mathbb{R}P^\infty$ and is therefore contractible. The zero section $\mathbb{R}P^\infty \longrightarrow D(\gamma_1)$ and the quotient map $D(\gamma_1) \longrightarrow T\gamma_1$ are homotopy equivalences, and their composite is the required homotopy equivalence j. □

Since $Sq^1(x) = x^2$ if $\deg x = 1$, the lemma implies that Sq^1 is non-zero on the Thom class of γ_1, verifying axiom 2. For axiom 3, we easily check that $T(\xi \oplus \varepsilon) \cong \Sigma T(\xi)$ for any vector bundle ξ and that the Thom class of $\xi \oplus \varepsilon$ is the suspension of the Thom class of ξ. Thus axiom 3 follows from the stability of the Steenrod operations. For axiom 4, we easily check that, for any vector bundles ζ and ξ, $T(\zeta \times \xi) \cong T\zeta \wedge T\xi$ and the Thom class of $\zeta \times \xi$ is the tensor product of the Thom classes of ζ and ξ. Interpreting the Cartan formula for the Steenrod operations externally in the cohomology of products and therefore of smash products, we see that it implies axiom 4. That is, the properties that axiomatize the Steenrod operations directly imply the properties that axiomatize the Stiefel-Whitney classes.

We next take the classifying space point of view. As we shall explain in §8, passage from topological groups to their classifying spaces is a product-preserving functor, at least up to homotopy. We may embed $(\mathbb{Z}_2)^n = O(1)^n$ in $O(n)$ as the subgroup of diagonal matrices. The classifying space $BO(1)$ is $\mathbb{R}P^\infty$, and we obtain a map

$$\omega : (\mathbb{R}P^\infty)^n \simeq B(O(1)^n) \longrightarrow BO(n)$$

upon passage to classifying spaces. The symmetric group Σ_n is contained in $O(n)$ as the subgroup of permutation matrices, and the diagonal subgroup $O(1)^n$ is closed under conjugation by symmetric matrices. Application of the classifying space functor to conjugation by permutation matrices induces the corresponding permutation of the factors of $BO(1)^n$, and it induces the identity map on $BO(n)$. Indeed, up to homotopy, inner conjugation by an element of G induces the identity map on BG for any topological group G.

By the Künneth theorem, we see that
$$H^*((\mathbb{R}P^\infty)^n;\mathbb{Z}_2) = \otimes_{i=1}^n H^*(\mathbb{R}P^\infty;\mathbb{Z}_2) = \mathbb{Z}_2[\alpha_1,\ldots,\alpha_n],$$
where the generators α_i are of degree one. The symmetric group Σ_n acts on this cohomology ring by permuting the variables α_i. The subring $H^*((\mathbb{R}P^\infty)^n;\mathbb{Z}_2)^{\Sigma_n}$ of elements invariant under the action is the polynomial algebra on the elementary symmetric functions σ_i, $1 \leq i \leq n$, in the variables α_i. Here
$$\sigma_i = \sum \alpha_{j_1} \cdots \alpha_{j_i}, \quad 1 \leq j_1 < \cdots < j_n,$$
has degree i. The induced map $\omega^* : H^*(BO(n);\mathbb{Z}_2) \longrightarrow H^*((\mathbb{R}P^\infty)^n;\mathbb{Z}_2)$ takes values in $H^*((\mathbb{R}P^\infty)^n;\mathbb{Z}_2)^{\Sigma_n}$. We shall give a general reason why this is so in §8. The resulting map
$$\omega^* : H^*(BO(n);\mathbb{Z}_2) \longrightarrow H^*((\mathbb{R}P^\infty)^n;\mathbb{Z}_2)^{\Sigma_n}$$
is a ring homomorphism between polynomial algebras on generators of the same degrees. It turns out to be a monomorphism and therefore an isomorphism. We redefine the Stiefel-Whitney classes by letting w_i be the unique element such that $\omega^*(w_i) = \sigma_i$ for $1 \leq i \leq n$ and defining $w_0 = 1$ and $w_i = 0$ for $i > n$. Then axioms 1 and 2 for the Stiefel-Whitney classes are obvious, and we derive axioms 3 and 4 from algebraic properties of elementary symmetric functions.

One advantage of this approach is that, since we know the Steenrod operations on $H^*(\mathbb{R}P^\infty;\mathbb{Z}_2)$ and can read them off on $H^*((\mathbb{R}P^\infty)^n;\mathbb{Z}_2)$ by the Cartan formula, it leads to a purely algebraic calculation of the Steenrod operations in $H^*(BO(n);\mathbb{Z}_2)$. Explicitly, the following "Wu formula" holds:
$$Sq^i(w_j) = \sum_{t=0}^{i} \binom{j+t-i-1}{t} w_{i-t}w_{j+t}.$$

7. Chern, Pontryagin, and Euler classes

The theory of the previous sections extends appropriately to complex vector bundles and to oriented real vector bundles. The proof of the classification theorem for complex n-plane bundles works in exactly the same way as for real n-plane bundles, using complex Grassmann varieties. For oriented real n-plane bundles, we use the Grassmann varieties of oriented n-planes, the points of which are planes x together with a chosen orientation. In fact, the fundamental groups of the real Grassmann varieties are $\mathbb{Z}_2$, and their universal covers are their orientation covers. These covers are the oriented Grassmann varieties $\tilde{G}_n(\mathbb{R}^q)$. We write $BU(n) = G_n(\mathbb{C}^\infty)$ and $BSO(n) = \tilde{G}_n(\mathbb{R}^\infty)$, and we construct universal complex n-plane bundles $\gamma_n : EU_n \longrightarrow BU(n)$ and oriented n-plane bundles $\tilde{\gamma}_n : \tilde{E}_n \longrightarrow BSO(n)$ as in the first section. Let $\mathscr{E}U_n(B)$ denote the set of equivalence classes of complex n-plane bundles over B and let $\tilde{\mathscr{E}}_n(B)$ denote the set of equivalence classes of oriented real n-plane bundles over B; it is required that bundle maps (g, f) be orientation preserving, in the sense that the induced map of Thom spaces carries the orientation of the target bundle to the orientation of the source bundle. The universal bundle $\tilde{\gamma}_n$ has a canonical orientation which determines an orientation on $f^*\tilde{E}_n$ for any map $f : B \longrightarrow BSO(n)$.

THEOREM. *The natural transformation* $\Phi : [-, BU(n)] \longrightarrow \mathscr{E}U_n(-)$ *obtained by sending the homotopy class of a map* $f : B \longrightarrow BU(n)$ *to the equivalence class of the n-plane bundle* f^*EU_n *is a natural isomorphism of functors.*

THEOREM. *The natural transformation* $\Phi : [-, BSO(n)] \longrightarrow \tilde{\mathscr{E}}_n(-)$ *obtained by sending the homotopy class of a map* $f : B \longrightarrow BSO(n)$ *to the equivalence class of the oriented n-plane bundle* $f^*\tilde{E}_n$ *is a natural isomorphism of functors.*

The definition of characteristic classes for complex n-plane bundles and for oriented real n-plane bundles in a cohomology theory k^* is the same as for real n-plane bundles, and the Yoneda lemma applies.

LEMMA. *Evaluation on* γ_n *specifies a canonical bijection between characteristic classes of complex n-plane bundles and elements of* $k^*(BU(n))$.

LEMMA. *Evaluation on* $\tilde{\gamma}_n$ *specifies a canonical bijection between characteristic classes of oriented n-plane bundles and elements of* $k^*(BSO(n))$.

Clearly we have a 2-fold cover $\pi_n : BSO(n) \longrightarrow BO(n)$. The mod 2 characteristic classes for oriented n-plane bundles are as one might expect from this. Continue to write w_i for $\pi^*(w_i) \in H^i(BSO(n); \mathbb{Z}_2)$; here $w_1 = 0$ since $BSO(n)$ is simply connected.

THEOREM. $H^*(BSO(n); \mathbb{Z}_2) \cong \mathbb{Z}_2[w_2, \ldots, w_n]$.

If we regard a complex n-plane bundle as a real $2n$-plane bundle, then the complex structure induces a canonical orientation. By the Yoneda lemma, the resulting natural transformation $r : \mathscr{E}U_n(-) \longrightarrow \tilde{\mathscr{E}}_n(-)$ is represented by a map $r : BU(n) \longrightarrow BSO(2n)$. Explicitly, ignoring its complex structure, we may identify $\mathbb{C}^\infty$ with $\mathbb{R}^\infty \oplus \mathbb{R}^\infty \cong \mathbb{R}^\infty$ and so regard a complex n-plane in $\mathbb{C}^\infty$ as an oriented n-plane in $\mathbb{R}^\infty$. Similarly, we may complexify real bundles fiberwise and so obtain a natural transformation $c : \mathscr{E}_n(-) \longrightarrow \mathscr{E}U_n(-)$. It is represented by a map $c : BO(n) \longrightarrow BU(n)$. Explicitly, identifying $\mathbb{C}^\infty$ with $\mathbb{R}^\infty \otimes_\mathbb{R} \mathbb{C}$, we may complexify an n-plane in $\mathbb{R}^\infty$ to obtain an n-plane in $\mathbb{C}^\infty$.

The Thom space of a complex or oriented real vector bundle is the Thom space of its underlying real vector bundle. We obtain characteristic classes in cohomology with any coefficients by applying cohomology operations to Thom classes, but it is rarely the case that the resulting characteristic classes generate all characteristic classes: the cases $H^*(BO(n); \mathbb{Z}_2)$ and $H^*(BSO(n); \mathbb{Z}_2)$ are exceptional. Characteristic classes constructed in this fashion satisfy homotopy invariance properties that fail for general characteristic classes.

In the complex case, with integral coefficients, we have a parallel to our second approach to Stiefel-Whitney classes that leads to a description of $H^*(BU(n); \mathbb{Z})$ in terms of Chern classes. We may embed $(S^1)^n = U(1)^n$ in $U(n)$ as the subgroup of diagonal matrices. The classifying space $BU(1)$ is $\mathbb{C}P^\infty$, and we obtain a map

$$\omega : (\mathbb{C}P^\infty)^n \simeq B(U(1)^n) \longrightarrow BU(n)$$

upon passage to classifying spaces. The symmetric group Σ_n is contained in $U(n)$ as the subgroup of permutation matrices, and the diagonal subgroup $U(1)^n$ is closed under conjugation by symmetric matrices. Application of the classifying space functor to conjugation by permutation matrices induces the corresponding permutation of the factors of $BU(1)^n$, and it induces the identity map on $BU(n)$.

By the Künneth theorem, we see that

$$H^*((\mathbb{C}P^\infty)^n; \mathbb{Z}) = \otimes_{i=1}^n H^*(\mathbb{C}P^\infty; \mathbb{Z}) = \mathbb{Z}[\beta_1, \ldots, \beta_n],$$

where the generators β_i are of degree two. The symmetric group Σ_n acts on this cohomology ring by permuting the variables β_i. The subring $H^*((\mathbb{C}P^\infty)^n; \mathbb{Z})^{\Sigma_n}$ of elements invariant under the action is the polynomial algebra on the elementary symmetric functions σ_i, $1 \leq i \leq n$, in the variables β_i. Here

$$\sigma_i = \sum \beta_{j_1} \cdots \beta_{j_i}, \quad 1 \leq j_1 < \cdots < j_n,$$

has degree $2i$. The induced map $\omega^* : H^*(BU(n); \mathbb{Z}) \longrightarrow H^*((\mathbb{C}P^\infty)^n; \mathbb{Z})$ takes values in $H^*((\mathbb{C}P^\infty)^n; \mathbb{Z})^{\Sigma_n}$. The resulting map

$$\omega^* : H^*(BU(n); \mathbb{Z}) \longrightarrow H^*((\mathbb{C}P^\infty)^n; \mathbb{Z})^{\Sigma_n}$$

is a ring homomorphism between polynomial algebras on generators of the same degrees. It turns out to be a monomorphism and thus an isomorphism when tensored with any field, and it is therefore an isomorphism. We define the Chern classes by letting c_i, $1 \leq i \leq n$, be the unique element such that $\omega^*(c_i) = \sigma_i$.

THEOREM. *For $n \geq 1$, there are elements $c_i \in H^{2i}(BU(n); \mathbb{Z})$, $i \geq 0$, called the Chern classes. They satisfy and are uniquely characterized by the following axioms.*

1. $c_0 = 1$ and $c_i = 0$ if $i > n$.
2. c_1 is the canonical generator of $H^2(BU(1); \mathbb{Z})$ when $n = 1$.
3. $i_n^*(c_i) = c_i$.
4. $p_{m,n}^*(c_i) = \sum_{j=0}^{i} c_j \otimes c_{i-j}$.

The integral cohomology $H^*(BU(n); \mathbb{Z})$ is the polynomial algebra $\mathbb{Z}[c_1, \ldots, c_n]$.

Here we take axiom 1 as a definition and we interpret axiom 2 as meaning that c_1 corresponds to the identity map of $\mathbb{C}P^\infty$ under the canonical identification of $[\mathbb{C}P^\infty, \mathbb{C}P^\infty]$ with $H^2(\mathbb{C}P^\infty; \mathbb{Z})$. Axioms 3 and 4 can be read off from algebraic properties of elementary symmetric functions. The theorem admits an immediate interpretation in terms of characteristic classes. Observe that, since $H^*(BU(n); \mathbb{Z})$ is a free Abelian group, the theorem remains true precisely as stated with $\mathbb{Z}$ replaced by any other commutative ring of coefficients R. We continue to write c_i for the image of c_i in $H^*(BU(n); R)$ under the homomorphism induced by the unit $\mathbb{Z} \longrightarrow R$ of the ring R.

The reader deserves to be warned about a basic inconsistency in the literature.

REMARK. With the discussion above, $c_1(\gamma_1^{n+1})$ is the canonical generator of $H^2(\mathbb{C}P^n; \mathbb{Z})$, where γ_1^{n+1} is the canonical line bundle of lines in $\mathbb{C}^{n+1}$ and points on the line. This is the standard convention in algebraic topology. In algebraic geometry, it is more usual to define Chern classes so that the first Chern class of the dual of γ_1^{n+1} is the canonical generator of $H^2(\mathbb{C}P^n; \mathbb{Z})$. With this convention, the nth Chern class would be $(-1)^n c_n$. It is often unclear in the literature which convention is being followed.

Turning to oriented real vector bundles, we define the Pontryagin and Euler classes as follows, taking cohomology with coefficients in any commutative ring R.

DEFINITION. Define the Pontryagin classes $p_i \in H^{4i}(BO(n); R)$ by

$$p_i = (-1)^i c^*(c_{2i}),$$

$c^* : H^{4i}(BU(n); R) \longrightarrow H^{4i}(BO(n); R)$; also write p_i for $\pi_n^*(p_i) \in H^{4i}(BSO(n); R)$.

DEFINITION. Define the Euler class $e(\xi) \in H^n(B; R)$ of an R-oriented n-plane bundle ξ over the base space B by $e(\xi) = \Phi^{-1}\mu^2$, where $\mu \in H^n(T\xi; R)$ is the Thom class. Giving the universal oriented n-plane bundle over $BSO(n)$ the R-orientation induced by its integral orientation, this defines the Euler class $e \in H^n(BSO(n); R)$.

If n is odd, then $2\mu^2 = 0$ and thus $2e = 0$. If $R = \mathbb{Z}_2$, then $Sq^n(\mu) = \mu^2$ and thus $e = w_n$. The name "Euler class" is justified by the following classical result, which well illustrates the kind of information that characteristic numbers can encode.[1]

THEOREM. *If M is a smooth closed oriented manifold, then the characteristic number $e[M] = \langle e(\tau(M)), z \rangle \in \mathbb{Z}$ is the Euler characteristic of M.*

The evident inclusion $T^n \cong SO(2)^n \longrightarrow SO(2n)$ is a maximal torus, and it induces a map $BT^n \longrightarrow BSO(2n)$. A calculation shows that e restricts to the nth elementary symmetric polynomial $\beta_1 \cdots \beta_n$. The cited inclusion factors through the homomorphism $U(n) \longrightarrow SO(2n)$, hence $BT^n \longrightarrow BSO(2n)$ factors through $r : BU(n) \longrightarrow BSO(2n)$. This implies another basic fact about the Euler class.

PROPOSITION. $r^* : H^*(BSO(2n); \mathbb{Z}) \longrightarrow H^*(BU(n); \mathbb{Z})$ *sends e to c_n.*

The presence of 2-torsion makes the description of the integral cohomology rings of $BO(n)$ and $BSO(n)$ quite complicated, and these rings are almost never used in applications. Rather, one uses the mod 2 cohomology rings and the following description of the cohomology rings that result by elimination of 2-torsion.

THEOREM. *Take coefficients in a ring R in which 2 is a unit. Then*

$$H^*(BO(2n)) \cong H^*(BO(2n+1)) \cong H^*(BSO(2n+1)) \cong R[p_1, \ldots, p_n]$$

and

$$H^*(BSO(2n)) \cong R[p_1, \ldots, p_{n-1}, e], \text{ with } e^2 = p_n.$$

8. A glimpse at the general theory

We should place the theory of vector bundles in a more general context. We have written $BO(n)$, $BU(n)$, and $BSO(n)$ for certain "classifying spaces" in this chapter, but we defined a classifying space BG for any topological group G in Chapter 16 §5. In fact, the spaces here are homotopy equivalent to the spaces of the same name that we defined there, and we here explain why.

Consider bundles $\xi : Y \longrightarrow B$ with fiber G. For spaces U in a numerable open cover $\mathscr{O}$ of B, there are homeomorphisms $\phi : U \times G \longrightarrow p^{-1}(U)$ such that $p \circ \phi = \pi_1$. We say that Y is a principal G-bundle if Y has a free right action by G, B is the orbit space Y/G, ξ is the quotient map, and the ϕ are maps of right G-spaces. We say that $\xi : Y \longrightarrow B$ is a universal principal G-bundle if Y is a contractible space. In particular, for any topological group G whose identity element is a nondegenerate basepoint, such as any Lie group G, the map $p : EG \longrightarrow BG$ constructed in Chapter 16 §5 is a universal principal G-bundle. The classification theorem below implies that the base spaces of any two universal principal G-bundles are homotopy equivalent, and it is usual to write BG for any space in this homotopy type. Observe that the long exact sequence of homotopy groups of a universal principal G-bundle gives isomorphisms $\pi_q(BG) \cong \pi_{q-1}(G)$ for $q \geq 1$.

We have implicitly constructed other examples of universal principal G-bundles when G is $O(n)$, $U(n)$, or $SO(n)$. To see this, consider $V_n(\mathbb{R}^q)$. Write $\mathbb{R}^q = \mathbb{R}^n \times \mathbb{R}^{q-n}$ and note that this fixes embeddings of $O(n)$ and $O(q-n)$ in the orthogonal group $O(q)$. Of course, $O(q)$ acts on vectors in $\mathbb{R}^q$ and thus on n-frames. Consider the fixed n-frame $x_0 = \{e_1, \ldots, e_n\}$. Any other n-frame can be obtained from

---
[1] See Corollary 11.12 of Milnor and Stasheff *Characteristic Classes* for a proof.

this one by the action of an element of $O(q)$, and the isotropy group of x_0 is $O(q-n)$. Thus the action of $O(q)$ is transitive, and evaluation on x_0 induces a homeomorphism $O(q)/O(q-n) \longrightarrow V_n(\mathbb{R}^q)$ of $O(q)$-spaces. The action of $O(n) \subset O(q)$ is free, and passage to orbits gives a homeomorphism $O(q)/O(n) \times O(q-n) \longrightarrow G_n(\mathbb{R}^q)$. It is intuitively clear and not hard to prove that the colimit over q of the inclusions $O(q-n) \longrightarrow O(q)$ is a homotopy equivalence and that this implies the contractibility of $V_n(\mathbb{R}^\infty)$. We deduce that $V_n(\mathbb{R}^\infty)$ is a universal principal $O(n)$-bundle. We have analogous universal principal $U(n)$-bundles and $SO(n)$-bundles.

There is a classification theorem for principal G-bundles. Let $\mathscr{P}G(B)$ denote the set of equivalence classes of principal G-bundles over B, where two principal G-bundles over B are equivalent if there is a G-homeomorphism over B between them. Via pullback of bundles, this is a contravariant set-valued functor on the homotopy category of spaces.

THEOREM. *Let $\gamma : Y \longrightarrow Y/G$ be any universal principal G-bundle. The natural transformation $\Phi : [-, Y/G] \longrightarrow \mathscr{P}G(-)$ obtained by sending the homotopy class of a map $f : B \longrightarrow Y/G$ to the equivalence class of the principal G-bundle f^*Y is a natural isomorphism of functors.*

Now let F be any space on which G acts effectively from the left. Here an action is effective if $gf = f$ for every $f \in F$ implies $g = e$. For a principal G-bundle Y, let G act on $Y \times F$ by $g(y, f) = (yg^{-1}, gf)$ and let $Y \times_G F$ be the orbit space $(Y \times F)/G$. With the correct formal definition of a fiber bundle with group G and fiber F, every such fiber bundle $p : E \longrightarrow B$ is equivalent to one of the form $Y \times_G F \longrightarrow Y/G \cong B$ for some principal G-bundle Y over B; moreover Y is uniquely determined up to equivalence.

In fact, the "associated principal G-bundle" Y can be constructed as the function space of all maps $\psi : F \longrightarrow E$ such that ψ is an admissible homeomorphism onto some fiber $F_b = p^{-1}(b)$. Here admissibility means that the composite of ψ with the homeomorphism $F_b \longrightarrow F$ determined by a coordinate chart $\phi : U \times F \xrightarrow{\cong} p^{-1}(U)$, $b \in U$, coincides with action by some element of G. The left action of G on F induces a right action of G on Y; this action is free because the given action on F is effective. The projection $Y \longrightarrow B$ sends ψ to b when $\psi : F \xrightarrow{\cong} F_b$, and it factors through a homeomorphism $Y/G \longrightarrow B$. Y inherits local triviality from p, and the evaluation map $Y \times F \longrightarrow E$ induces an equivalence of bundles $Y \times_G F \longrightarrow E$.

We conclude that, for any F, $\mathscr{P}G(B)$ is naturally isomorphic to the set of equivalence classes of bundles with group G and fiber F over B. Fiber bundles with group $O(n)$ and fiber $\mathbb{R}^n$ are real n-plane bundles, fiber bundles with group $U(n)$ and fiber $\mathbb{C}^n$ are complex n-plane bundles, and fiber bundles with group $SO(n)$ and fiber $\mathbb{R}^n$ are oriented real n-plane bundles. Thus the classification theorems of the previous sections could all be rederived as special cases of the general classification theorem for principal G-bundles stated in this section.

In our discussion of Stiefel-Whitney and Chern classes, we used that passage to classifying spaces is a product-preserving functor, at least up to homotopy. For the functoriality, if $f : G \longrightarrow H$ is a homomorphism of topological groups, then consideration of the way bundles are constructed by gluing together coordinate charts shows that a principal G-bundle $\xi : Y \longrightarrow B$ naturally gives rise to a principal H-bundle $f_*Y \longrightarrow B$. This construction is represented on the classifying space level by a map $Bf : BG \longrightarrow BH$.

In fact, if $EG \longrightarrow BG$ and $EH \longrightarrow BH$ are universal principal bundles, then any map $\tilde{f} : EG \longrightarrow EH$ such that $\tilde{f}(xg) = \tilde{f}(x)f(g)$ for all $x \in EG$ and $g \in G$ induces a map in the homotopy class Bf on passage to orbits. For example, if $f : G \longrightarrow G$ is given by conjugation by $\gamma \in G$, $f(g) = \gamma^{-1}g\gamma$, then $\tilde{f}(x) = x\gamma$ satisfies this equivariance property and therefore Bf is homotopic to the identity. This explains why inner conjugations induce the identity map on passage to classifying spaces, as we used in our discussion of Stiefel-Whitney and Chern classes.

If $EG \longrightarrow BG$ and $EG' \longrightarrow BG'$ are universal principal G and G' bundles, then $EG \times EG'$ is a contractible space with a free action by $G \times G'$. The orbit space is $BG \times BG'$, and this shows that $BG \times BG'$ is a choice for the classifying space $B(G \times G')$ and is therefore homotopy equivalent to any other choice.

The explicit construction of BG given in Chapter 16 §5 is functorial in G on the point-set level and not just up to homotopy, and it is product preserving in the strong sense that the projections induce a homeomorphism $B(H \times G) \cong BH \times BG$.

PROBLEMS

1. Verify that $w(\mathbb{R}P^q) = 1$ if and only if $q = 2^k - 1$ for some k.
2. Prove that $\mathbb{R}P^{2^k}$ cannot immerse in $\mathbb{R}^{2^{k+1}-2}$. (By the Whitney embedding theorem, any smooth closed n-manifold immerses in $\mathbb{R}^{2n-1}$, so this is a best possible non-immersion result.)
3. Prove that all tangential Stiefel-Whitney numbers of $\mathbb{R}P^q$ are zero if and only if q is odd.
4. * Try to construct a smooth compact manifold whose boundary is $\mathbb{R}P^3$.
5. Prove that a smooth closed n-manifold M is R-orientable if and only its tangent bundle is R-orientable.

CHAPTER 24

An introduction to K-theory

The first generalized cohomology theory to be discovered was K-theory, and it plays a vital role in the connection of algebraic topology to analysis and algebraic geometry. The fact that it is a generalized cohomology theory is a consequence of the Bott periodicity theorem, which is one of the most important and influential theorems in all of topology. We give some basic information about K-theory and, following Adams and Atiyah, we explain how the Adams operations in K-theory allow a quick solution to the "Hopf invariant one problem." One implication is the purely algebraic theorem that the only possible dimensions of a real (not necessarily associative) division algebra are 1, 2, 4, and 8. We shall only discuss complex K-theory, although there is a precisely analogous construction of real K-theory KO. From the point of view of algebraic topology, real K-theory is a substantially more powerful invariant, but complex K-theory is usually more relevant to applications in other fields.

1. The definition of K-theory

Except where otherwise noted, we work with complex vector bundles throughout this chapter. Dimension will mean complex dimension and line bundles will mean complex line bundles. We consider the set $Vect(X)$ of equivalence classes of vector bundles over a space X. We assume unless otherwise specified that X is compact. We remind the reader that vector bundles can have different dimension over different components of X. The set $Vect(X)$ forms an Abelian monoid (= semi-group) under Whitney sum, and it forms a semi-ring with multiplication given by the (internal) tensor product of vector bundles over X.

There is a standard construction, called the Grothendieck construction, of an Abelian group $G(M)$ associated to an Abelian monoid M: one takes the quotient of the free Abelian group generated by the elements of M by the subgroup generated by the set of elements of the form $m + n - m \oplus n$, where $\oplus$ is the sum in M. The evident morphism of Abelian monoids $i : M \longrightarrow G(M)$ is universal: for any homomorphism of monoids $f : M \longrightarrow G$, where G is an Abelian group, there is a unique homomorphism of groups $\tilde{f} : G(M) \longrightarrow G$ such that $\tilde{f} \circ i = f$. If M is a semi-ring, then its multiplication induces a multiplication on $G(M)$ such that $G(M)$ is a ring, called the Grothendieck ring of M. If the semi-ring M is commutative, then the ring $G(M)$ is commutative.

DEFINITION. The K-theory of X, denoted $K(X)$, is the Grothendieck ring of the semi-ring $Vect(X)$. An element of $K(X)$ is called a virtual bundle over X. We write $[\xi]$ for the element of $K(X)$ determined by a vector bundle ξ.

Since ε is the identity element for the product in $K(X)$, it is standard to write $q = [\varepsilon^q]$, where ε^q is the q-dimensional trivial bundle. For vector bundles over a based space X, we have the function $d : Vect(X) \longrightarrow \mathbb{Z}$ that sends a vector bundle to the dimension of its restriction to the component of the basepoint $*$. Since d is a homomorphism of semi-rings, it induces a dimension function $d : K(X) \longrightarrow \mathbb{Z}$, which is a homomorphism of rings. Since d is an isomorphism when X is a point, d can be identified with the induced map $K(X) \longrightarrow K(*)$.

DEFINITION. The reduced K-theory $\tilde{K}(X)$ of a based space X is the kernel of $d : K(X) \longrightarrow \mathbb{Z}$. It is an ideal of $K(X)$ and thus a ring without identity. Clearly $K(X) \cong \tilde{K}(X) \times \mathbb{Z}$.

We have a homotopical interpretation of these definitions, and it is for this that we need X to be compact. By the classification theorem, we know that $\mathscr{E}U_n(X)$ is naturally isomorphic to $[X_+, BU(n)]$; we have adjoined a disjoint basepoint because we are thinking cohomologically and want the brackets to denote based homotopy classes of maps. We have maps $i_n : BU(n) \longrightarrow BU(n+1)$. With our construction of classifying spaces via Grassmannians, these maps are inclusions, and we define BU to be the colimit of the $BU(n)$, with the topology of the union.

We say that bundles ζ and ξ are stably equivalent if, for a sufficiently large q, the bundles $\zeta \oplus \varepsilon^{q-m}$ and $\xi \oplus \varepsilon^{q-n}$ are equivalent, where $m = d(\zeta)$ and $n = d(\xi)$. Let $\mathscr{E}U(X)$ be the set of stable equivalence classes of vector bundles over X. If X is connected, or if we restrict attention to vector bundles that are n-plane bundles for some n, then $\mathscr{E}U$ is isomorphic to $\operatorname{colim} \mathscr{E}U_n(X)$, where the colimit is taken over the maps $\mathscr{E}U_n(X) \longrightarrow \mathscr{E}U_{n+1}(X)$ obtained by sending a bundle ξ to $\xi \oplus \varepsilon$. Since a map from a compact space X into BU has image in one of the $BU(n)$, and similarly for homotopies, we see that in this case $[X_+, BU] \cong \operatorname{colim}[X_+, BU(n)]$ and therefore
$$\mathscr{E}U(X) \cong [X_+, BU].$$
A deeper use of compactness gives the following basic fact.

PROPOSITION. *If* $\xi : E \longrightarrow X$ *is a vector bundle over* X, *then there is a bundle* η *over* X *such that* $\xi \oplus \eta$ *is equivalent to* ε^q *for some* q.

SKETCH PROOF. The space ΓE of sections of E is a vector space under fiberwise addition and scalar multiplication. Using a partition of unity argument, one can show that there is a finite dimensional vector subspace V of $\Gamma(E)$ such that the map $g : X \times V \longrightarrow E$ specified by $g(x, s) = s(x)$ is an epimorphism of bundles over X. The resulting short exact sequence of vector bundles, like any other short exact sequence of vector bundles, splits as a direct sum, and the conclusion follows. □

COROLLARY. *Every virtual bundle over* X *can be written in the form* $[\xi] - q$ *for some bundle* ξ *and non-negative integer* q.

PROOF. Given a virtual bundle $[\omega] - [\zeta]$, where ω and ζ are bundles, choose η such that $\zeta \oplus \eta \cong \varepsilon^q$ and let $\xi = \omega \oplus \eta$. Then $[\omega] - [\zeta] = [\xi] - q$ in $K(X)$. □

COROLLARY. *There is a natural isomorphism* $\mathscr{E}U(X) \longrightarrow \tilde{K}(X)$.

PROOF. Writing $\{\xi\}$ for the stable equivalence class of a bundle ξ, the required isomorphism is given by the correspondence $\{\xi\} \leftrightarrow [\xi] - d(\xi)$. □

COROLLARY. *Give* $\mathbb{Z}$ *the discrete topology. For compact spaces* X, *there is a natural isomorphism*
$$K(X) \cong [X_+, BU \times \mathbb{Z}].$$

For nondegenerately based compact spaces X, there is a natural isomorphism
$$\tilde{K}(X) \cong [X, BU \times \mathbb{Z}].$$

PROOF. When X is connected, the first isomorphism sends $[\xi] - q$ to $(f, n - q)$, where ξ is an n-plane bundle with classifying map $f : X \longrightarrow BU(n) \subset BU$. The isomorphism for non-connected spaces follows since both functors send disjoint unions to Cartesian products. The second isomorphism follows from the first since $d : K(X) \longrightarrow \mathbb{Z}$ can be identified with the map $[X_+, BU \times \mathbb{Z}] \longrightarrow [S^0, BU \times \mathbb{Z}]$ induced by the cofibration $S^0 \longrightarrow X_+$, and the latter has kernel $[X, BU \times \mathbb{Z}]$ since $X_+/S^0 = X$. □

For general, non-compact, spaces X, it is best to define K-theory to mean represented K-theory. Here we implicitly apply CW approximation, or else use the definition in the following form.

DEFINITION. For a space X of the homotopy type of a CW complex, define
$$K(X) = [X_+, BU \times \mathbb{Z}].$$
For a nondegenerately based space of the homotopy type of a CW complex, define
$$\tilde{K}(X) = [X, BU \times \mathbb{Z}].$$

When X is compact, we know that $K(X)$ is a ring. It is natural to expect this to remain true for general X. That this is the case is a direct consequence of the following result, which the reader should regard as an aside.

PROPOSITION. *The space $BU \times \mathbb{Z}$ is a ring space up to homotopy. That is, there are additive and multiplicative H-space structures on $BU \times \mathbb{Z}$ such that the associativity, commutativity, and distributivity diagrams required of a ring commute up to homotopy.*

INDICATIONS OF PROOF. By passage to colimits over m and n, the maps $p_{m,n} : BU(m) \times BU(n) \longrightarrow BU(m+n)$ induce an "addition" $\oplus : BU \times BU \longrightarrow BU$. In fact, we can define BU in terms of planes in any copy of $\mathbb{C}^\infty$, and the explicit maps $p_{m,n}$ of Chapter 23 §2 pass to colimits to give
$$G_\infty(\mathbb{C}^\infty) \times G_\infty(\mathbb{C}^\infty) \longrightarrow G_\infty(\mathbb{C}^\infty \oplus \mathbb{C}^\infty);$$
use of an isomorphism $\mathbb{C}^\infty \oplus \mathbb{C}^\infty \cong \mathbb{C}^\infty$ gives the required map $\oplus$, which is well defined, associative, and commutative up to homotopy; the zero-dimensional plane provides a convenient basepoint 0 with which to check that we have a zero element up to homotopy. Using ordinary addition on $\mathbb{Z}$, we obtain the additive H-space structure on $BU \times \mathbb{Z}$. Tensor products of universal bundles give rise to classifying maps $q_{m,n} : BU(m) \times BU(n) \longrightarrow BU(mn)$. These do not pass to colimits so readily, since one must take into account the bilinearity of the tensor product, for example the relation $(\gamma_m \oplus \varepsilon) \otimes \gamma_n \cong (\gamma_m \otimes \gamma_n) \oplus \gamma_n$, and we merely affirm that, by fairly elaborate arguments, one can pass to colimits to obtain a product on $BU \times \mathbb{Z}$. It actually factors through the smash product with respect to the basepoint 0, since that acts as zero for the tensor product, and it restricts to an H-space structure on $BO \times \{1\}$ with basepoint $(0,1)$. □

The study of ring spaces such as this is a relatively new, and quite deep, part of algebraic topology. However, the reader should feel reasonably comfortable with the additive H-space structure on BU.

2. The Bott periodicity theorem

There are various ways to state, and various ways to prove, this basic result. We describe several versions and implications. One starting point is the following calculation. We have a canonical line bundle γ_1^2 over $S^2 \cong \mathbb{C}P^1$; its points are pairs (L, x), where L is a line in $\mathbb{C}^2$ and x is a point on that line. We let $H = \mathrm{Hom}(\gamma_1^2, \varepsilon)$ denote its dual.

THEOREM. $K(S^2)$ *is generated as a ring by* $[H]$ *subject to the single relation* $([H] - 1)^2 = 0$. *Therefore, as Abelian groups,* $K(S^2)$ *is free on the basis* $\{1, [H]\}$ *and* $\tilde{K}(S^2)$ *is free on the basis* $\{1 - [H]\}$.

INDICATION OF PROOF. We think of S^2 as the one-point compactification of $\mathbb{C}$ decomposed as the union of the unit disk D and the complement D' of the interior of D, so that $D \cap D' = S^1$. Any n-plane bundle over S^2 restricts to a trivial bundle over D and D', and these trivial bundles restrict to the same bundle over S^1. Conversely, an isomorphism f from the trivial bundle over S^1 to itself gives a way to glue together the trivial bundles over D and D' to reconstruct a bundle over S^2. Say that two such "clutching functions" f are equivalent if the bundles they give rise to are equivalent. A careful analysis of the form of the possible clutching functions f leads to a canonical example in each equivalence class and thus to the required calculation. □

For any pair of spaces X and Y, we have a Künneth-type ring homomorphism
$$\alpha : K(X) \otimes K(Y) \longrightarrow K(X \times Y)$$
specified by $\alpha(x \otimes y) = \pi_1^*(x)\pi_2^*(y)$.

THEOREM (Bott periodicity). *For compact spaces* X,
$$\alpha : K(X) \otimes K(S^2) \longrightarrow K(X \times S^2)$$
is an isomorphism.

INDICATION OF PROOF. The restrictions to $X \times D$ and $X \times D'$ of a bundle over $X \times S^2$ are equivalent to pullbacks of bundles over X, and their further restrictions to S^1 are equivalent. Conversely, bundles ζ and ξ over X together with an equivalence f between the restrictions to $X \times S^1$ of the pullbacks of ζ and ξ to $X \times D$ and $X \times D'$ determine a bundle over $X \times S^2$. Again, a careful analysis, which is similar to that in the special case when $X = pt$, of the equivalence classes of the possible clutching data (ζ, f, ξ) leads to the conclusion. □

The following useful observation applies to any representable functor, not just K-theory.

LEMMA. *For nondegenerately based spaces X and Y, the projections of $X \times Y$ on X and on Y and the quotient map $X \times Y \longrightarrow X \wedge Y$ induce a natural isomorphism*
$$\tilde{K}(X \wedge Y) \oplus \tilde{K}(X) \oplus \tilde{K}(Y) \cong \tilde{K}(X \times Y),$$
and $\tilde{K}(X \wedge Y)$ is the kernel of the map $\tilde{K}(X \times Y) \longrightarrow \tilde{K}(X) \oplus \tilde{K}(Y)$ induced by the inclusions of X and Y in $X \times Y$.

PROOF. The inclusion $X \vee Y \longrightarrow X \times Y$ is a cofibration with quotient $X \wedge Y$, and X and Y are retracts of $X \times Y$ via the inclusions and projections. □

2. THE BOTT PERIODICITY THEOREM

It follows easily that the Künneth map $\alpha : K(X) \otimes K(Y) \longrightarrow K(X \times Y)$ induces a reduced Künneth map $\beta : \tilde{K}(X) \otimes \tilde{K}(Y) \longrightarrow \tilde{K}(X \wedge Y)$. We have a splitting

$$\tilde{K}(X) \otimes \tilde{K}(Y) \oplus \tilde{K}(X) \oplus \tilde{K}(Y) \oplus \mathbb{Z} \cong K(X) \otimes K(Y)$$

that is compatible with the splitting of the lemma. Therefore the following reduced form of the Bott periodicity theorem is equivalent to the unreduced form that we have already stated.

THEOREM (Bott periodicity). *For nondegenerately based compact spaces X,*

$$\beta : \tilde{K}(X) \otimes \tilde{K}(S^2) \longrightarrow \tilde{K}(X \wedge S^2) = \tilde{K}(\Sigma^2 X)$$

is an isomorphism.

Write $b = 1 - [H] \in \tilde{K}(S^2)$. Since $\tilde{K}(S^2) \cong \mathbb{Z}$ with generator b, the theorem implies that multiplication by the "Bott element" b specifies an isomorphism

$$[X, BU \times \mathbb{Z}] \cong \tilde{K}(X) \longrightarrow \tilde{K}(\Sigma^2 X) \cong [X, \Omega^2(BU \times \mathbb{Z})]$$

for nondegenerately based compact spaces X. Here the addition in the source and target is derived from the natural additive H-space structure on $BU \times \mathbb{Z}$ on the left and the displayed double loop space on the right. If we had this isomorphism for general non-compact spaces X, we could apply it with $X = BU \times \mathbb{Z}$ and see that it is induced by a homotopy equivalence of H-spaces

$$\beta : BU \times \mathbb{Z} \longrightarrow \Omega^2(BU \times \mathbb{Z}).$$

In fact, one can deduce such a homotopy equivalence from the Bott periodicity theorem as just stated, but there are more direct proofs. On the right, the double loop space obviously depends only on the basepoint component $BU = BU \times \{0\}$. Since $\pi_2(BU) = \mathbb{Z}$, a little argument with H-spaces shows that $\Omega^2(BU \times \mathbb{Z})$ is equivalent as an H-space to $(\Omega_0^2 BU) \times \mathbb{Z}$, where $\Omega_0^2 BU$ denotes the component of the basepoint in $\Omega^2 BU$. Using the identity function on the factor $\mathbb{Z}$, we see that what is needed is an equivalence of H-spaces $\beta : BU \longrightarrow \Omega_0^2 BU$. In fact, it is easily deduced from the form of Bott periodicity that, up to homotopy, β must be the adjoint of the composite

$$\Sigma^2 BU = BU \wedge S^2 \xrightarrow{\text{id} \wedge b} BU \wedge BU \xrightarrow{\otimes} BU.$$

The infinite unitary group U is defined to be the union of the unitary groups $U(n)$, where $U(n)$ is embedded in $U(n+1)$ as matrices with last row and column zero except for 1 on the diagonal. Then ΩBU is homotopy equivalent as an H-space to U. Since $\pi_1(U) = \mathbb{Z}$ and the universal cover of U is the infinite special unitary group SU, ΩU is equivalent as an H-space to $(\Omega SU) \times \mathbb{Z}$. Therefore β may be viewed as a map $BU \longrightarrow \Omega SU$. Bott's original proof of the Bott periodicity theorem used the Grassmannian model for BU to write down an explicit map β in the required homotopy class and then used Morse theory to prove that β is a homotopy equivalence.

Bott's map β can also be proved to be a homotopy equivalence using only basic algebraic topology. Since BU and ΩSU are simply connected spaces of the homotopy types of CW complexes, a relative version of the Hurewicz theorem called the Whitehead theorem shows that β will be a weak equivalence and therefore a homotopy equivalence if it induces an isomorphism on integral homology. Since $H^*(BU(n)) = \mathbb{Z}[c_1, \ldots, c_n]$, $H^*(BU) \cong \mathbb{Z}[c_i | i \geq 1]$. The H-space structure on BU is induced by the maps $p_{m,n}$, and we find that the map $\psi : H^*(BU) \longrightarrow$

$H^*(BU \times BU) \cong H^*(BU) \otimes H^*(BU)$ induced by the product is given by $\psi(c_k) = \sum_{i+j=k} c_i \otimes c_j$. A purely algebraic dualization argument proves that, as a ring,

$$H_*(BU) \cong \mathbb{Z}[\gamma_i | i \geq 1],$$

where γ_i is the image of a generator of $H_{2i}(\mathbb{C}P^\infty)$ under the map induced by the inclusion of $\mathbb{C}P^\infty = BU(1)$ in BU. One can calculate $H_*(\Omega SU)$ and see that it too is a polynomial algebra with an explicitly given generator in each even degree. A direct inspection of the map β shows that it carries generators to generators.

In any case, it should now be clear that we have a periodic Ω-prespectrum and therefore a generalized cohomology theory represented by it.

DEFINITION. The K-theory Ω-prespectrum KU has spaces $KU_{2i} = BU \times \mathbb{Z}$ and $KU_{2i+1} = U$ for all $i \geq 0$. The structure maps are given by the canonical homotopy equivalence $U \simeq \Omega BU = \Omega(BU \times \mathbb{Z})$ and the Bott equivalence $BU \times \mathbb{Z} \simeq \Omega U$.

We have a resulting reduced cohomology theory on based spaces such that $\tilde{K}^{2i}(X) = \tilde{K}(X)$ and $\tilde{K}^{2i+1}(X) = \tilde{K}(\Sigma X)$ for all integers i. This theory has products that are induced by tensor products of bundles over compact spaces and that are induced by suitable maps $\phi : KU_i \wedge KU_j \longrightarrow KU_{i+j}$ in general, just as for the cup product in ordinary cohomology. It is standard to view this simply as a $\mathbb{Z}_2$-graded theory with groups $\tilde{K}^0(X)$ and $\tilde{K}^1(X)$.

3. The splitting principle and the Thom isomorphism

Returning to our bundle theoretic construction of K-theory, with X compact, we describe briefly some important generalizations of the Bott periodicity theorem. The reader should recall the Thom isomorphism theorem in ordinary cohomology from Chapter 23 §5. We let $\xi : E \longrightarrow X$ be an n-plane bundle over X, fixed throughout this section. (We shall use the letters E and ξ more or less interchangeably.) Results for general vector bundles over non-connected spaces X can be deduced by applying the results to follow to one component of X at a time.

DEFINITION. Let E_0 be the zero section of E. Define the projective bundle $\pi : P(E) \longrightarrow X$ by letting the non-zero complex numbers act on $E - E_0$ by scalar multiplication on fibers and taking the orbit space under this action. Equivalently, the fiber $\pi^{-1}(x) \subset P(E)$ is the complex projective space of lines through the origin in the fiber $\xi^{-1}(x) \subset E$. Define the canonical line bundle $L(E)$ over $P(E)$ to be the subbundle of the pullback π^*E of ξ along π whose points are the pairs consisting of a line in a fiber of E and a point on that line. Let $Q(E)$ be the quotient bundle $\pi^*E/L(E)$ and let $H(E)$ denote the dual of $L(E)$.

Observe that $P(\varepsilon^2) = X \times \mathbb{C}P^1$ is the trivial bundle over X with fiber $\mathbb{C}P^1 \cong S^2$. The first version of Bott periodicity generalizes, with essentially the same proof by analysis of clutching data, to the following version. Regard $K(P(E))$ as a $K(X)$-algebra via $\pi^* : K(X) \longrightarrow K(P(E))$.

THEOREM (Bott periodicity). *Let L be a line bundle over X and let $H = H(L \oplus \varepsilon)$. Then the $K(X)$-algebra $K(P(L \oplus \varepsilon))$ is generated by the single element $[H]$ subject to the single relation $([H] - 1)([L][H] - 1) = 0$.*

There is a further generalization to arbitrary bundles E. To place it in context, we shall first explain a cohomological analogue that expresses a different approach to the Chern classes than the one that we sketched before. It will be based on a

3. THE SPLITTING PRINCIPLE AND THE THOM ISOMORPHISM

generalization to projective bundles of the calculation of $H^*(\mathbb{C}P^n)$. The proofs of both results are intertwined with the proof of the following "splitting principle," which allows the deduction of explicit formulas about general bundles from formulas about sums of line bundles.

THEOREM (Splitting principle). *There is a compact space $F(E)$ and a map $p : F(E) \longrightarrow X$ such that p^*E is a sum of line bundles over $F(E)$ and both $p^* : H^*(X; \mathbb{Z}) \longrightarrow H^*(F(E); \mathbb{Z})$ and $p^* : K(X) \longrightarrow K(F(E))$ are monomorphisms.*

This is an easy inductive consequence of the following result, which we shall refer to as the "splitting lemma."

LEMMA (Splitting lemma). *Both $\pi^* : H^*(X; \mathbb{Z}) \longrightarrow H^*(P(E); \mathbb{Z})$ and $\pi^* : K(X) \longrightarrow K(P(E))$ are monomorphisms.*

PROOF OF THE SPLITTING PRINCIPLE. The pullback π^*E splits as the sum $L(E) \oplus Q(E)$. (The splitting is canonically determined by a choice of a Hermitian metric on E.) Applying this construction to the bundle $Q(E)$ over $P(E)$, we obtain a map $\pi : P(Q(E)) \longrightarrow P(E)$ with similar properties. We obtain the desired map $p : F(E) \longrightarrow X$ by so reapplying the projective bundle construction n times. Explicitly, using a Hermitian metric on E, we find that the fiber $F(E)_x$ is the space of splittings of the fiber E_x as a sum of n lines, and the points of the bundle p^*E are n-tuples of vectors in given lines. The splitting lemma implies the desired monomorphisms on cohomology and K-theory. □

THEOREM. *Let $x = c_1(L(E)) \in H^2(P(E); \mathbb{Z})$. Then $H^*(P(E); \mathbb{Z})$ is the free $H^*(X; \mathbb{Z})$-module on the basis $\{1, x, \ldots, x^{n-1}\}$, and the Chern classes of ξ are characterized by $c_0(\xi) = 1$ and the formula*

$$\sum_{k=0}^{n}(-1)^k c_k(E) x^{n-k} = 0.$$

SKETCH PROOF. This is another case where the Serre spectral sequence shows that the bundle behaves cohomologically as if it were trivial and the Künneth theorem applied. This gives the structure of $H^*(P(E))$ as an $H^*(X)$-module. In particular, it implies the splitting lemma and thus the splitting principle in ordinary cohomology. It also implies that there must be some description of x^n as a linear combination of the x^k for $k < n$, and the splitting principle may now be used to help determine that description. Write

$$x^n = \sum_{k=1}^{n}(-1)^{k+1} c'_k(E) x^{n-k}.$$

This defines characteristic classes $c'_k(E)$. One deduces that $c'_k(E) = c_k(E)$ by verifying that the c'_k satisfy the axioms that characterize the Chern classes. For a line bundle E, $L(E) = E$ and $c_1(E) = c'_1(E)$ by the definition of x. One first verifies by direct calculation that if $E = L_1 \oplus \cdots \oplus L_n$ is a sum of line bundles, then $\prod_{1 \leq k \leq n}(x - c_1(L_k)) = 0$. This implies that $c'_k(E)$ is the kth elementary symmetric polynomial in the $c_1(L_k)$. By the Whitney sum formula for the Chern classes, this implies that $c'_k(E) = c_k(E)$ in this case. The general case follows from the splitting principle. Indeed, we have a map $P(p^*E) \longrightarrow P(E)$ of projective bundles whose

induced map on base spaces is $p : F(E) \longrightarrow X$. Writing $p^*E \cong L_1 \oplus \cdots \oplus L_n$ and using the naturality of the classes c'_k, we have
$$p^*(c'_k(E)) = c'_k(L_1 \oplus \cdots \oplus L_n) = \sigma_k(c_1(L_1), \ldots, c_k(L_n)).$$
It follows easily that the c'_k satisfy the Whitney sum axiom for the Chern classes. Since the remaining axioms are clear, this implies that $c'_k = c_k$. □

The following analogue in K-theory of the previous theorem holds. Observe that, since they are continuous operations on complex vector spaces, the exterior powers λ^k can be applied fiberwise to give natural operations on vector bundles.

THEOREM. *Let $H = H(E)$. Then $K(P(E))$ is the free $K(X)$-module on the basis $\{1, [H], \ldots, [H]^{n-1}\}$, and the following formula holds:*
$$\sum_{k=0}^n (-1)^k [H]^k [\lambda^k E] = 0.$$

SKETCH PROOF. Suppose first that E is the sum of n line bundles. Using the fact that if E is an n-plane bundle and L is a line bundle, then $P(E)$ is canonically isomorphic to $P(E \otimes L)$, one can reduce to the case when the last line bundle is trivial. One can then argue by induction from the previous form of the Bott periodicity theorem. For a general bundle E, one then deduces the structure of $K(P(E))$ as a $K(X)$-module by a patching argument from coordinate charts and the case of trivial bundles. This implies the splitting lemma and thus the splitting principle in K-theory. It also implies that there must be some formula describing $[H]^n$ as a polynomial in the $[H]^k$ for $k < n$. One reason that the given formula holds will be indicated shortly. □

Projective bundles are closely related to Thom spaces. The inclusion of vector bundles $\xi \subset \xi \oplus \varepsilon$ induces an inclusion of projective bundles $P(E) \subset P(E \oplus \varepsilon)$. We give E a Hermitian metric and regard the Thom space $T\xi$ as the quotient $D(E)/S(E)$ of the unit disk bundle by the unit sphere bundle. The total space of ε is $X \times \mathbb{C}$ and we write $1_x = (x, 1)$. Define a map $\eta : D(E) \longrightarrow P(E \oplus \varepsilon)$ by sending a point e_x in the fiber over x to the line generated by $e_x - (1 - |e_x|^2)1_x$. Then η maps $D(E) - S(E)$ homeomorphically onto $P(E \oplus \varepsilon) - P(E)$ and maps $S(E)$ onto $P(E)$ by the evident Hopf map. Therefore η induces a homeomorphism
$$T(\xi) \cong D(E)/S(E) \cong P(E \oplus \varepsilon)/P(E).$$
Just as in ordinary cohomology, the Thom diagonal gives rise to a product
$$K(X) \otimes \tilde{K}(T\xi) \longrightarrow \tilde{K}(T\xi).$$
The description of $K(P(E))$ and the exact sequence in K-theory induced by the cofibering
$$P(E) \longrightarrow P(E \oplus \varepsilon) \longrightarrow T(\xi)$$
lead to the Thom isomorphism in K-theory. There is a natural way to associate elements of $K(X)$ to complexes of vector bundles over X, and the exterior algebra of the bundle E gives rise to an element $\lambda_E \in \tilde{K}(T\xi)$. This element restricts to a generator of $\tilde{K}(S_x^n)$ for each $x \in X$, and these Thom classes are compatible with Whitney sum, in the sense that $\lambda_{E \oplus E'} = \lambda_E \cdot \lambda_{E'}$. Moreover, the image of λ_E in $K(P(E \oplus \varepsilon))$ is $\sum_{k=0}^n (-1)^k [H]^k [\lambda^k E]$. Therefore this element maps to zero in $K(P(E))$, and this gives the formula in the previous theorem.

THEOREM (Thom isomorphism theorem). *Define* $\Phi : K(X) \longrightarrow \tilde{K}(T(\xi))$ *by*
$$\Phi(x) = x \cdot \lambda_E.$$
Then Φ is an isomorphism.

4. The Chern character; almost complex structures on spheres

We have seen above that ordinary cohomology and K-theory enjoy similar properties. The splitting theorem implies a direct connection between them. Let R be any commutative ring and consider a formal power series $f(t) = \sum a_i t^i \in R[[t]]$. Given an element $x \in H^n(X; R)$, we let $f(x) = \sum a_i x^i \in H^{**}(X; R)$. The sums will be finite in our applications of this formula. Via the splitting principle, we can use f to construct a natural homomorphism of Abelian monoids $\hat{f} : Vect(X) \longrightarrow H^{**}(X; R)$, where X is any compact space. For a line bundle over X, we set
$$\hat{f}(L) = f(c_1(L)).$$
For a sum $E = L_1 \oplus \cdots \oplus L_n$ of line bundles over X, we set
$$\hat{f}(E) = \sum_{i=1}^{n} f(c_1(L_i)).$$
For a general n-plane bundle E over X, we let $\hat{f}(E)$ be the unique element of $H^{**}(X; R)$ such that $p^*(\hat{f}(E)) = \hat{f}(p^*(E)) \in H^{**}(F(E))$. More explicitly, writing $p^*E = L_1 \oplus \cdots \oplus L_n$, we see that $\hat{f}(p^*(E))$ is a symmetric polynomial in the $c_1(L_i)$ and can therefore be written as a polynomial in the elementary symmetric polynomials $p^*(c_k(E))$. Application of this polynomial to the $c_k(E)$ gives $\hat{f}(E)$. (For vector bundles E over non-connected spaces X, we add the elements obtained by restricting E to the components of X.) By the universal property of $K(X)$, $\hat{f}$ extends to a homomorphism $\hat{f} : K(X) \longrightarrow H^{**}(X; R)$.

There is an analogous multiplicative extension $\bar{f}$ of f that starts from the definition
$$\bar{f}(E) = \prod_{i=1}^{n} f(c_1(L_i))$$
on a sum $E = L_1 \oplus \cdots \oplus L_n$ of line bundles L_i.

EXAMPLE. For any R, if $f(t) = 1 + t$, then $\bar{f}(E) = c(E)$ is the total Chern class of E.

The example we are interested in is the "Chern character," which gives rise to an isomorphism between rationalized K-theory and rational cohomology.

EXAMPLE. Taking $R = \mathbb{Q}$, define the Chern character $ch(E) \in H^{**}(X; \mathbb{Q})$ by $ch(E) = \hat{f}(E)$, where $f(t) = e^t = \sum t^i/i!$.

For line bundles L and L', we have $c_1(L \otimes L') = c_1(L) + c_1(L')$. One way to see this is to recall that $BU(1) \simeq K(\mathbb{Z}, 2)$ and that line bundles are classified by their Chern classes regarded as elements of
$$[X_+, BU(1)] \cong H^2(BU; \mathbb{Z}).$$
The tensor product is represented by a product $\phi : BU(1) \times BU(1) \longrightarrow BU(1)$ that gives $BU(1)$ an H-space structure. We may think of ϕ as an element of
$$H^2(BU \times BU; \mathbb{Z}) \cong H^2(BU; \mathbb{Z}) \oplus H^2(BU; \mathbb{Z}).$$

and this element is the sum of the Chern classes in the two copies of $H^2(BU;\mathbb{Z})$ (since a basepoint of BU is a homotopy identity element for ϕ). This has the following implication.

LEMMA. *The Chern character specifies a ring homomorphism*
$$ch : K(X) \longrightarrow H^{**}(X;Q).$$

PROOF. We must check that $ch(E \otimes E) = ch(E) \cdot ch(E')$ for bundles E and E' over X. It suffices to check this when E and E' are sums of line bundles, in which case the result follows directly from the bilinearity of the tensor product and the relation $e^{t+t'} = e^t e^{t'}$. □

This leads to the following calculation.

LEMMA. *For $n \geq 1$, the Chern character maps $\tilde{K}(S^{2n})$ isomorphically onto the image of $H^{2n}(S^{2n};\mathbb{Z})$ in $H^{2n}(S^{2n};\mathbb{Q})$. Therefore $c_n : \tilde{K}(S^{2n}) \longrightarrow H^{2n}(S^{2n};\mathbb{Z})$ is a monomorphism with cokernel $\mathbb{Z}_{(n-1)!}$.*

PROOF. The first statement is clear for $n = 1$, when $ch = c_1$, and follows by compatibility with external products for $n > 1$. The definition of ch implies that the component ch_n of ch in degree $2n$ is $c_n/(n-1)!$ plus terms decomposable in terms of the c_i for $i < n$, and the second statement follows. □

Together with some of the facts given in Chapter 23 §7, this has a remarkable application to the study of almost complex structures on spheres. Recall that a smooth manifold of even dimension admits an almost complex structure if its tangent bundle is the underlying real vector bundle of a complex bundle.

THEOREM. *S^2 and S^6 are the only spheres that admit an almost complex structure.*

PROOF. It is classical that S^2 and S^6 admit almost complex structures and that S^4 does not. Assume that S^{2n} admits an almost complex structure. We shall show that $n \leq 3$. We are given that the tangent bundle τ is the realification of a complex bundle. Its nth Chern class is its Euler class: $c_n(\tau) = \chi(\tau)$. Since the Euler characteristic of S^{2n} is 2, $\chi(\tau) = 2\iota_{2n}$, where $\iota_{2n} \in H^{2n}(S^{2n},\mathbb{Z})$ is the canonical generator. However, $c_n(\tau)$ must be divisible by $(n-1)!$. This can only happen if $n \leq 3$. □

Obviously the image of ch lies in the sum of the even degree elements in $H^{**}(X;\mathbb{Q})$, which we denote by $H^{even}(X;\mathbb{Q})$. We define $H^{odd}(X;\mathbb{Q})$ similarly, and we extend ch to $\mathbb{Z}_2$-graded reduced cohomology by defining ch on $\tilde{K}^1(X)$ to be the composite
$$\tilde{K}^1(X) \cong \tilde{K}(\Sigma X) \xrightarrow{ch} \tilde{H}^{even}(\Sigma X;\mathbb{Q}) \cong \tilde{H}^{odd}(X;\mathbb{Q}).$$

We then have the following basic result, which actually holds for general compact spaces X provided that we replace singular cohomology by Čech cohomology.

THEOREM. *For any finite based CW complex X, ch induces an isomorphism*
$$\tilde{K}^*(X) \otimes \mathbb{Q} \longrightarrow \tilde{H}^{**}(X;\mathbb{Q}).$$

SKETCH PROOF. We think of both the source and target as $\mathbb{Z}_2$-graded. The lemma above implies the conclusion when $X = S^n$ for any n. One can check that the displayed maps for varying X give a map of $\mathbb{Z}_2$-graded cohomology theories. The conclusion then follows from the five lemma and induction on the number of cells of X. □

5. The Adams operations

There are natural operations in K-theory, called the Adams operations, that are somewhat analogous to the Steenrod operations in mod 2 cohomology. In fact, the analogy can be given content by establishing a precise relationship between the Adams and Steenrod operations, but we will not go into that here.

THEOREM. *For each non-zero integer k, there is a natural homomorphism of rings $\psi^k : K(X) \longrightarrow K(X)$. These operations satisfy the following properties.*
1. $\psi^1 = \text{id}$ *and* ψ^{-1} *is induced by complex conjugation of bundles.*
2. $\psi^k \psi^\ell = \psi^{k\ell} = \psi^\ell \psi^k$.
3. $\psi^p(x) \equiv x^p \mod p$ *for any prime p.*
4. $\psi^k(\xi) = \xi^k$ *if ξ is a line bundle.*
5. $\psi^k(x) = k^n x$ *if $x \in \tilde{K}(S^{2n})$.*

We explain the construction. By property 2, $\psi^{-k} = \psi^k \psi^{-1}$, hence by property 1 we can concentrate on the case $k > 1$. The exterior powers of bundles satisfy the relation
$$\lambda^k(\xi \oplus \eta) = \oplus_{i+j=k} \lambda^i(\xi) \otimes \lambda^j(\eta).$$
It follows formally that the λ^k extend to operations $K(X) \longrightarrow K(X)$. Indeed, form the group G of power series with constant coefficient 1 in the ring $K(X)[[t]]$ of formal power series in the variable t. We define a function from (equivalence classes of) vector bundles to this Abelian group by setting
$$\Lambda(\xi) = 1 + \lambda^1(\xi)t + \cdots + \lambda^k(\xi)t^k + \cdots.$$
Visibly, this is a morphism of monoids,
$$\Lambda(\xi \oplus \eta) = \Lambda(\xi)\Lambda(\eta).$$
It therefore extend to a homomorphism of groups $\Lambda : K(X) \longrightarrow G$, and we let $\lambda^k(x)$ be the coefficient of t^k in $\Lambda(x)$.

We define the ψ^k as suitable polynomials in the λ^k. Recall that the subring of symmetric polynomials in the polynomial algebra $\mathbb{Z}[x_1, \ldots, x_n]$ is the polynomial algebra $\mathbb{Z}[\sigma_1, \ldots, \sigma_n]$, where $\sigma_i = x_1 x_2 \cdots x_i + \cdots$ is the ith elementary symmetric function. We may write the power sum $\pi_k = x_1^k + \cdots + x_n^k$ as a polynomial
$$\pi_k = Q_k(\sigma_1, \ldots, \sigma_k)$$
in the first k elementary symmetric functions. Provided $n \geq k$, Q_k does not depend on n. We define
$$\psi^k(x) = Q_k(\lambda^1(x), \ldots, \lambda^k(x)).$$
For example, $\pi_2 = \sigma_1^2 - 2\sigma_2$, hence $\psi^2(x) = x^2 - 2\lambda^2(x)$. The naturality of the ψ^k is clear from the naturality of the λ^k.

If ξ is a line bundle, then $\lambda^1(\xi) = \xi$ and $\lambda^k(\xi) = 0$ for $k \geq 2$. Clearly $\sigma_1^k = \pi_k +$ other terms and π_k does not occur as a summand of any other monomial in the σ_i. Therefore $Q_k \equiv \sigma_1^k$ modulo terms in the ideal generated by the σ_i for $i > 1$. This immediately implies property 4. Moreover, if $\xi_1, \ldots, \xi_n$ are line bundles, then

$$\begin{aligned}\Lambda(\xi_1 \oplus \cdots \oplus \xi_n) &= (1+\xi_1)\cdots(1+\xi_n) \\ &= 1 + \sigma_1(\xi_1,\ldots,\xi_n)t + \sigma_2(\xi_1,\ldots,\xi_n)t^2 + \cdots.\end{aligned}$$

This implies the generalization of property 4 to sums of line bundles:

4' $\psi^k(\xi_1 \oplus \cdots \oplus \xi_n) = \pi_k(\xi_1,\ldots,\xi_n)$ for line bundles ξ_i.

Now, if x and y are sums of line bundles, the following formulas are immediate:

$$\psi^k(x+y) = \psi^k(x) + \psi^k(y), \quad \psi^k(xy) = \psi^k(x)\psi^k(y), \quad \psi^k\psi^\ell(x) = \psi^{k\ell}(x)$$

and $\psi^p(x) \equiv x^p$ mod p for a prime p.

For arbitrary bundles, these formulas follow directly from the splitting principle and naturality, and they then follow formally for arbitrary virtual bundles. This completes the proof of all properties except 5. We have that $\tilde{K}(S^2)$ is generated by $1 - [H]$, where $(1 - [H])^2 = 0$. Clearly $\psi^k(1 - [H]) = 1 - [H]^k$. By induction on k, $1 - [H]^k = k(1 - [H])$. Since $S^{2n} = S^2 \wedge \cdots \wedge S^2$ and $\tilde{K}(S^{2n})$ is generated by the k-fold external tensor power $(1 - [H]) \otimes \cdots \otimes (1 - [H])$, property 5 follows from the fact that ψ^k preserves products.

REMARK. By the splitting principle, it is clear that the ψ^k are the unique natural and additive operations with the specified behavior on line bundles.

Two further properties of the ψ^k should be mentioned. The first is a direct consequence of the multiplicativity of the ψ^k and their behavior on spheres.

PROPOSITION. *The following diagram does not commute for based spaces X, where β is the periodicity isomorphism:*

$$\begin{CD} \tilde{K}(X) @>\beta>> \tilde{K}(\Sigma^2 X) \\ @V\psi^k VV @VV\psi^k V \\ \tilde{K}(X) @>>\beta> \tilde{K}(\Sigma^2 X). \end{CD}$$

Rather, $\psi^k \beta = k\beta\psi^k$.

Therefore the ψ^k do not give stable operations on the $\mathbb{Z}$-graded theory K^*.

PROPOSITION. *Define ψ_H^k on $H^{even}(X;\mathbb{Z})$ by letting $\psi_H^k(x) = k^r x$ for $x \in H^{2r}(X;\mathbb{Z})$. Then the following diagram commutes:*

$$\begin{CD} K(X) @>ch>> H^{even}(X;\mathbb{Q}) \\ @V\psi^k VV @VV\psi_H^k V \\ K(X) @>>ch> H^{even}(X;\mathbb{Q}). \end{CD}$$

PROOF. It suffices to prove this on vector bundles E. By the splitting principle in K-theory and cohomology, we may assume that E is a sum of line bundles. By additivity, we may then assume that E is a line bundle. Here $\psi^k(E) = E^k$ and $c_1(E^k) = kc_1(E)$. The conclusion follows readily from the definition of ch in terms of e^t. □

REMARK. The observant reader will have noticed that, by analogy with the definition of the Stiefel-Whitney classes, we can define characteristic classes in K-theory by use of the Adams operations and the Thom isomorphism, setting $\rho^k(E) = \Phi^{-1}\psi^k\Phi(1)$ for n-plane bundles E.

6. The Hopf invariant one problem and its applications

We give one of the most beautiful and impressive illustrations of the philosophy described in the first chapter. We define a numerical invariant, called the "Hopf invariant," of maps $f : S^{2n-1} \longrightarrow S^n$ and show that it can only rarely take the value one. We then indicate several problems whose solution can be reduced to the question of when such maps f take the value one. Adams' original solution to the Hopf invariant one problem used secondary cohomology operations in ordinary cohomology and was a critical starting point of modern algebraic topology. The later realization that a problem that required secondary operations in ordinary cohomology could be solved much more simply using primary operations in K-theory had a profound impact on the further development of the subject.

Take cohomology with integer coefficients unless otherwise specified.

DEFINITION. Let X be the cofiber of a based map $f : S^{2n-1} \longrightarrow S^n$, where $n \geq 2$. Then X is a CW complex with a single vertex, a single n-cell i, and a single $2n$-cell j. The differential in the cellular chain complex of X is zero for obvious dimensional reasons, hence $\tilde{H}^*(X)$ is free Abelian on generators $x = [i]$ and $y = [j]$. Define an integer $h(f)$, the Hopf invariant of f, by $x^2 = h(f)y$. We usually regard $h(f)$ as defined only up to sign (thus ignoring problems of orientations of cells). Note that $h(f)$ depends only on the homotopy class of f.

If n is odd, then $2x^2 = 0$ and thus $x^2 = 0$. We assume from now on that n is even. Although not essential to the main point of this section, we record the following basic properties of the Hopf invariant.

PROPOSITION. *The Hopf invariant enjoys the following properties.*
1. *If $g : S^{2n-1} \longrightarrow S^{2n-1}$ has degree d, then $h(f \circ g) = dh(f)$.*
2. *If $e : S^n \longrightarrow S^n$ has degree d, then $h(e \circ f) = d^2 h(f)$.*
3. *The Hopf invariant defines a homomorphism $\pi_{2n-1}(S^n) \longrightarrow \mathbb{Z}$.*
4. *There is a map $f : S^{2n-1} \longrightarrow S^n$ such that $h(f) = 2$.*

PROOF. We leave the first three statements to the reader. For property 4, let $\pi : D^n \longrightarrow D^n/S^{n-1} \cong S^n$ be the quotient map and define
$$f : S^{2n-1} \cong (D^n \times S^{n-1}) \cup (S^{n-1} \times D^n) \longrightarrow S^n$$
by $f(x,y) = \pi(x)$ and $f(y,x) = \pi(x)$ for $x \in D^n$ and $y \in S^{n-1}$. We leave it to the reader to verify that $h(f) = 2$. □

We have adopted the standard definition of $h(f)$, but we could just as well have defined it in terms of K-theory. To see this, consider the cofiber sequence
$$S^{2n-1} \xrightarrow{f} S^n \xrightarrow{i} X \xrightarrow{\pi} S^{2n} \xrightarrow{\Sigma f} S^{n+1}.$$

Obviously $i^* : H^n(X) \longrightarrow H^n(S^n)$ and $\pi^* : H^{2n}(S^{2n}) \longrightarrow H^{2n}(X)$ are isomorphisms. We have the commutative diagram with exact rows

$$\begin{array}{ccccccccc}
0 & \longrightarrow & \tilde{K}(S^{2n}) & \xrightarrow{\pi^*} & \tilde{K}(X) & \xrightarrow{i^*} & \tilde{K}(S^n) & \longrightarrow & 0 \\
& & \downarrow ch & & \downarrow ch & & \downarrow ch & & \\
0 & \longrightarrow & \tilde{H}^{**}(S^{2n};\mathbb{Q}) & \xrightarrow{\pi^*} & \tilde{H}^{**}(X;\mathbb{Q}) & \xrightarrow{i^*} & \tilde{H}^{**}(S^n;\mathbb{Q}) & \longrightarrow & 0.
\end{array}$$

Here the top row is exact since $\tilde{K}^1(S^n) = 0$ and $\tilde{K}^1(S^{2n}) = 0$. The vertical arrows are monomorphisms since they are rational isomorphisms. By a lemma in the previous section, generators i_n of $\tilde{K}(S^n)$ and i_{2n} of $\tilde{K}(S^{2n})$ map under ch to generators of $H^n(S^n)$ and $H^{2n}(S^{2n})$. Choose $a \in \tilde{K}(X)$ such that $i^*(a) = i_n$ and let $b = \pi^*(i_{2n})$. Then $\tilde{K}(X)$ is the free Abelian group on the basis $\{a, b\}$. Since $i_n^2 = 0$, we have $a^2 = h'(f)b$ for some integer $h'(f)$. The diagram implies that, up to sign, $ch(b) = y$ and $ch(a) = x + qy$ for some rational number q. Since ch is a ring homomorphism and since $y^2 = 0$ and $xy = 0$, we conclude that $h'(f) = h(f)$.

THEOREM. *If $h(f) = \pm 1$, then $n = 2$, 4, or 8.*

PROOF. Write $n = 2m$. Since $\psi^k(i_{2n}) = k^{2m}i_{2n}$ and $\psi^k(i_n) = k^m i_n$, we have

$$\psi^k(b) = k^{2m} b \quad \text{and} \quad \psi^k(a) = k^m a + \mu_k b$$

for some integer μ_k. Since $\psi^2(a) \equiv a^2 \bmod 2$, $h(f) = \pm 1$ implies that μ_2 is odd. Now, for any odd k,

$$\begin{aligned}
\psi^k \psi^2(a) &= \psi^k(2^m a + \mu_2 b) \\
&= k^m 2^m a + (2^m \mu_k + k^{2m}\mu_2)b
\end{aligned}$$

while

$$\begin{aligned}
\psi^2 \psi^k(a) &= \psi^2(k^m a + \mu_k b) \\
&= 2^m k^m a + (k^m \mu_2 + 2^{2m}\mu_k)b.
\end{aligned}$$

Since these must be equal, we find upon equating the coefficients of b that

$$2^m(2^m - 1)\mu_k = k^m(k^m - 1)\mu_2.$$

If μ_2 is odd, this implies that 2^m divides $k^m - 1$. Already with $k = 3$, an elementary number theoretic argument shows that this implies $m = 1$, 2, or 4. □

This allows us to determine which spheres can admit an H-space structure. Recall from a problem in Chapter 18 that S^{2m} cannot be an H-space. Clearly S^n is an H-space for $n = 0$, 1, 3, and 7: view S^n as the unit sphere in the space of real numbers, complex numbers, quaternions, or Cayley numbers.

THEOREM. *If S^{n-1} is an H-space, then $n = 1$, 2, 4, or 8.*

The strategy of proof is clear: given an H-space structure on S^{n-1}, we construct from it a map $f : S^{2n-1} \longrightarrow S^n$ of Hopf invariant one. The following construction and lemma do this and more.

CONSTRUCTION (Hopf construction). Let $\phi : S^{n-1} \times S^{n-1} \longrightarrow S^{n-1}$ be a map. Let $CX = (X \times I)/(X \times \{1\})$ be the unreduced cone functor and note that we have canonical homeomorphisms of pairs

$$(D^n, S^{n-1}) \cong (CS^{n-1}, S^{n-1})$$

6. THE HOPF INVARIANT ONE PROBLEM AND ITS APPLICATIONS

and
$$(D^{2n}, S^{2n-1}) \cong (D^n \times D^n, (D^n \times S^{n-1}) \cup (S^{n-1} \times D^n))$$
$$\cong (CS^{n-1} \times CS^{n-1}, (CS^{n-1} \times S^{n-1}) \cup (S^{n-1} \times CS^{n-1})).$$

Take S^n to be the unreduced suspension of S^{n-1}, with the upper and lower hemispheres D^n_+ and D^n_- corresponding to the points with suspension coordinate $1/2 \leq t \leq 1$ and $0 \leq t \leq 1/2$, respectively. Define
$$f : S^{2n-1} \cong (CS^{n-1} \times S^{n-1}) \cup (S^{n-1} \times CS^{n-1}) \longrightarrow S^n$$
as follows. Let $x, y \in S^{n-1}$ and $t \in I$. On $CS^{n-1} \times S^{n-1}$, f is the composite
$$CS^{n-1} \times S^{n-1} \xrightarrow{\alpha} C(S^{n-1} \times S^{n-1}) \xrightarrow{C\phi} CS^{n-1} \xrightarrow{\beta} D^n_-,$$
where $\alpha([x,t], y) = [(x, y), t]$ and $\beta([x, t]) = [x, (1-t)/2]$. On $S^{n-1} \times CS^{n-1}$, f is the composite
$$S^{n-1} \times CS^{n-1} \xrightarrow{\alpha'} C(S^{n-1} \times S^{n-1}) \xrightarrow{C\phi} CS^{n-1} \xrightarrow{\beta'} D^n_+,$$
where $\alpha'(x, [y, t]) = [(x, y), t]$ and $\beta'([x, t]) = [x, (1+t)/2]$. The map f, or rather the resulting 2-cell complex $X = S^n \cup_f D^{2n}$, is called the Hopf construction on ϕ.

Giving S^{n-1} a basepoint, we obtain inclusions of S^{n-1} onto the first and second copies of S^{n-1} in $S^{n-1} \times S^{n-1}$. The bidegree of a map $\phi : S^{n-1} \times S^{n-1} \longrightarrow S^{n-1}$ is the pair of integers given by the two resulting composite maps $S^{n-1} \longrightarrow S^{n-1}$. Thus ϕ gives S^{n-1} an H-space structure if its bidegree is $(1, 1)$.

LEMMA. *If the bidegree of $\phi : S^{n-1} \times S^{n-1} \longrightarrow S^{n-1}$ is (d_1, d_2), then the Hopf invariant of the Hopf construction on ϕ is $\pm d_1 d_2$.*

PROOF. Making free use of the homeomorphisms of pairs specified in the construction, we see that the diagonal map of X, its top cell j, evident quotient maps, and projections π_i onto first and second coordinates give rise to a commutative diagram in which the maps marked $\simeq$ are homotopy equivalences and those marked $\cong$ are homeomorphisms:

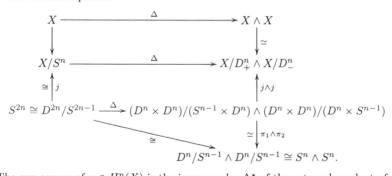

The cup square of $x \in H^n(X)$ is the image under Δ^* of the external product of x with itself. The maps on the left induce isomorphisms on H^{2n}. The inclusions of D^n in the ith factor of $D^n \times D^n$ induce homotopy inverses
$$\iota_1 : D^n/S^{n-1} \longrightarrow (D^n \times D^n)/(S^{n-1} \times D^n)$$
and
$$\iota_2 : D^n/S^{n-1} \longrightarrow (D^n \times D^n)/(D^n \times S^{n-1})$$

to the projections π_i in the diagram, and it suffices to prove that, up to sign, the composites

$$j \circ \iota_1 : D^n/S^{n-1} \longrightarrow X/D^n_+ \quad \text{and} \quad j \circ \iota_2 : D^n/S^{n-1} \longrightarrow X/D^n_-$$

induce multiplication by d_1 and by d_2 on H^n. However, by construction, these maps factor as composites

$$D^n/S^{n-1} \xrightarrow{\gamma_1} S^n/D^n_+ \longrightarrow X/D^n_+ \quad \text{and} \quad D^n/S^{n-1} \xrightarrow{\gamma_2} S^n/D^n_- \longrightarrow X/D^n_-,$$

where, up to signs and identifications of spheres, γ_1 and γ_2 are the suspensions of the restrictions of ϕ to the two copies of S^{n-1} in $S^{n-1} \times S^{n-1}$. □

The determination of which spheres are H-spaces has the following implications.

THEOREM. *Let $\omega : \mathbb{R}^n \times \mathbb{R}^n \longrightarrow \mathbb{R}^n$ be a map with a two-sided identity element $e \neq 0$ and no zero divisors. Then $n = 1, 2, 4$, or 8.*

PROOF. The product restricts to give $\mathbb{R}^n - \{0\}$ an H-space structure. Since S^{n-1} is homotopy equivalent to $\mathbb{R}^n - \{0\}$, it inherits an H-space structure. Explicitly, we may assume that $e \in S^{n-1}$, by rescaling the metric, and we give S^{n-1} the product $\phi : S^{n-1} \times S^{n-1} \longrightarrow S^{n-1}$ specified by $\phi(x, y) = \omega(x, y)/|\omega(x, y)|$. □

Note that ω need not be bilinear, just continuous. Also, it need not have a strict unit; all that is required is that e be a two-sided unit up to homotopy for the restriction of ω to $\mathbb{R}^n - \{0\}$.

THEOREM. *If S^n is parallelizable, then $n = 0, 1, 3$, or 7.*

PROOF. Exclude the trivial case $n = 0$ and suppose that S^n is parallelizable, so that its tangent bundle τ is trivial. We will show that S^n is an H-space. Define a map $\mu : \tau \longrightarrow S^n$ as follows. Think of the tangent plane τ_x as affinely embedded in $\mathbb{R}^{n+1}$ with origin at x. We have a parallel translate of this plane to an affine plane with origin at $-x$. Define μ by sending a tangent vector $y \in \tau_x$ to the intersection with S^n of the line from x to the translate of y. Composing with a trivialization $S^n \times \mathbb{R}^n \cong \tau$, this gives a map $\mu : S^n \times \mathbb{R}^n \longrightarrow S^n$. Let S^n_∞ be the one-point compactification of $\mathbb{R}^n$. Extend μ to a map $\phi : S^n \times S^n_\infty \longrightarrow S^n$ by letting $\phi(x, \infty) = x$; ϕ is continuous since $\mu(x, y)$ approaches x as y approaches ∞. By construction, ∞ is a right unit for this product. For a fixed x, $y \longrightarrow \phi(x, y)$ is a degree one homeomorphism $S^n_\infty \longrightarrow S^n_\infty$. The conclusion follows. □

CHAPTER 25

An introduction to cobordism

Cobordism theories were introduced shortly after K-theory, and their use pervades modern algebraic topology. We shall describe the cobordism of smooth closed manifolds, but this is in fact a particularly elementary example. Other examples include smooth closed manifolds with extra structure on their stable normal bundles: orientation, complex structure, Spin structure, or symplectic structure for example. All of these except the symplectic case have been computed completely. The complex case is particularly important since complex cobordism and theories constructed from it have been of central importance in algebraic topology for the last few decades, quite apart from their geometric origins in the classification of manifolds. The area is pervaded by insights from algebraic topology that are quite mysterious geometrically. For example, the complex cobordism groups turn out to be concentrated in even degrees: every smooth closed manifold of odd dimension with a complex structure on its stable normal bundle is the boundary of a compact manifold (with compatible bundle information). However, there is no geometric understanding of why this should be the case. The analogue with "complex" replaced by "symplectic" is false.

1. The cobordism groups of smooth closed manifolds

We consider the problem of classifying smooth closed n-manifolds M. One's first thought is to try to classify them up to diffeomorphism, but that problem is in principle unsolvable. Thom's discovery that one can classify such manifolds up to the weaker equivalence relation of "cobordism" is one of the most beautiful advances of twentieth century mathematics. We say that two smooth closed n-manifolds M and N are cobordant if there is a smooth compact manifold W whose boundary is the disjoint union of M and N, $\partial W = M \amalg N$. We write $\mathcal{N}_n$ for the set of cobordism classes of smooth closed n-manifolds. It is convenient to allow the empty set $\emptyset$ as an n-manifold for every n. Disjoint union gives an addition on the set $\mathcal{N}_n$. This operation is clearly associative and commutative and it has $\emptyset$ as a zero element. Since

$$\partial(M \times I) = M \amalg M,$$

$M \amalg M$ is cobordant to $\emptyset$. Thus $M = -M$ and $\mathcal{N}_n$ is a vector space over $\mathbb{Z}_2$. Cartesian product of manifolds defines a multiplication $\mathcal{N}_m \times \mathcal{N}_n \longrightarrow \mathcal{N}_{m+n}$. This operation is bilinear, associative, and commutative, and the zero dimensional manifold with a single point provides an identity element. We conclude that $\mathcal{N}_*$ is a graded $\mathbb{Z}_2$-algebra.

THEOREM (Thom). $\mathcal{N}_*$ is a polynomial algebra over $\mathbb{Z}_2$ on generators u_i of dimension i for $i > 1$ and not of the form $2^r - 1$.

As already stated in our discussion of Stiefel-Whitney numbers, it follows from the proof of the theorem that a manifold is a boundary if and only if its normal Stiefel-Whitney numbers are zero. We can restate this as follows.

THEOREM. *Two smooth closed n-manifolds are cobordant if and only if their normal Stiefel-Whitney numbers, or equivalently their tangential Stiefel-Whitney numbers, are equal.*

Explicit generators u_i are known. Write $[M]$ for the cobordism class of a manifold M. Then we can take $u_{2i} = [\mathbb{R}P^{2i}]$. We have seen that the Stiefel-Whitney numbers of $\mathbb{R}P^{2i-1}$ are zero, so we need different generators in odd dimensions. For $m < n$, define $H_{n,m}$ to be the hypersurface in $\mathbb{R}P^n \times \mathbb{R}P^m$ consisting of those pairs $([x_0,\ldots,x_n],[y_0,\ldots,y_m])$ such that $x_0 y_0 + \cdots + x_m y_m = 0$; here $(x_0,\ldots,x_n) \in S^n$ and $[x_0,\ldots,x_n]$ denotes its image in $\mathbb{R}P^n$. We may write an odd number i not of the form $2^r - 1$ in the form $i = 2^p(2q+1) - 1 = 2^{p+1}q + 2^p - 1$, where $p \geq 1$ and $q \geq 1$. Then we can take $u_i = [H_{2^{p+1}q, 2^p}]$.

The strategy for the proof of Thom's theorem is to describe $\mathcal{N}_n$ as a homotopy group of a certain Thom space. The homotopy group is a stable one, and it turns out to be computable by the methods of generalized homology theory.

Consider the universal q-plane bundle $\gamma_q : E_q \longrightarrow G_q(\mathbb{R}^\infty) = BO(q)$. Let $TO(q)$ be its Thom space. Recall that we have maps $i_q : BO(q) \longrightarrow BO(q+1)$ such that $i_q^*(\gamma_{q+1}) = \gamma_q \oplus \varepsilon$. The Thom space $T(\gamma_q \oplus \varepsilon)$ is canonically homeomorphic to the suspension $\Sigma TO(q)$, and the bundle map $\gamma_q \oplus \varepsilon \longrightarrow \gamma_{q+1}$ induces a map $\sigma_q : \Sigma TO(q) \longrightarrow TO(q+1)$. Thus the spaces $TO(q)$ and maps σ_q constitute a prespectrum TO. By definition, the homotopy groups of a prespectrum $T = \{T_q\}$ are

$$\pi_n(T) = \operatorname{colim} \pi_{n+q}(T_q),$$

where the colimit is taken over the maps

$$\pi_{n+q}(T_q) \xrightarrow{\Sigma} \pi_{n+q+1}(\Sigma T_q) \xrightarrow{\sigma_{q*}} \pi_{n+q+1}(T_{q+1}).$$

In the case of TO, it turns out that these maps are isomorphisms if q is sufficiently large, and we have the following translation of our problem in manifold theory to a problem in homotopy theory. We shall sketch the proof in the next section, where we shall also explain the ring structure on $\pi_*(TO)$ that makes it a $\mathbb{Z}_2$-algebra.

THEOREM (Thom). *For sufficiently large q, $\mathcal{N}_n$ is isomorphic to $\pi_{n+q}(TO(q))$. Therefore*

$$\mathcal{N}_n \cong \pi_n(TO).$$

Moreover, $\mathcal{N}_*$ and $\pi_*(TO)$ are isomorphic as $\mathbb{Z}_2$-algebras.

2. Sketch proof that $\mathcal{N}_*$ is isomorphic to $\pi_*(TO)$

Given a smooth closed n-manifold M, we may embed it in $\mathbb{R}^{n+q}$ for q sufficiently large, and we let ν be the normal bundle of the embedding. (By the Whitney embedding theorem, $q = n$ suffices, but the precise estimate is not important to us.) Embed M as the zero section of the total space $E(\nu)$. Then a standard result in differential topology known as the tubular neighborhood theorem implies that the identity map of M extends to an embedding of $E(\nu)$ onto an open neighborhood U of M in $\mathbb{R}^{n+q}$.

2. SKETCH PROOF THAT $\mathscr{N}_*$ IS ISOMORPHIC TO $\pi_*(TO)$

Think of S^{n+q} as the one-point compactification of $\mathbb{R}^{n+q}$. The "Pontryagin-Thom construction" associates a map $t : S^{n+q} \longrightarrow T(\nu)$ to our tubular neighborhood U. Observing that $T\nu - \{\infty\} = E(\nu)$, we let t restrict on U to the identification $U \cong E(\nu)$ and let t send all points of $\mathbb{R}^{n+q} - U$ to the point at infinity. The Thom space was tailor made to allow this construction. For q large enough, any two embeddings of M in $\mathbb{R}^{n+q}$ are isotopic, and the homotopy class of t is independent of the choice of the embedding of M in $\mathbb{R}^{n+q}$. Now choose a classifying map $f : M \longrightarrow BO(q)$ for ν. The composite $Tf \circ t : S^{n+q} \longrightarrow TO(q)$ represents an element of $\pi_{n+q}(TO(q))$.

As the reader should think through, it is intuitively plausible that cobordant manifolds induce homotopic maps $S^{n+q} \longrightarrow TO(q)$, so that this construction gives a well defined function $\alpha : \mathscr{N}_n \longrightarrow \pi_{n+q}(TO(q))$. However, technically, one can arrange the argument so that this fact drops out without explicit verification. Given two n-manifolds, we can embed them and their tubular neighborhoods disjointly in $\mathbb{R}^{n+q}$, and it follows easily that α is a homomorphism.

We construct an inverse β to α. Any map $g : S^{n+q} \longrightarrow TO(q)$ has image contained in $T(\gamma_q^r)$ for a sufficiently large $r > q$, where γ_q^r is the restriction of the universal bundle γ_q to the compact manifold $G_q(\mathbb{R}^r)$. By an implication of Sard's theorem known as the transversality theorem, we can deform the restriction of g to $g^{-1}(T\gamma_q^r - \{\infty\}) = g^{-1}(E(\gamma_q^r))$ so as to obtain a homotopic map that is smooth and transverse to the zero section. This use of transversality is the crux of the proof of the theorem. It follows that the inverse image $g^{-1}(G_q(\mathbb{R}^r))$ is a smooth closed n-manifold embedded in $\mathbb{R}^{n+q} = S^{n+q} - \{\infty\}$. It is intuitively plausible that homotopic maps $g_i : S^{n+q} \longrightarrow TO(q)$, $i = 0, 1$, give rise to cobordant n-manifolds by this construction. Indeed, with the g_i smooth and transverse to the zero section, we can approximate a homotopy between them by a homotopy h which is smooth on $h^{-1}(T(\gamma_q^r) - \{\infty\})$ and transverse to the zero section. Then $h^{-1}(G_q(\mathbb{R}^r))$ is a manifold whose boundary is $g_0^{-1}(G_q(\mathbb{R}^r)) \amalg g_1^{-1}(G_q(\mathbb{R}^r))$. It is easy to verify that the resulting function $\beta : \pi_{n+q}(TO(q)) \longrightarrow \mathscr{N}_n$ is a homomorphism.

If we start with a manifold M embedded in $\mathbb{R}^{n+q}$ and construct the classifying map f for its normal bundle to be the Gauss map described in our sketch proof of the classification theorem in Chapter 23 §1, then the composite $Tf \circ t$ is smooth and transverse to the zero section, and the inverse image of the zero section is exactly M. This proves that β is an epimorphism. To complete the proof, it suffices to show that β is a monomorphism. It will follow formally that α is well defined and inverse to β.

Thus suppose given $g : S^{n+q} \longrightarrow T\gamma_q^r$ such that $g^{-1}(E(\nu))$ is smooth and transverse to the zero section and suppose that $M = g^{-1}(G_q(\mathbb{R}^r))$ is a boundary, say $M = \partial W$. The inclusion of M in S^{n+q} extends to a embedding of W in D^{n+q+1}, by the Whitney embedding theorem for manifolds with boundary (assuming as always that q is sufficiently large). We may assume that $U = g^{-1}(T\gamma_q^r - \{\infty\})$ is a tubular neighborhood and that $g : U \longrightarrow E(\gamma_q^r)$ is a map of vector bundles. A relative version of the tubular neighborhood theorem then shows that U can be extended to a tubular neighborhood V of W in D^{n+q+1} and that g extends to a map of vector bundles $h : V \longrightarrow E(\gamma_q^r)$. We can then extend h to a map $D^{n+q+1} \longrightarrow T(\gamma_q^r)$ by mapping $D^{n+q+1} - V$ to ∞. This extension of g to the disk implies that g is null homotopic.

218   AN INTRODUCTION TO COBORDISM

We must still define the ring structure on $\pi_*(TO)$ and prove that we have an isomorphism of rings and therefore of $\mathbb{Z}_2$-algebras. Recall that we have maps $p_{m,n} : BO(m) \times BO(n) \longrightarrow BO(m+n)$ such that $p^*_{m,n}(\gamma_{m+n}) = \gamma_m \times \gamma_n$. The Thom space $T(\gamma_m \times \gamma_n)$ is canonically homeomorphic to the smash product $TO(m) \wedge TO(n)$, and the bundle map $\gamma_m \times \gamma_n \longrightarrow \gamma_{m+n}$ induces a map $\phi_{m,n} : TO(m) \wedge TO(n) \longrightarrow TO(m+n)$. If we have maps $f : S^{m+q} \longrightarrow TO(m)$ and $g : S^{n+q} \longrightarrow TO(n)$, then we can compose their smash product with $\phi_{m,n}$ to obtain a composite map

$$S^{m+n+q+r} \cong S^{m+q} \wedge S^{n+r} \xrightarrow{f \wedge g} TO(m) \wedge TO(n) \xrightarrow{\phi_{m,n}} TO(m+n).$$

We can relate the maps $\phi_{m,n}$ to the maps σ_n. In fact, TO is a commutative and associative ring prespectrum in the sense of the following definition.

DEFINITION. Let T be a prespectrum. Then T is a ring prespectrum if there are maps $\eta : S^0 \longrightarrow T_0$ and $\phi_{m,n} : T_m \wedge T_n \longrightarrow T_{m+n}$ such that the following diagrams are homotopy commutative:

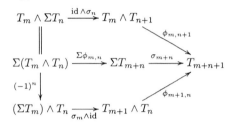

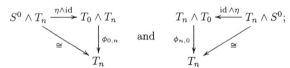

T is associative if the following diagrams are homotopy commutative:

$$\begin{array}{ccc} T_m \wedge T_n \wedge T_p & \xrightarrow{\phi_{m,n} \wedge \mathrm{id}} & T_{m+n} \wedge T_p \\ \mathrm{id} \wedge \phi_{n,p} \downarrow & & \downarrow \phi_{m+n,p} \\ T_m \wedge T_{n+p} & \xrightarrow{\phi_{m,n+p}} & T_{m+n+p}; \end{array}$$

T is commutative if there are equivalences $(-1)^{mn} : T_{m+n} \longrightarrow T_{m+n}$ that suspend to $(-1)^{mn}$ on ΣT_{m+n} and if the following diagrams are homotopy commutative:

$$\begin{array}{ccc} T_m \wedge T_n & \xrightarrow{t} & T_n \wedge T_m \\ \phi_{m,n} \downarrow & & \downarrow \phi_{n,m} \\ T_{m+n} & \xrightarrow{(-1)^{mn}} & T_{m+n}. \end{array}$$

When T is an Ω-prespectrum, we can restate this as $\phi_{m,n} \simeq (-1)^{mn} \phi_{n,m} t$.

For example, the Eilenberg-Mac Lane Ω-prespectrum of a commutative ring R is an associative and commutative ring prespectrum by the arguments in Chapter 22 §3. It is denoted HR or sometimes, by abuse, $K(R, 0)$. Similarly, the K-theory

Ω-prespectrum is an associative and commutative ring prespectrum. The sphere prespectrum, whose nth space is S^n, is another example. For TO, the required maps $(-1)^{mn} : TO(m+n) \longrightarrow TO(m+n)$ are obtained by passage to Thom complexes from a map $\gamma_{m+n} \longrightarrow \gamma_{m+n}$ of universal bundles given on the domains of coordinate charts by the evident interchange isomorphism $\mathbb{R}^{m+n} \longrightarrow \mathbb{R}^{m+n}$. The following lemma is immediate by passage to colimits.

LEMMA. *If T is an associative ring prespectrum, then $\pi_*(T)$ is a graded ring. If T is commutative, then $\pi_*(T)$ is commutative in the graded sense.*

Returning to the case at hand, we show that the maps α for varying n transport products of manifolds to products in $\pi_*(TO)$. Thus let M be an m-manifold embedded in $\mathbb{R}^{m+q}$ with tubular neighborhood $U \cong E(\nu_M)$ and N be an n-manifold embedded in $\mathbb{R}^{n+r}$ with tubular neighborhood $V \cong E(\nu_N)$. Then $M \times N$ is embedded in $\mathbb{R}^{m+q+n+r}$ with tubular neighborhood $U \times V \cong E(\nu_{M \times N})$. Identifying $S^{m+q+n+r}$ with $S^{m+q} \wedge S^{n+r}$, we find that the Pontryagin-Thom construction for $M \times N$ is the smash product of the Pontryagin-Thom constructions for M and N. That is, the left square in the following diagram commutes. The right square commutes up to homotopy by the definition of $\phi_{q,r}$.

$$\begin{array}{ccccc} S^{m+q} \wedge S^{n+r} & \xrightarrow{t \wedge t} & T\nu_m \wedge T\nu_N & \longrightarrow & TO(q) \wedge TO(r) \\ \| & & \downarrow \cong & & \downarrow \phi_{q,r} \\ S^{m+q+n+r} & \xrightarrow{t} & T(\nu_{M \times N}) & \longrightarrow & TO(q+r). \end{array}$$

This implies the claimed multiplicativity of the maps α.

3. Prespectra and the algebra $H_*(TO;\mathbb{Z}_2)$

Calculation of the homotopy groups $\pi_*(TO)$ proceeds by first computing the homology groups $H_*(TO;\mathbb{Z}_2)$ and then showing that the stable Hurewicz homomorphism maps $\pi_*(TO)$ monomorphically onto an identifiable part of $H_*(TO;\mathbb{Z}_2)$. We explain the calculation of homology groups in this section and the next, connect the calculation with Stiefel-Whitney numbers in §5, and describe how to complete the desired calculation of homotopy groups in §6.

We must first define the homology groups of prespectra and the stable Hurewicz homomorphism. Just as we defined the homotopy groups of a prespectrum T by the formula

$$\pi_n(T) = \operatorname{colim} \pi_{n+q}(T_q),$$

we define the homology and cohomology groups of T with respect to a homology theory k_* and cohomology theory k^* on spaces by the formulas

$$k_n(T) = \operatorname{colim} \tilde{k}_{n+q}(T_q),$$

where the colimit is taken over the maps

$$\tilde{k}_{n+q}(T_q) \xrightarrow{\Sigma_*} \tilde{k}_{n+q+1}(\Sigma T_q) \xrightarrow{\sigma_{q*}} \tilde{k}_{n+q+1}(T_{q+1}),$$

and

$$k^n(T) = \lim \tilde{k}^{n+q}(T_q),$$

where the limit is taken over the maps

$$\tilde{k}^{n+q+1}(T_{q+1}) \xrightarrow{\sigma_q^*} \tilde{k}^{n+q+1}(\Sigma T_q) \xrightarrow{\Sigma^{-1}} \tilde{k}^{n+q}(T_q).$$

In fact, this definition of cohomology is inappropriate in general, differing from the appropriate definition by a $\lim^1$ error term. However, the definition is correct when k^* is ordinary cohomology with coefficients in a field R and each $\tilde{H}^{n+q}(T_q; R)$ is a finite dimensional vector space over R. This is the only case that we will need in the work of this chapter. In this case, it is clear that $H^n(T; R)$ is the vector space dual of $H_n(T; R)$, a fact that we shall use repeatedly.

Observe that there is no cup product in $H^*(T; R)$: the maps in the limit system factor through the reduced cohomologies of suspensions, in which cup products are identically zero (see Problem 5 at the end of Chapter 19). However, if T is an associative and commutative ring prespectrum, then the homology groups $H_*(T; R)$ form a graded commutative R-algebra.

The Hurewicz homomorphisms $\pi_{n+q}(T_q) \longrightarrow \tilde{H}_{n+q}(T_q; \mathbb{Z})$ pass to colimits to give the stable Hurewicz homomorphism

$$h : \pi_n(T) \longrightarrow H_n(T; \mathbb{Z}).$$

We may compose this with the map $H_n(T; \mathbb{Z}) \longrightarrow H_n(T; R)$ induced by the unit of a ring R, and we continue to denote the composite by h. If T is an associative and commutative ring prespectrum, then $h : \pi_*(T) \longrightarrow H_*(T; R)$ is a map of graded commutative rings.

We shall write H_* and H^* for homology and cohomology with coefficients in $\mathbb{Z}_2$ throughout §§3–6, and we tacitly assume that all homology and cohomology groups in sight are finite dimensional $\mathbb{Z}_2$-vector spaces. Recall that we have Thom isomorphisms

$$\Phi_q : H^n(BO(q)) \longrightarrow \tilde{H}^{n+q}(TO(q))$$

obtained by cupping with the Thom class $\mu_q \in \tilde{H}^q(TO(q))$. Naturality of the Thom diagonal applied to the map of bundles $\gamma_q \oplus \varepsilon \longrightarrow \gamma_{q+1}$ gives the commutative diagram

$$\begin{array}{ccc} \Sigma TO(q) & \xrightarrow{\Delta} & BO(q)_+ \wedge \Sigma TO(q) \\ \sigma_q \downarrow & & \downarrow i_q \wedge \sigma_q \\ TO(q+1) & \xrightarrow{\Delta} & BO(q+1)_+ \wedge TO(q+1). \end{array}$$

This implies that the following diagram is commutative:

$$\begin{array}{ccc} H^n(BO(q+1)) & \xrightarrow{i_q^*} & H^n(BO(q)) \\ \Phi_{q+1} \downarrow & & \downarrow \Phi_q \\ \tilde{H}^{n+q+1}(TO(q+1)) & \xrightarrow[\sigma_q^*]{} \tilde{H}^{n+q+1}(\Sigma TO(q)) \xrightarrow[\Sigma^{-1}]{} & \tilde{H}^{n+q}(TO(q)). \end{array}$$

We therefore obtain a "stable Thom isomorphism"

$$\Phi : H^n(BO) \longrightarrow H^n(TO)$$

on passage to limits. We have dual homology Thom isomorphisms

$$\Phi_n : \tilde{H}_{n+q}(TO(q)) \longrightarrow H_n(BO(q))$$

that pass to colimits to give a stable Thom isomorphism

$$\Phi : H_n(T) \longrightarrow H_n(BO).$$

Naturality of the Thom diagonal applied to the map of bundles $\gamma_q \oplus \gamma_r \longrightarrow \gamma_{q+r}$ gives the commutative diagram

$$\begin{array}{ccc} TO(q) \wedge TO(r) & \xrightarrow{\Delta \wedge \Delta} & BO(q)_+ \wedge TO(q) \wedge BO(r)_+ \wedge TO(r) \\ & & \downarrow \text{id} \wedge t \wedge \text{id} \\ \phi_{q,r} \downarrow & & (BO(q) \times BO(r))_+ \wedge TO(q) \wedge TO(r) \\ & & \downarrow (p_{q,r})_+ \wedge \phi_{q,r} \\ TO(q+r) & \xrightarrow{\Delta} & BO(q+r)_+ \wedge TO(q+r). \end{array}$$

As we observed for BU in the previous chapter, the maps $p_{q,r}$ pass to colimits to give BO an H-space structure, and it follows that $H_*(BO)$ is a $\mathbb{Z}_2$-algebra. On passage to homology and colimits, these diagrams imply the following conclusion.

PROPOSITION. *The Thom isomorphism* $\Phi : H_*(TO) \longrightarrow H_*(BO)$ *is an isomorphism of* $\mathbb{Z}_2$-*algebras.*

The description of the $H^*(BO(n))$ and the maps i_q^* in Chapter 23 §2 implies that

$$H^*(BO) = \mathbb{Z}_2[w_i | i \geq 1]$$

as an algebra. However, we are more interested in its "coalgebra" structure, which is given by the vector space dual

$$\psi : H^*(BO) \longrightarrow H^*(BO) \otimes H^*(BO)$$

of its product in homology. It is clear from the description of the $p_{q,r}^*$ that

$$\psi(w_k) = \sum_{i+j=k} w_i \otimes w_j.$$

From here, determination of $H_*(BO)$ and therefore $H_*(TO)$ as an algebra is a purely algebraic, but non-trivial, problem in dualization. Let $i : \mathbb{R}P^\infty = BO(1) \longrightarrow BO$ be the inclusion. Let $x_i \in H_i(\mathbb{R}P^\infty)$ be the unique non-zero element and let $b_i = i_*(x_i)$. Then the solution of our dualization problem takes the following form.

THEOREM. $H_*(BO)$ *is the polynomial algebra* $\mathbb{Z}_2[b_i | i \geq 1]$.

Let $a_i \in H_i(TO)$ be the element characterized by $\Phi(a_i) = b_i$.

COROLLARY. $H_*(TO)$ *is the polynomial algebra* $\mathbb{Z}_2[a_i | i \geq 1]$.

Using the compatibility of the Thom isomorphisms for $BO(1)$ and BO, we see that the a_i come from $H_*(TO(1))$. Remember that elements of $H_{i+1}(TO(1))$ map to elements of $H_i(TO)$ in the colimit; in particular, the non-zero element of $H_1(TO(1))$ maps to the identity element $1 \in H_0(TO)$. Recall from Chapter 23 §6 that we have a homotopy equivalence $j : \mathbb{R}P^\infty \longrightarrow TO(1)$.

COROLLARY. *For* $i \geq 0$, $j_*(x_{i+1})$ *maps to* a_i *in* $H_*(TO)$, *where* $a_0 = 1$.

4. The Steenrod algebra and its coaction on $H_*(TO)$

Since the Steenrod operations are stable and natural, they pass to limits to define natural operations $Sq^i : H^n(T) \longrightarrow H^{n+i}(T)$ for $i \geq 0$ and prespectra T. Here $Sq^0 = \mathrm{id}$, but it is not true that $Sq^i(x) = 0$ for $i > \deg x$. For example, we have the "stable Thom class" $\Phi(1) = \mu \in H^0(TO)$, and it is immediate from the definition of the Stiefel-Whitney classes that $\Phi(w_i) = Sq^i(\mu)$. Of course, $Sq^i(1) = 0$ for $i > 0$, so that Φ does not commute with Steenrod operations. The homology and cohomology of TO are built up from $\pi_*(TO)$ and Steenrod operations. We need to make this statement algebraically precise to determine $\pi_*(TO)$, and we need to assemble the Steenrod operations into an algebra to do this.

DEFINITION. The mod 2 Steenrod algebra A is the quotient of the free associative $\mathbb{Z}_2$-algebra generated by elements Sq^i, $i \geq 1$, by the ideal generated by the Adem relations (which are stated in Chapter 22 §5).

The following lemmas should be clear.

LEMMA. For spaces X, $H^*(X)$ has a natural A-module structure.

LEMMA. For prespectra T, $H^*(T)$ has a natural A-module structure.

The elements of A are stable mod 2 cohomology operations, and our description of the cohomology of $K(\mathbb{Z}_2, q)$s in Chapter 22 §5 implies that A is in fact the algebra of all stable mod 2 cohomology operations, with multiplication given by composition. Passage to limits over q leads to the following lemma. Alternatively, with the more formal general definitions of the next section, it will become yet another application of the Yoneda lemma. Recall that $H\mathbb{Z}_2$ denotes the Eilenberg-Mac Lane Ω-prespectrum $\{K(\mathbb{Z}_2, q)\}$.

LEMMA. As a vector space, A is isomorphic to $H^*(H\mathbb{Z}_2)$.

We shall see how to describe the composition in A homotopically in the next section. What is more important at the moment is that the lemma allows us to read off a basis for A.

THEOREM. A has a basis consisting of the operations $Sq^I = Sq^{i_1} \cdots Sq^{i_j}$, where I runs over the sequences $\{i_1, \ldots, i_j\}$ of positive integers such that $i_r \geq 2i_{r+1}$ for $1 \leq r < j$.

What is still more important to us is that A not only has the composition product $A \otimes A \longrightarrow A$, it also has a coproduct $\psi : A \longrightarrow A \otimes A$. Giving $A \otimes A$ its natural structure as an algebra, ψ is the unique map of algebras specified on generators by $\psi(Sq^k) = \sum_{i+j=k} Sq^i \otimes Sq^j$. The fact that ψ is a well defined map of algebras is a formal consequence of the Cartan formula. Algebraic structures like this, with compatible products and coproducts, are called "Hopf algebras."

We write A_* for the vector space dual of A, and we give it the dual basis to the basis just specified on A. While A_* is again a Hopf algebra, we are only interested in its algebra structure at the moment. In contrast with A, the algebra A_* is commutative, as is apparent from the form of the coproduct on the generators of A. Recall that $H\mathbb{Z}_2$ is an associative and commutative ring prespectrum, so that $H_*(H\mathbb{Z}_2)$ is a commutative $\mathbb{Z}_2$-algebra. The definition of the product on $H\mathbb{Z}_2$ (in Chapter 22 §3) and the Cartan formula directly imply the following observation.

LEMMA. A_* is isomorphic as an algebra to $H_*(H\mathbb{Z}_2)$.

4. THE STEENROD ALGEBRA AND ITS COACTION ON $H_*(TO)$

We need an explicit description of this algebra. In principle, this is a matter of pure algebra from the results already stated, but the algebraic work is non-trivial.

THEOREM. *For $r \geq 1$, define $I_r = (2^{r-1}, 2^{r-2}, \ldots, 2, 1)$ and define ξ_r to be the basis element of A_* dual to Sq^{I_r}. Then A_* is the polynomial algebra $\mathbb{Z}_2[\xi_r | r \geq 1]$.*

We need a bit of space level motivation for the particular relevance of the elements ξ_r. We left the computation of the Steenrod operations in $H^*(\mathbb{R}P^\infty)$ as an exercise, and the reader should follow up by proving the following result.

LEMMA. *In $H^*(\mathbb{R}P^\infty) = \mathbb{Z}_2[\alpha]$, $Sq^{I_r}(\alpha) = \alpha^{2^r}$ for $r \geq 1$ and $Sq^I(\alpha) = 0$ for all other basis elements sq^I of A.*

The A-module structure maps

$$A \otimes H^*(X) \longrightarrow H^*(X) \quad \text{and} \quad A \otimes H^*(T) \longrightarrow H^*(T)$$

for spaces X and prespectra T dualize to give "A_*-comodule" structure maps

$$\gamma : H_*(X) \longrightarrow A_* \otimes H_*(X) \quad \text{and} \quad \gamma : H_*(T) \longrightarrow A_* \otimes H_*(T).$$

We remind the reader that we are implicitly assuming that all homology and cohomology groups in sight are finitely generated $\mathbb{Z}_2$-vector spaces, although these "coactions" can in fact be defined without this assumption.

Formally, the notion of a comodule N over a coalgebra C is defined by reversing the direction of arrows in a diagrammatic definition of a module over an algebra. For example, for any vector space V, $C \otimes V$ is a comodule with action

$$\psi \otimes \mathrm{id} : C \otimes V \longrightarrow C \otimes C \otimes V.$$

Note that, dualizing the unit of an algebra, a $\mathbb{Z}_2$-coalgebra is required to have a counit $\varepsilon : C \longrightarrow \mathbb{Z}_2$. We understand all of these algebraic structures to be graded, and we say that a coalgebra is connected if $C_i = 0$ for $i < 0$ and $\varepsilon : C_0 \longrightarrow \mathbb{Z}_2$ is an isomorphism. When considering the Hurewicz homomorphism of $\pi_*(TO)$, we shall need the following observation.

LEMMA. *Let C be a connected coalgebra and V be a vector space. An element $y \in C \otimes V$ satisfies $(\psi \otimes \mathrm{id})(y) = 1 \otimes y$ if and only if $y \in C_0 \otimes V \cong V$.*

If V is a C-comodule with coaction $\nu : V \longrightarrow C \otimes V$, then ν is a morphism of C-comodules. Therefore the coaction maps γ above are maps of A_*-comodules for any space X or prespectrum T. We also need the following observation, which is implied by the Cartan formula.

LEMMA. *If T is an associative ring prespectrum, then $\gamma : H_*(T) \longrightarrow A_* \otimes H_*(T)$ is a homomorphism of algebras.*

The lemma above on Steenrod operations in $H^*(\mathbb{R}P^\infty)$ dualizes as follows.

LEMMA. *Write the coaction $\gamma : H_*(\mathbb{R}P^\infty) \longrightarrow A_* \otimes H_*(\mathbb{R}P^\infty)$ in the form $\gamma(x_i) = \sum_j a_{i,j} \otimes x_j$. Then*

$$a_{i,1} = \begin{cases} \xi_r & \text{if } i = 2^r \text{ for some } r \geq 1 \\ 0 & \text{otherwise.} \end{cases}$$

Note that $a_{i,i} = 1$, dualizing $Sq^0(\alpha^i) = \alpha^i$.

Armed with this information, we return to the study of the algebra $H_*(TO)$. We know that it is isomorphic to $H_*(BO)$, but the crux of the matter is to redescribe it in terms of A_*.

THEOREM. *Let N_* be the algebra defined abstractly by*
$$N_* = \mathbb{Z}_2[u_i | i > 1 \text{ and } i \neq 2^r - 1],$$
where $\deg u_i = i$. *Define a homomorphism of algebras* $f : H_*(TO) \longrightarrow N_*$ *by*
$$f(a_i) = \begin{cases} u_i & \text{if } i \text{ is not of the form } 2^r - 1 \\ 0 & \text{if } i = 2^r - 1. \end{cases}$$
Then the composite
$$g : H_*(TO) \xrightarrow{\gamma} A_* \otimes H_*(TO) \xrightarrow{\text{id} \otimes f} A_* \otimes N_*$$
is an isomorphism of both A-comodules and $\mathbb{Z}_2$-algebras.

PROOF. It is clear from things already stated that g is a map of both A-comodules and $\mathbb{Z}_2$-algebras. We must prove that it is an isomorphism. Its source and target are both polynomial algebras with one generator of degree i for each $i \geq 1$, hence it suffices to show that g takes generators to generators. Recall that $a_i = j_*(x_{i+1})$. This allows us to compute $\gamma(a_i)$. Modulo terms that are decomposable in the algebra $A_* \otimes H_*(TO)$, we find
$$\gamma(a_i) \equiv \begin{cases} 1 \otimes a_i & \text{if } i \text{ is not of the form } 2^r - 1 \\ \xi_r \otimes 1 + 1 \otimes a_{2^r - 1} & \text{if } i = 2^r - 1. \end{cases}$$
Applying $\text{id} \otimes f$ to these elements, we obtain $1 \otimes u_i$ in the first case and $\xi_r \otimes 1$ in the second case. □

Now consider the Hurewicz homomorphism $h : \pi_*(T) \longrightarrow H_*(T)$ of a prespectrum T. We have the following observation, which is a direct consequence of the definition of the Hurewicz homomorphism and the fact that $Sq^i = 0$ for $i > 0$ in the cohomology of spheres.

LEMMA. *For $x \in \pi_*(T)$, $\gamma(h(x)) = 1 \otimes h(x)$.*

Therefore, identifying N_* as the subalgebra $\mathbb{Z}_2 \otimes N_*$ of $A_* \otimes N_*$, we see that $g \circ h$ maps $\pi_*(TO)$ to N_*. We shall prove the following result in §6 and so complete the proof of Thom's theorem.

THEOREM. $h : \pi_*(TO) \longrightarrow H_*(TO)$ *is a monomorphism and $g \circ h$ maps $\pi_*(TO)$ isomorphically onto N_*.*

5. The relationship to Stiefel-Whitney numbers

We shall prove that a smooth closed n-manifold M is a boundary if and only if all of its normal Stiefel-Whitney numbers are zero. Polynomials in the Stiefel-Whitney classes are elements of $H^*(BO)$. We have seen that the normal Stiefel-Whitney numbers of a boundary are zero, and it follows that cobordant manifolds have the same normal Stiefel-Whitney numbers. The assignment of Stiefel-Whitney numbers to corbordism classes of n-manifolds specifies a homomorphism
$$\# : H^n(BO) \otimes \mathcal{N}_n \longrightarrow \mathbb{Z}_2.$$
We claim that the following diagram is commutative:

$$\begin{array}{ccccc} H^n(BO) \otimes \mathcal{N}_n & \xrightarrow{\text{id} \otimes \alpha} & H^n(BO) \otimes \pi_n(TO) & \xrightarrow{\text{id} \otimes h} & H^n(BO) \otimes H_n(TO) \\ \scriptstyle{\#} \downarrow & & & & \downarrow \scriptstyle{\text{id} \otimes \Phi} \\ \mathbb{Z}_2 & & \xleftarrow{\langle , \rangle} & & H^n(BO) \otimes H_n(BO). \end{array}$$

5. THE RELATIONSHIP TO STIEFEL-WHITNEY NUMBERS

To say that all normal Stiefel-Whitney numbers of M are zero is to say that $w\#[M] = 0$ for all $w \in H^n(BO)$. Granted the commutativity of the diagram, this is the same as to say that $\langle w, (\Phi \circ h \circ \alpha)([M]) \rangle = 0$ for all $w \in H^n(BO)$. Since $\langle \, , \, \rangle$ is the evaluation pairing of dual vector spaces, this implies that $(\Phi \circ h \circ \alpha)([M]) = 0$. Since Φ and α are isomorphisms and h is a monomorphism, this implies that $[M] = 0$ and thus that M is a boundary.

Thus we need only prove that the diagram is commutative. Embed M in $\mathbb{R}^{n+q}$ with normal bundle ν and let $f : M \longrightarrow BO(q)$ classify ν. Then $\alpha([M])$ is represented by the composite $S^{n+q} \xrightarrow{t} T\nu \xrightarrow{Tf} TO(q)$. In homology, we have the commutative diagram

$$\begin{array}{ccc} \tilde{H}_{n+q}(S^{n+q}) & \xrightarrow{t_*} \tilde{H}_{n+q}(T\nu) \xrightarrow{(Tf)_*} & \tilde{H}_{n+q}(TO(q)) \\ & \downarrow \Phi & \downarrow \Phi \\ & H_n(M) \xrightarrow{f_*} & H_n(BO(q)). \end{array}$$

Let $i_{n+q} \in \tilde{H}_{n+q}(S^{n+q})$ be the fundamental class. By the diagram and the definitions of α and the Hurewicz homomorphism,

$$(f_* \circ \Phi \circ t_*)(i_{n+q}) = (\Phi \circ (Tf)_* \circ t_*)(i_{n+q}) = (\Phi \circ h \circ \alpha)([M]) \in H_n(BO(q)).$$

Let $z = (\Phi \circ t_*)(i_{n+q}) \in H_n(M)$. We claim that z is the fundamental class. Granting the claim, it follows immediately that, for $w \in H^n(BO(q))$,

$$\begin{aligned} w\#[M] = \langle w(\nu), z \rangle & = \langle (f^*w(\gamma_q)), (\Phi \circ t_*)(i_{n+q}) \rangle \\ & = \langle w(\gamma_q), (f_* \circ \Phi \circ t_*)(i_{n+q}) \rangle \\ & = \langle w(\gamma_q), (\Phi \circ h \circ \alpha)([M]) \rangle. \end{aligned}$$

Thus we are reduced to proving the claim. It suffices to show that z maps to a generator of $H_n(M, M - x)$ for each $x \in M$. Since we must deal with pairs, it is convenient to use the homeomorphism between $T\nu$ and the quotient $D(\nu)/S(\nu)$ of the unit disk bundle by the unit sphere bundle. Recall that we have a relative cap product

$$\cap : H^q(D(\nu), S(\nu)) \otimes H_{i+q}(D(\nu), S(\nu)) \longrightarrow H_i(D(\nu)).$$

Letting $p : D(\nu) \longrightarrow M$ be the projection, which of course is a homotopy equivalence, we find that the homology Thom isomorphism

$$\Phi : H_{i+q}(D(\nu), S(\nu)) \longrightarrow H_i(M)$$

is given by the explicit formula

$$\Phi(a) = p_*(\mu \cap a).$$

Let $x \in U \subset M$, where $U \cong \mathbb{R}^n$. Let $D(U)$ and $S(U)$ be the inverse images in U of the unit disk and unit sphere in $\mathbb{R}^n$ and let $V = D(U) - S(U)$. Since $D(U)$ is contractible, $\nu|_{D(U)}$ is trivial and thus isomorphic to $D(U) \times D^q$. Write

$$\partial(D(U) \times D^q) = (D(U) \times S^{q-1}) \cup (S(U) \times D^q)$$

and observe that we obtain a homotopy equivalence

$$t : S^{n+q} \longrightarrow (D(U) \times D^q)/\partial(D(U) \times D^q) \cong S^{n+q}$$

by letting t be the quotient map on the restriction of the tubular neighborhood of ν to $D(\nu|_{D(U)})$ and letting t send the complement of this restriction to the basepoint.

Interpreting $t: S^{n+q} \longrightarrow D(\nu)/S(\nu)$ similarly, we obtain the following commutative diagram:

$$\begin{array}{ccccc}
\tilde{H}_{n+q}(S^{n+q}) & \xrightarrow[\cong]{t_*} & H_{n+q}(D(U) \times D^q, \partial(D(U) \times D^q)) & \xrightarrow[\cong]{\Phi} & H_n(D(U), S(U)) \\
\downarrow t_* & & \downarrow & & \downarrow \cong \\
& & H_{n+q}(D(\nu), S(\nu) \cup D(\nu|_{M-V})) & \xrightarrow{\Phi} & H_n(M, M-V) \\
\downarrow & & & & \downarrow \cong \\
H_{n+q}(D(\nu), S(\nu)) & \xrightarrow{\Phi} & H_n(M) & \longrightarrow & H_n(M, M-x).
\end{array}$$

The unlabeled arrows are induced by inclusions, and the right vertical arrows are excision isomorphisms. The maps Φ are of the general form $\Phi(a) = p_*(\mu \cap a)$. For the top map Φ, $\mu \in H_{n+q}(D(\nu|_{D(U)}), S(\nu|_{D(U)})) \cong H_{n+q}(S^{n+q})$, and, up to evident isomorphisms, Φ is just the inverse of the suspension isomorphism $\tilde{H}_n(S^n) \longrightarrow \tilde{H}_{n+q}(S^{n+q})$. The diagram shows that z maps to a generator of $H_n(M, M-x)$, as claimed.

6. Spectra and the computation of $\pi_*(TO) = \pi_*(MO)$

We must still prove that $h: \pi_*(TO) \longrightarrow H_*(TO)$ is a monomorphism and that $g \circ h$ maps $\pi_*(TO)$ isomorphically onto N_*. Write N for the dual vector space of N_*. (Of course, N is a coalgebra, but that is not important for this part of our work.) Remember that the Steenrod algebra A is dual to A_* and that $A \cong H^*(H\mathbb{Z}_2)$. The dual of $g: H_*(TO) \longrightarrow A_* \otimes N_*$ is an isomorphism of A-modules (and of coalgebras) $g^*: A \otimes N \longrightarrow H^*(TO)$. Thus, if we choose a basis $\{y_i\}$ for N, where $\deg y_i = n_i$ say, then $H^*(TO)$ is the free graded A-module on the basis $\{y_i\}$.

At this point, we engage in a conceptual thought exercise. We think of prespectra as "stable objects" that have associated homotopy, homology, and cohomology groups. Imagine that we have a good category of stable objects, analogous to the category of based spaces, that is equipped with all of the constructions that we have on based spaces: wedges (= coproducts), colimits, products, limits, suspensions, loops, homotopies, cofiber sequences, fiber sequences, smash products, function objects, and so forth. Let us call the stable objects in our imagined category "spectra" and call the category of such objects $\mathscr{S}$. We have in mind an analogy with the notions of presheaf and sheaf.

Whatever spectra are, there must be a way of constructing a spectrum from a prespectrum without changing its homotopy, homology, and cohomology groups. In turn, a based space X determines the prespectrum $\Sigma^\infty X = \{\Sigma^n X\}$. The homology and cohomology groups of $\Sigma^\infty X$ are the (reduced) homology and cohomology groups of X; the homotopy groups of $\Sigma^\infty X$ are the stable homotopy groups of X.

Because homotopy groups, homology groups, and cohomology groups on based spaces satisfy the weak equivalence axiom, the real domain of definition of these invariants is the category $\bar{h}\mathscr{T}$ that is obtained from the homotopy category $h\mathscr{T}$ of based spaces by adjoining inverses to the weak equivalences. This category is equivalent to the homotopy category $h\mathscr{C}$ of based CW complexes. Explicitly, the morphisms from X to Y in $\bar{h}\mathscr{T}$ can be defined to be the based homotopy classes of maps $\Gamma X \longrightarrow \Gamma Y$, where ΓX and ΓY are CW approximations of X and Y. Composition is defined in the evident way.

6. SPECTRA AND THE COMPUTATION OF $\pi_*(TO) = \pi_*(MO)$ 227

Continuing our thought exercise, we can form the homotopy category $h\mathscr{S}$ of spectra and can define homotopy groups in terms of homotopy classes of maps from sphere spectra to spectra. Reflection on the periodic nature of K-theory suggests that we should define sphere spectra of negative dimension and define homotopy groups $\pi_q(X)$ for all integers q. We say that a map of spectra is a weak equivalence if it induces an isomorphism on homotopy groups. We can form the "stable category" $\bar{h}\mathscr{S}$ from $h\mathscr{S}$ exactly as we formed the category $\bar{h}\mathscr{T}$ from $h\mathscr{T}$. That is, we develop a theory of CW spectra using sphere spectra as the domains of attaching maps. The Whitehead and cellular approximation theorems hold, and every spectrum X admits a CW approximation $\Gamma X \longrightarrow X$. We define the set $[X, Y]$ of morphisms $X \longrightarrow Y$ in $\bar{h}\mathscr{S}$ to be the set of homotopy classes of maps $\Gamma X \longrightarrow \Gamma Y$. This is a *stable* category in the sense that the functor $\Sigma : \bar{h}\mathscr{S} \longrightarrow \bar{h}\mathscr{S}$ is an equivalence of categories. More explicitly, the natural maps $X \longrightarrow \Omega\Sigma X$ and $\Sigma\Omega X \longrightarrow X$ are isomorphisms in $\bar{h}\mathscr{S}$.

In particular, up to isomorphism, every object in the category $\bar{h}\mathscr{S}$ is a suspension, hence a double suspension. This implies that each $[X, Y]$ is an Abelian group and composition is bilinear. Moreover, for any map $f : X \longrightarrow Y$, the canonical map $Ff \longrightarrow \Omega Cf$ and its adjoint $\Sigma Ff \longrightarrow Cf$ (see Chapter 8 §7) are also isomorphisms in $\bar{h}\mathscr{S}$, so that cofiber sequences and fiber sequences are equivalent. Therefore cofiber sequences give rise to long exact sequences of homotopy groups.

The homotopy groups of wedges and products of spectra are given by

$$\pi_*(\bigvee_i X_i) = \sum_i \pi_*(X_i) \quad \text{and} \quad \pi_*(\prod_i X_i) = \prod_i \pi_*(X_i).$$

Therefore, if only finitely many $\pi_q(X_i)$ are non-zero for each q, then the natural map $\bigvee_i X_i \longrightarrow \prod_i X_i$ is an isomorphism.

We have homology groups and cohomology groups defined on $\bar{h}\mathscr{S}$. A spectrum E represents a homology theory E_* and a cohomology theory E^* specified in terms of smash products and function spectra by

$$E_q(X) = \pi_q(X \wedge E) \quad \text{and} \quad E^q(X) = \pi_{-q}F(X, E) \cong [X, \Sigma^q E].$$

Verifications of the exactness, suspension, additivity, and weak equivalence axioms are immediate from the properties of the category $\bar{h}\mathscr{S}$. Moreover, every homology or cohomology theory on $\bar{h}\mathscr{S}$ is so represented by some spectrum E.

As will become clear later, Ω-prespectra are more like spectra than general prespectra, and we continue to write $H\pi$ for the "Eilenberg-Mac Lane spectrum" that represents ordinary cohomology with coefficients in π. Its only non-zero homotopy group is $\pi_0(H\pi) = \pi$, and the Hurewicz homomorphism maps this group isomorphically onto $H_0(H\pi;\mathbb{Z})$. When $\pi = \mathbb{Z}_2$, the natural map $H_0(H\mathbb{Z}_2;\mathbb{Z}) \longrightarrow H_0(H\mathbb{Z}_2;\mathbb{Z}_2)$ is also an isomorphism.

Returning to our motivating example, we write MO for the "Thom spectrum" that arises from the Thom prespectrum TO. The reader may sympathize with a student who claimed that MO stands for "Mythical Object."

We may choose a map $\bar{y}_i : MO \longrightarrow \Sigma^{n_i} H\mathbb{Z}_2$ that represents the element y_i. Define $K(N_*)$ to be the wedge of a copy of $\Sigma^{n_i} H\mathbb{Z}_2$ for each basis element y_i and note that $K(N_*)$ is isomorphic in $\bar{h}\mathscr{S}$ to the product of a copy of $\Sigma^{n_i} H\mathbb{Z}_2$ for each y_i. We think of $K(N_*)$ as a "generalized Eilenberg-Mac Lane spectrum." It satisfies $\pi_*(K(N_*)) \cong N_*$ (as Abelian groups and so as $\mathbb{Z}_2$-vector spaces), and the mod 2 Hurewicz homomorphism $h : \pi_*(K(N_*)) \longrightarrow H_*(K(N_*))$ is a monomorphism.

Using the $\bar{y}_i$ as coordinates, we obtain a map
$$\omega : MO \longrightarrow \prod_i \Sigma^{n_i} H\mathbb{Z}_2 \simeq K(N_*).$$
The induced map ω^* on mod 2 cohomology is an isomorphism of A-modules: $H^*(MO)$ and $H^*(K(N_*))$ are free A-modules, and we have defined ω so that ω^* sends basis elements to basis elements. Therefore the induced map on homology groups is an isomorphism. Here we are using mod 2 homology, but it can be deduced from the fact that both $\pi_*(MO)$ and $\pi_*(K(N_*))$ are $\mathbb{Z}_2$-vector spaces that ω induces an isomorphism on integral homology groups. Therefore the integral homology groups of $C\omega$ are zero. By the Hurewicz theorem in $\bar{h}\mathscr{S}$, the homotopy groups of $C\omega$ are also zero. Therefore ω induces an isomorphism of homotopy groups. That is, ω is an isomorphism in $\bar{h}\mathscr{S}$. Therefore $\pi_*(MO) \cong N_*$ and the Hurewicz homomorphism $h : \pi_*(MO) \longrightarrow H_*(MO)$ is a monomorphism. It follows that $g \circ h : \pi_*(MO) \longrightarrow N_*$ is an isomorphism since it is a monomorphism between vector spaces of the same finite dimension in each degree.

7. An introduction to the stable category

To give content to the argument just sketched, we should construct a good category of spectra. In fact, no such category was available when Thom first proved his theorem in 1960. With motivation from the introduction of K-theory and cobordism, a good stable category was constructed by Boardman (unpublished) around 1964 and an exposition of his category was given by Adams soon after. However, these early constructions were far more primitive than our outline suggests. While they gave a satisfactory stable category, the underlying category of spectra did not have products, limits, and function objects, and its smash product was not associative, commutative, or unital. In fact, a fully satisfactory category of spectra was not constructed until 1995.

We give a few definitions to indicate what is involved.

DEFINITION. A spectrum E is a prespectrum such that the adjoints $\tilde{\sigma} : E_n \longrightarrow \Omega E_{n+1}$ of the structure maps $\sigma : \Sigma E_n \longrightarrow E_{n+1}$ are *homeomorphisms*. A map $f : T \longrightarrow T'$ of prespectra is a sequence of maps $f_n : T_n \longrightarrow T'_n$ such that $\sigma'_n \circ \Sigma T_n = f_{n+1} \circ \sigma_n$ for all n. A map $f : E \longrightarrow E'$ of spectra is a map between E and E' regarded as prespectra.

We have a forgetful functor from the category $\mathscr{S}$ of spectra to the category $\mathscr{P}$ of prespectra. It has a left adjoint $L : \mathscr{P} \longrightarrow \mathscr{S}$. In $\mathscr{P}$, we define wedges, colimits, products, and limits spacewise. For example, $(T \vee T')_n = T_n \vee T'_n$, with the evident structure maps. We define wedges and colimits of spectra by first performing the construction on the prespectrum level and then applying the functor L. If we start with spectra and construct products or limits spacewise, then the result is again a spectrum; that is, limits of spectra are the limits of their underlying prespectra. Thus the category $\mathscr{S}$ is complete and cocomplete.

Similarly, we define the smash product $T \wedge X$ and function prespectrum $F(X,T)$ of a based space X and a prespectrum T spacewise. For a spectrum E, we define $E \wedge X$ by applying L to the prespectrum level construction; the prespectrum $F(X,E)$ is already a spectrum. We now have cylinders $E \wedge I_+$ and thus can define homotopies between maps of spectra. Similarly we have cones $CE = E \wedge I$ (where I has basepoint 1), suspensions $\Sigma E = E \wedge S^1$, path spectra $F(I,E)$ (where I has basepoint 0), and loop spectra $\Omega E = F(S^1, E)$. The development of cofiber and fiber sequences proceeds exactly as for based spaces.

The left adjoint L can easily be described explicitly on those prespectra T whose adjoint structure maps $\tilde{\sigma}_n : T_n \longrightarrow \Omega T_{n+1}$ are inclusions: we define $(LT)_n$ to be the union of the expanding sequence

$$T_n \xrightarrow{\tilde{\sigma}_n} \Omega T_{n+1} \xrightarrow{\Omega \tilde{\sigma}_{n+1}} \Omega^2 T_{n+2} \longrightarrow \cdots.$$

We then have

$$\Omega(LT)_{n+1} = \Omega(\bigcup \Omega^q T_{n+1+q}) \cong \bigcup \Omega^{q+1} T_{n+q+1} \cong (LT)_n.$$

We have an evident map of prespectra $\lambda : T \longrightarrow LT$, and a comparison of colimits shows (by a cofinality argument) that λ induces isomorphisms on homotopy and homology groups. The essential point is that homotopy and homology commute with colimits. It is not true that cohomology converts colimits to limits in general, because of $\lim^1$ error terms, and this is one reason that our definition of the cohomology of prespectra via limits is inappropriate except under restrictions that guarantee the vanishing of $\lim^1$ terms. Observe that there is no problem in the case of Ω-prespectra, for which λ is a spacewise weak equivalence.

For a based space X, we define the suspension spectrum $\Sigma^\infty X$ by applying L to the suspension prespectrum $\Sigma^\infty X = \{\Sigma^n X\}$. The inclusion condition is satisfied in this case. We define $QX = \cup \Omega^q \Sigma^q X$, and we find that the nth space of $\Sigma^\infty X$ is $Q\Sigma^n X$. It should be apparent that the homotopy groups of the space QX are the stable homotopy groups of X.

The adjoint structure maps of the Thom prespectrum TO are also inclusions, and our mythical object is $MO = LTO$.

In general, for a prespectrum T, we can apply an iterated mapping cylinder construction to define a spacewise equivalent prespectrum KT whose adjoint structure maps are inclusions. The prespectrum level homotopy, homology, and cohomology groups of KT are isomorphic to those of T. Thus, if we have a prespectrum T whose invariants we are interested in, such as an Eilenberg-Mac Lane Ω-prespectrum or the K-theory Ω-prespectrum, then we can construct a spectrum LKT that has the same invariants.

For a based space X and $q \geq 0$, we construct a prespectrum $\Sigma_q^\infty X$ whose nth space is a point for $n < q$ and is $\Sigma^{n-q} X$ for $n \geq q$; its structure maps for $n \geq q$ are identity maps. We continue to write $\Sigma_q^\infty X$ for the spectrum obtained by applying L to this prespectrum. We then define sphere spectra S^q for all integers q by letting $S^q = \Sigma^\infty S^q$ for $q \geq 0$ and $S^{-q} = \Sigma_q^\infty S^0$ for $q > 0$. The definition is appropriate since $\Sigma S^q \cong S^{q+1}$ for all integers q. We can now define homotopy groups in the obvious way. For example, the homotopy groups of the K-theory spectrum are $\mathbb{Z}$ for every even integer and zero for every odd integer.

From here, we can go on to define CW spectra in very much the same way that we defined CW complexes, and we can fill in the rest of the outline in the previous section. The real work involves the smash product of spectra, but this does not belong in our rapid course. While there is a good deal of foundational work involved, there is also considerable payoff in explicit concrete calculations, as the computation of $\pi_*(MO)$ well illustrates.

With the hope that this glimpse into the world of stable homotopy theory has whetted the reader's appetite for more, we will end at this starting point.

Suggestions for further reading

Rather than attempt a complete bibliography, I will give a number of basic references. I will begin with historical references and textbooks. I will then give references for specific topics, more or less in the order in which topics appear in the text. Where material has been collected in one or another book, I have often referred to such books rather than to original articles. However, the importance and quality of exposition of some of the original sources often make them still to be preferred today. The subject in its earlier days was blessed with some of the finest expositors of mathematics, for example Steenrod, Serre, Milnor, and Adams. Some of the references are intended to give historical perspective, some are classical papers in the subject, some are follow-ups to material in the text, and some give an idea of the current state of the subject. In fact, many major parts of algebraic topology are nowhere mentioned in any of the existing textbooks, although several were well established by the mid-1970s. I will indicate particularly accessible references for some of them; the reader can find more of the original references in the sources given.

1. A classic book and historical references

The axioms for homology and cohomology theories were set out in the classic:

S. Eilenberg and N. Steenrod. *Foundations of algebraic topology.* Princeton University Press. 1952.

I believe the only historical monograph on the subject is:

J. Dieudonné. *A history of algebraic and differential topology, 1900–1960.* Birkhäuser. 1989.

A large collection of historical essays will appear soon:

I.M. James, editor. *The history of topology.* Elsevier Science. To appear.

Among the contributions, I will advertise one of my own, available on the web:

J.P. May. *Stable algebraic topology, 1945–1966.* http://hopf.math.purdue.edu

2. Textbooks in algebraic topology and homotopy theory

These are ordered roughly chronologically (although this is obscured by the fact that the most recent editions or versions are cited). I have included only those texts that I have looked at myself, that are at least at the level of the more elementary chapters here, and that offer significant individuality of treatment. There are many other textbooks in algebraic topology.

Two classic early textbooks:

P.J. Hilton and S. Wylie. *Homology theory.* Cambridge University Press. 1960.

E. Spanier. *Algebraic topology*. McGraw-Hill. 1966.

An idiosyncratic pre-homology level book giving much material about groupoids:

R. Brown. *Topology. A geometric account of general topology, homotopy types, and the fundamental groupoid*. Second edition. Ellis Horwood. 1988.

A homotopical introduction close to the spirit of this book:

B. Gray. *Homotopy theory, an introduction to algebraic topology*. Academic Press. 1975.

The standard current textbooks in basic algebraic topology:

M.J. Greenberg and J. R. Harper. *Algebraic topology, a first course*. Benjamin/Cummings. 1981.

W.S. Massey. *A basic course in algebraic topology*. Springer-Verlag. 1991.

A. Dold. *Lectures on algebraic topology*. Reprint of the 1972 edition. Springer-Verlag. 1995.

J.W. Vick. *Homology theory; an introduction to algebraic topology*. Second edition. Springer-Verlag. 1994.

J.R. Munkres. *Elements of algebraic topology*. Addison Wesley. 1984.

J.J. Rotman. *An introduction to algebraic topology*. Springer-Verlag. 1986.

G.E. Bredon. *Topology and geometry*. Springer-Verlag. 1993.

Sadly, the following are still the only more advanced textbooks in the subject:

R.M. Switzer. *Algebraic topology. Homotopy and homology*. Springer-Verlag. 1975.

G.W. Whitehead. *Elements of homotopy theory*. Springer-Verlag. 1978.

3. Books on CW complexes

Two books giving more detailed studies of CW complexes than are found in textbooks (the second giving a little of the theory of compactly generated spaces):

A.T. Lundell and S. Weingram *The topology of CW complexes*. Van Nostrand Reinhold. 1969.

R. Fritsch and R.A. Piccinini. *Cellular structures in topology*. Cambridge University Press. 1990.

4. Differential forms and Morse theory

Two introductions to algebraic topology starting from de Rham cohomology:

R. Bott and L.W. Tu. *Differential forms in algebraic topology*. Springer-Verlag. 1982.

I. Madsen and J. Tornehave. *From calculus to cohomology. de Rham cohomology and characteristic classes*. Cambridge University Press. 1997.

The classic reference on Morse theory, with an exposition of the Bott periodicity theorem:

J. Milnor. *Morse theory*. Annals of Math. Studies No. 51. Princeton University Press. 1963.

A modern use of Morse theory for the analytic construction of homology:

M. Schwarz. *Morse homology*. Progress in Math. Vol. 111. Birkhäuser. 1993.

5. Equivariant algebraic topology

Two good basic references on equivariant algebraic topology, classically called the theory of transformation groups (see also §§16, 21 below):

G. Bredon. *Introduction to compact transformation groups.* Academic Press. 1972.

T. tom Dieck. *Transformation groups.* Walter de Gruyter. 1987.

A more advanced book, a precursor to much recent work in the area:

T. tom Dieck. *Transformation groups and representation theory.* Lecture Notes in Mathematics Vol. 766. Springer-Verlag. 1979.

6. Category theory and homological algebra

A revision of the following classic on basic category theory is in preparation:

S. Mac Lane. *Categories for the working mathematician.* Springer-Verlag. 1971.

Two classical treatments and a good modern treatment of homological algebra:

H. Cartan and S. Eilenberg. *Homological algebra.* Princeton University Press. 1956.

S. MacLane. *Homology.* Springer-Verlag. 1963.

C.A. Weibel. *An introduction to homological algebra.* Cambridge University Press. 1994.

7. Simplicial sets in algebraic topology

Two older treatments and a comprehensive modern treatment:

P. Gabriel and M. Zisman. *Calculus of fractions and homotopy theory.* Springer-Verlag. 1967.

J.P. May. *Simplicial objects in algebraic topology.* D. Van Nostrand 1967; reprinted by the University of Chicago Press 1982 and 1992.

P.G. Goerss and J.F. Jardine. *Simplicial homotopy theory.* Birkhäuser. To appear.

8. The Serre spectral sequence and Serre class theory

Two classic papers of Serre:

J.-P. Serre. *Homologie singuliére des espaces fibrés. Applications.* Annals of Math. (2)54(1951), 425–505.

J.-P. Serre. *Groupes d'homotopie et classes de groupes abéliens.* Annals of Math. (2)58(1953), 198–232.

A nice exposition of some basic homotopy theory and of Serre's work:

S.-T. Hu. *Homotopy theory.* Academic Press. 1959.

Many of the textbooks cited in §2 also treat the Serre spectral sequence.

9. The Eilenberg-Moore spectral sequence

There are other important spectral sequences in the context of fibrations, mainly due to Eilenberg and Moore. Three references:

S. Eilenberg and J.C. Moore. *Homology and fibrations, I.* Comm. Math. Helv. 40(1966), 199–236.

L. Smith. *Homological algebra and the Eilenberg-Moore spectral sequences.* Trans. Amer. Math. Soc. 129(1967), 58–93.

V.K.A.M. Gugenheim and J.P. May. *On the theory and applications of differential torsion products.* Memoirs Amer. Math. Soc. No. 142. 1974.

There is a useful guidebook to spectral sequences:

J. McCleary. *User's guide to spectral sequences.* Publish or Perish. 1985.

10. Cohomology operations

A compendium of the work of Steenrod and others on the construction and analysis of the Steenrod operations:

N.E. Steenrod and D.B.A. Epstein. *Cohomology operations.* Annals of Math. Studies No. 50. Princeton University Press. 1962.

A classic paper that first formalized cohomology operations, among other things:

J.-P. Serre. *Cohomologie modulo 2 des complexes d'Eilenberg-Mac Lane.* Comm. Math. Helv. 27(1953), 198–232.

A general treatment of Steenrod-like operations:

J.P. May. *A general algebraic approach to Steenrod operations.* In Lecture Notes in Mathematics Vol. 168, 153–231. Springer-Verlag. 1970.

A nice book on mod 2 Steenrod operations and the Adams spectral sequence:

R. Mosher and M. Tangora. *Cohomology operations and applications in homotopy theory.* Harper and Row. 1968.

11. Vector bundles

A classic and a more recent standard treatment that includes K-theory:

N.E. Steenrod. *Topology of fibre bundles.* Princeton University Press. 1951. Fifth printing, 1965.

D. Husemoller. *Fibre bundles.* Springer-Verlag. 1966. Third edition, 1994.

A general treatment of classification theorems for bundles and fibrations:

J.P. May. *Classifying spaces and fibrations.* Memoirs Amer. Math. Soc. No. 155. 1975.

12. Characteristic classes

The classic introduction to characteristic classes:

J. Milnor and J.D. Stasheff. *Characteristic classes.* Annals of Math. Studies No. 76. Princeton University Press. 1974.

A good reference for the basic calculations of characteristic classes:

A. Borel. *Topology of Lie groups and characteristic classes.* Bull. Amer. Math. Soc. 61(1955), 297–432.

Two proofs of the Bott periodicity theorem that only use standard techniques of algebraic topology, starting from characteristic class calculations:

H. Cartan et al. *Périodicité des groupes d'homotopie stables des groupes classiques, d'après Bott.* Séminaire Henri Cartan, 1959/60. Ecole Normale Supérieure. Paris.

E. Dyer and R.K. Lashof. *A topological proof of the Bott periodicity theorems.* Ann. Mat. Pure Appl. (4)54(1961), 231–254.

13. K-theory

Two classical lecture notes on K-theory:

R. Bott. *Lectures on $K(X)$.* W.A. Benjamin. 1969.

This includes a reprint of perhaps the most accessible proof of the complex case of the Bott periodicity theorem, namely:

M.F. Atiyah and R. Bott. *On the periodicity theorem for complex vector bundles.* Acta Math. 112(1994), 229–247.

M.F. Atiyah. *K-theory.* Notes by D.W. Anderson. Second Edition. Addison-Wesley. 1967.

This includes reprints of two classic papers of Atiyah, one that relates Adams operations in K-theory to Steenrod operations in cohomology and another that sheds insight on the relationship between real and complex K-theory:

M.F. Atiyah. *Power operations in K-theory.* Quart. J. Math. (Oxford) (2)17(1966), 165–193.

M.F. Atiyah. *K-theory and reality.* Quart. J. Math. (Oxford) (2)17(1966), 367–386.

Another classic paper that greatly illuminates real K-theory:

M.F. Atiyah, R. Bott, and A. Shapiro. *Clifford algebras.* Topology 3(1964), suppl. 1, 3–38.

A more recent book on K-theory:

M. Karoubi. *K-theory.* Springer-Verlag. 1978.

Some basic papers of Adams and Adams and Atiyah giving applications of K-theory:

J.F. Adams. *Vector fields on spheres.* Annals of Math. 75(1962), 603–632.

J.F. Adams. *On the groups $J(X)$ I, II, III, and IV.* Topology 2(1963), 181–195; 3(1965), 137-171 and 193–222; 5(1966), 21–71.

J.F. Adams and M.F. Atiyah. *K-theory and the Hopf invariant.* Quart. J. Math. (Oxford) (2)17(1966), 31–38.

14. Hopf algebras; the Steenrod algebra, Adams spectral sequence

The basic source for the structure theory of (connected) Hopf algebras:

J. Milnor and J.C. Moore. *On the structure of Hopf algebras.* Annals of Math. 81(1965), 211–264.

The classic analysis of the structure of the Steenrod algebra as a Hopf algebra:

J. Milnor. *The Steenrod algebra and its dual.* Annals of Math. 67(1958), 150–171.

Two classic papers of Adams; the first constructs the Adams spectral sequence relating the Steenrod algebra to stable homotopy groups and the second uses secondary cohomology operations to solve the Hopf invariant one problem:

J.F. Adams. *On the structure and applications of the Steenrod algebra.* Comm. Math. Helv. 32(1958), 180–214.

J.F. Adams. *On the non-existence of elements of Hopf invariant one*. Annals of Math. 72(1960), 20–104.

15. Cobordism

The beautiful classic paper of Thom is still highly recommended:

R. Thom. *Quelques propriétés globals des variétés différentiables*. Comm. Math. Helv. 28(1954), 17–86.

Thom computed unoriented cobordism. Oriented and complex cobordism came later. In simplest form, the calculations use the Adams spectral sequence:

J. Milnor. *On the cobordism ring Ω^* and a complex analogue*. Amer. J. Math. 82(1960), 505–521.

C.T.C. Wall. *A characterization of simple modules over the Steenrod algebra mod 2*. Topology 1(1962), 249–254.

A. Liulevicius. *A proof of Thom's theorem*. Comm. Math. Helv. 37(1962), 121–131.

A. Liulevicius. *Notes on homotopy of Thom spectra*. Amer. J. Math. 86(1964), 1–16.

A very useful compendium of calculations of cobordism groups:

R. Stong. *Notes on cobordism theory*. Princeton University Press. 1968.

16. Generalized homology theory and stable homotopy theory

Two classical references, the second of which also gives detailed information about complex cobordism that is of fundamental importance to the subject.

G.W. Whitehead. *Generalized homology theories*. Trans. Amer. Math. Soc. 102(1962), 227–283.

J.F. Adams. *Stable homotopy and generalised homology*. Chicago Lectures in Mathematics. University of Chicago Press. 1974. Reprinted in 1995.

An often overlooked but interesting book on the subject:

H.R. Margolis. *Spectra and the Steenrod algebra. Modules over the Steenrod algebra and the stable homotopy category*. North-Holland. 1983.

Foundations for equivariant stable homotopy theory are established in:

L.G. Lewis, Jr., J.P. May, and M.Steinberger (with contributions by J.E. McClure). *Equivariant stable homotopy theory*. Lecture Notes in Mathematics Vol. 1213. Springer-Verlag. 1986.

17. Quillen model categories

In the introduction, I alluded to axiomatic treatments of "homotopy theory." Here are the original and two more recent references:

D.G. Quillen. *Homotopical algebra*. Lecture Notes in Mathematics Vol. 43. Springer-Verlag. 1967.

W.G. Dwyer and J. Spalinski. *Homotopy theories and model categories*. In A handbook of algebraic topology, edited by I.M. James. North-Holland. 1995.

The cited "*Handbook*" (over 1300 pages) contains an uneven but very interesting collection of expository articles on a wide variety of topics in algebraic topology.

M. Hovey. *Model categories.* Amer. Math. Soc. Surveys and Monographs No. 63. 1998.

18. Localization and completion; rational homotopy theory

Since the early 1970s, it has been standard practice in algebraic topology to localize and complete topological spaces, and not just their algebraic invariants, at sets of primes and then to study the subject one prime at a time, or rationally. Two of the basic original references are:

D. Sullivan. *The genetics of homotopy theory and the Adams conjecture.* Annals of Math. 100(1974), 1–79.

A.K. Bousfield and D.M. Kan. *Homotopy limits, completions, and localizations.* Lecture Notes in Mathematics Vol. 304. Springer-Verlag. 1972.

A more accessible introduction to localization and a readable recent paper on completion are:

P. Hilton, G. Mislin, and J. Roitberg. *Localization of nilpotent groups and spaces.* North-Holland. 1975.

F. Morel. *Quelques remarques sur la cohomologie modulo p continue des pro-p-espaces et les resultats de J. Lannes concernent les espaces fonctionnel Hom(BV, X).* Ann. Sci. Ecole Norm. Sup. (4)26(1993), 309–360.

When spaces are rationalized, there is a completely algebraic description of the result. The main original reference and a more accessible source are:

D. Sullivan. *Infinitesimal computations in topology.* Publ. Math. IHES 47(1978), 269–332.

A.K. Bousfield and V.K.A.M. Gugenheim. *On PL de Rham theory and rational homotopy type.* Memoirs Amer. Math. Soc. No. 179. 1976.

19. Infinite loop space theory

Another area well established by the mid-1970s. The following book is a delightful read, with capsule introductions of many topics other than infinite loop space theory, a very pleasant starting place for learning modern algebraic topology:

J.F. Adams. *Infinite loop spaces.* Annals of Math. Studies No. 90. Princeton University Press. 1978.

The following survey article is less easy going, but gives an indication of the applications to high dimensional geometric topology and to algebraic K-theory:

J.P. May. *Infinite loop space theory.* Bull. Amer. Math. Soc. 83(1977), 456–494.

Five monographs, each containing a good deal of expository material, that give a variety of theoretical and calculational developments and applications in this area:

J.P. May. *The geometry of iterated loop spaces.* Lecture Notes in Mathematics Vol. 271. Springer-Verlag. 1972.

J.M. Boardman and R.M. Vogt. *Homotopy invariant algebraic structures on topological spaces.* Lecture Notes in Mathematics Vol. 347. Springer-Verlag. 1973.

F.R. Cohen, T.J. Lada, and J.P. May. *The homology of iterated loop spaces.* Lecture Notes in Mathematics Vol. 533. Springer-Verlag. 1976.

J.P. May (with contributions by F. Quinn, N. Ray, and J. Tornehave). E_∞ ring spaces and E_∞ ring spectra. Lecture Notes in Mathematics Vol. 577. Springer-Verlag. 1977.

R. Bruner, J.P. May, J.E. McClure, and M. Steinberger. H_∞ ring spectra and their applications. Lecture Notes in Mathematics Vol. 1176. Springer-Verlag. 1986.

20. Complex cobordism and stable homotopy theory

Adams' book cited in §16 gives a spectral sequence for the computation of stable homotopy groups in terms of generalized cohomology theories. Starting from complex cobordism and related theories, its use has been central to two waves of major developments in stable homotopy theory. A good exposition for the first wave:

D.C. Ravenel. Complex cobordism and stable homotopy groups of spheres. Academic Press. 1986.

The essential original paper and a very nice survey article on the second wave:

E. Devinatz, M.J. Hopkins, and J.H. Smith. Nilpotence and stable homotopy theory. Annals of Math. 128(1988), 207–242.

M.J. Hopkins. Global methods in homotopy theory. In Proceedings of the 1985 LMS Symposium on homotopy theory, edited by J.D.S. Jones and E. Rees. London Mathematical Society. 1987.

The cited Proceedings contain good introductory survey articles on several other topics in algebraic topology. A larger scale exposition of the second wave is:

D.C. Ravenel. Nilpotence and periodicity in stable homotopy theory. Annals of Math. Studies No. 128. Princeton University Press. 1992.

21. Follow-ups to this book

There is a leap from the level of this introductory book to that of the most recent work in the subject. One recent book that helps fill the gap is:

P. Selick. Introduction to homotopy theory. Fields Institute Monographs No. 9. American Mathematical Society. 1997.

There is a recent expository book for the reader who would like to jump right in and see the current state of algebraic topology; although it focuses on equivariant theory, it contains introductions and discussions of many non-equivariant topics:

J.P. May et al. Equivariant homotopy and cohomology theory. NSF-CBMS Regional Conference Monograph. 1996.

For the reader of the last section of this book whose appetite has been whetted for more stable homotopy theory, there is an expository article that motivates and explains the properties that a satisfactory category of spectra should have:

J.P. May. Stable algebraic topology and stable topological algebra. Bulletin London Math. Soc. 30(1998), 225–234.

The following monograph gives such a category, with many applications; more readable accounts appear in the Handbook cited in §16 and in the book just cited:

A. Elmendorf, I. Kriz, M.A. Mandell, and J.P. May, with an appendix by M. Cole. Rings, modules, and algebras in stable homotopy theory. Amer. Math. Soc. Surveys and Monographs No. 47. 1997.

Index

Adams operations, 209
additivity axiom, 93, 94, 108, 109, 135, 136, 144, 145
Adem relations, 181
adjoint functors, 14
almost complex structure, 208
antipodal map, 101, 139
associated principal G-bundle, 197
attaching map, 71
Aut($\mathscr{E}$), 26, 29
Aut$_G(S)$, 24

base space, 21
bidegree of a map, 213
Bockstein operation, 91, 181
$BO(n)$, 184
Borsuk-Ulam theorem, 139
Bott periodicity, 202–204
boundary, 89
boundary of a manifold, 166
Brouwer fixed point theorem, 10
Brown representability theorem, 175
$BSO(n)$, 193
$BU(n)$, 193
bundle, 49

canonical line bundle, 184, 188, 195
cap product, 150, 151, 176
  relative, 153
Cartan formula, 181
category, 13
  cocomplete, 17
  complete, 17
  connected, 15
  discrete, 16
  homotopy, 14
  of canonical orbits, 24
  opposite, 13
  small, 13
cell, 71
cellular approximation theorem, 74
cellular chain complex, 95–98, 117
  of a product, 99
  reduced, 98
cellular cochains, 136, 147
cellular homotopy, 73
cellular map, 71

chain complex, 89
chain homotopy, 90
chain map, 89
characteristic class, 185, 189
characteristic classes
  in K-theory, 211
characteristic number
  normal, 189
  tangential, 189
Chern character, 207
Chern classes, 195, 205
CHP, 47
classification theorem
  for complex n-plane bundles, 193
  for oriented n-plane bundles, 194
  for principal G-bundles, 197
  for real n-plane bundles, 185
classifying space, 126, 184, 196
closed manifold, 166
clutching function, 202
coalgebra, 221
cobordant manifolds, 215
cobordism, 215
cochain complex, 89
  dual, 131
coequalizer, 16
cofiber homotopy equivalence, 44
cofiber sequence, 57
cofibration, 41, 106, 143
  based, 56
cohomology operation, 180
  stable, 180
cohomology theory, 135
  generalized, 143
  multiplicative, 148
  on spectra, 227
  ordinary, 147, 173
  reduced, 144
cohomology with compact supports, 158
colimit, 16, 38
  cohomology of, 146
  homology of, 113
commutativity
  graded, 137, 141
comodule, 223
compact space, 37

239

INDEX

compact-open topology, 39
compactly closed subspace, 37
compactly generated space, 37
compactly supported cohomology, 158
compactly supported homology, 155
cone, 55
 reduced, 55, 106
 unreduced, 67, 106
connecting homomorphism, 91
contractible space, 15
coproduct, 16
$\mathrm{Cov}(B)$, 28
$\mathrm{Cov}(\mathscr{B})$, 26
cover, 21
covering, 21
 of a groupoid, 23
 regular, 22, 25, 26, 29
 universal, 22, 25, 26, 29, 30
covering homotopy property, 47
covering space, 21
cup product, 137, 175
 relative, 148
cup product pairing, 149, 164
CW complex, 71
 relative, 71
cycle, 89
cylinder
 reduced, 56

deformation, 33
deformation cochain, 140
deformation retract, 33
degeneracy map, 121, 125
degree of a map, 10
diagram
 $\mathscr{D}$-shaped, 16
dimension axiom, 93, 94, 135, 136
dimension function, 200
disk, 71
double of a manifold, 168
DR-pair, 43

$\mathscr{E}_n(-)$, 183
$\tilde{\mathscr{E}}_n(B)$, 193
Eckmann-Hilton duality, 41
edge, 33
edge path, 33
 closed, 33
 reduced, 33
effective group action, 197
Eilenberg-MacLane space, 119, 126, 171
Eilenberg-MacLane spectrum, 227
equalizer, 17
equivalent
 categories, 14
 paths, 6
$\mathscr{E}U(X)$, 200
$\mathscr{E}U_n(B)$, 193
Euler characteristic, 196

 of a finite graph, 35
 of a CW complex, 80
 of a graded vector space, 91
 of a space, 163
Euler class, 195
evaluation pairing, 149, 177
exact sequence, 91
exactness axiom, 93, 94, 108, 109, 135, 136, 144
excision axiom, 93, 94, 135, 136
Ext, 132
exterior powers, 206, 209

face map, 121, 125
fiber, 21
fiber homotopy equivalence, 50
fiber translation functor, 23
fibration, 47
 based, 59
 Serre, 47
first axiom of countability, 37
fixed point space, 32
free group, 35
free product, 19
Freudenthal suspension theorem, 83
full subcategory, 15
function space, 39
 based, 55
functor, 13
 adjoint, 14
 contravariant, 13
 covariant, 13
 homotopy invariant, 14
fundamental class, 150
fundamental group, 6–7, 35
fundamental groupoid, 15
fundamental neighborhood, 21
fundamental theorem
 of algebra, 10
 of covering groupoid theory, 25
 of covering space theory, 28

Gauss map, 185
geometric realization, 122, 124
graded R-module, 89
graph, 33
Grassmann variety, 184
 of oriented n-planes, 193
Grothendieck construction, 199
Grothendieck ring, 199
group action, 23
 free, 24
 transitive, 24
groupoid, 15
groupoid action, 24
 transitive, 24

HELP, 73
HEP, 41

Hom, 131
homologous cycles, 89
homology group
  of a chain complex, 89
homology theory, 93, 94
  generalized, 105
  on spectra, 227
  ordinary, 115, 117, 172
  reduced, 108, 109, 172
homotopy, 5
  based, 56, 59
  relative to a subspace, 67
homotopy category, 14
homotopy cofiber, 36, 57
homotopy colimit, 114
homotopy equivalence, 14
  of pairs, 45
homotopy excision theorem, 81
homotopy extension and lifting property, 73
homotopy extension property, 41
homotopy fiber, 59
homotopy groups, 63
  relative, 63
  stable, 84, 173
Hopf algebra, 222
Hopf bundles, 65
Hopf construction, 212
Hopf invariant, 211
H-space, 141, 174, 212
Hurewicz homomorphism, 115
  stable, 220
Hurewicz theorem, 116
hypersurface $H_{n,m}$, 216

index, 164
invariance of domain, 166
isotropy group, 24

$K(X)$, 199
$\tilde{K}(X)$, 200
K-theory, 199
  periodic, 204
  reduced, 200
  represented, 201
k-invariants, 179
Klein bottle, 20, 71, 101
k-space, 37
KU, 204
Künneth map, 202, 203
Künneth theorem, 130
  relative, 165

$\lim^1$, 146
limit, 16, 38
local homeomorphism, 32
locally contractible space, 33
locally path connected space, 21
long exact sequence
  of based spaces, 57, 59

of pointed sets, 57, 59
loop, 6
loop space, 56

manifold with boundary, 166
mapping cylinder, 42
mapping path space, 47, 59
Mayer-Vietoris sequence, 110, 145
  relative, 111, 145
Mittag-Leffler condition, 147
MO, 227, 229
Moore space, 119
Mythical Object, 227

$\mathcal{N}_n$, 215
natural transformation, 14
n-connected space, 74
NDR-pair, 43
n-equivalence, 67, 81
nondegenerately based space, 56
n-plane bundle, 183
n-simple space, 140
n-simplex
  singular, 122
  topological, 121
numerable open cover, 49

obstruction cocycle, 140
orbit category, 24
orbit space, 30
orientable, 155, 161, 167
orientable manifold, 154
orientation, 154
orientation cover, 161

$\mathcal{P}$, 228
parallelizable manifold, 187
parallelizable spheres, 214
path lifting function, 48
  based, 59
path space, 56
$\mathcal{P}G(B)$, 197
Poincaré duality theorem, 150, 159
  relative, 168
Pontryagin classes, 195
Pontryagin-Thom construction, 217
Postnikov system, 178
prespectrum, 172
  cohomology groups of, 219
  homology groups of, 219
  homotopy groups of, 216
  Ω, 174
  ring, 218
  Steenrod operations of, 222
principal G-bundle, 196
product, 16
products on $\mathbb{R}^n$, 214
projective bundle, 204
projective plane, 20, 71, 101

projective space
  complex, 72, 151
  real, 72, 101, 138, 151
pullback, 17, 47
pushout, 16, 41

quotient space, 38
QX, 229

reduced cochains, 147
reduced cohomology, 143
reduced homology, 105
  provisional definition, 97, 117
represented functor, 173, 180
R-fundamental class, 149, 154, 168
ring prespectrum, 218
ring space, 201
R-orientable, 149, 154, 167
R-orientation, 150, 154, 167, 191
R-orientation cover, 162
R-oriented manifold, 150

$\mathscr{S}$, 226, 228
semi-locally simply connected space, 30
Serre spectral sequence, 191
short exact sequence
  of chain complexes, 91
signature, 164
simple space, 140, 178
simplicial object, 125
simplicial set, 124
singular chain complex, 122
singular cochains, 136
singular homology, 122
skeleton
  of a category, 15
  of a CW complex, 71
smash product, 55
spectrum, 226, 228
$Sph(E)$, 190
sphere prespectrum, 172
splitting lemma, 205
splitting principle, 205
stable, 93, 148
stable category, 227
stable homotopy groups, 84
stable homotopy theory, 173
stable invariants, 93
stable objects, 226
stably equivalent bundles, 200
star, 22
Steenrod algebra, 222
Steenrod operations, 181
Stiefel variety, 184
Stiefel-Whitney classes, 187, 192
Stiefel-Whitney numbers, 189, 216
  normal, 189, 216, 224
  tangential, 189, 216
subcomplex

  of a CW complex, 71
suspension, 55
  reduced, 55, 106
  unreduced, 106
suspension axiom, 108, 109, 144
suspension homomorphism, 83
suspension prespectrum, 172
suspension spectrum, 229

$\mathscr{T}$, 14, 38
telescope, 113
tensor product
  of chain complexes, 90
Thom class, 191, 206
  stable, 222
Thom cobordism theorem, 215, 216
Thom diagonal, 191
Thom isomorphism, 191, 207
  stable, 220
Thom prespectrum, 216
Thom space, 190, 206, 216
  of a complex bundle, 194
Thom spectrum, 227
TO, 216
$TO(q)$, 216
topological boundary map, 96, 107
topological collar, 167
topological group, 8, 32
  commutative, 127
topology of the union, 38
Tor, 129
torsion product, 129
torus, 20, 71, 101
total Chern class, 207
total space, 21
total Steenrod operation, 188
total Stiefel-Whitney class, 187
transfer homomorphism, 127, 141
transversality, 217
tree, 33
  maximal, 34
triad, 77
  CW, 77
  excisive, 77
triad homotopy groups, 84
triple, 63, 77
  exact sequence of, 84, 110, 145
tubular neighborhood theorem, 216
Tychonoff space, 39

$\mathscr{U}$, 13, 38
unique path lifting theorem, 22
universal coefficient theorem, 130, 132
universal n-plane bundle, 184
  complex, 193
  oriented, 193
universal principal G-bundle, 196
universal property, 14

van Kampen theorem, 17–18
vanishing theorem, 154
$Vect(X)$, 199
vector bundle, 183
vertex, 33, 71
virtual bundle, 199

weak equivalence, 67, 93
weak equivalence axiom, 93, 108, 135, 144
weak Hausdorff space, 37
weak product, 172
wedge, 16
well pointed space, 56
Weyl group, 24
Whitehead theorem, 73, 74, 203
Whitney duality formula, 187
Whitney sum, 186
Wu formula, 188, 193

Yoneda lemma, 180, 186